EX LIBRIS

UNIVERSITATIS SANCTI JOANNIS

Synchrotron Radiation Applied to Biophysical and Biochemical Research

NATO ADVANCED STUDY INSTITUTES SERIES

A series of edited volumes comprising multifaceted studies of contemporary scientific issues by some of the best scientific minds in the world, assembled in cooperation with NATO Scientific Affairs Division.

Series A: Life Sciences

Recent Volumes in this Series

The series is published by an international board of publishers in conjunction with NATO Scientific Affairs Division

A Life Sciences	Plenum Publishing Corporation
B Physics	New York and London
C Mathematical and Physical Sciences	D. Reidel Publishing Company Dordrecht and Boston
D Behavioral and Social Sciences	Sijthoff International Publishing Company Leiden
E Applied Sciences	Noordhoff International Publishing Leiden

Synchrotron Radiation Applied to Biophysical and Biochemical Research

Edited by

A. Castellani

Comitato Nazionale Energia Nucleare
Rome, Italy

and

I. F. Quercia

Istituto Nazionale Fisica Nucleare
Rome, Italy

PLENUM PRESS • NEW YORK AND LONDON
Published in cooperation with NATO Scientific Affairs Division

Library of Congress Cataloging in Publication Data

Nato Advanced Study Institute on Synchrotron Radiation Applied to Biophysical
and Biochemical Research, Frascati, 1978.
Synchrotron radiation applied to biophysical and biochemical research.   .

(NATO advanced study institutes series: Series A, Life sciences; v. 25)
Includes index.
1. Synchrotron radiation–Congresses. 2. Biological physics–Technique–Congresses.
3. Biological chemistry–Technique–Congresses. I. Castellani, Amleto. II. Quercia,
Italo Federico. III. North Atlantic Treaty Organization. IV. Title. V. Series.
QH324.9.S95N37 1978   574′.1′92   79-16560
ISBN 978-1-4684-3592-4   ISBN 978-1-4684-3590-0 (eBook)
DOI 10.1007/978-1-4684-3590-0

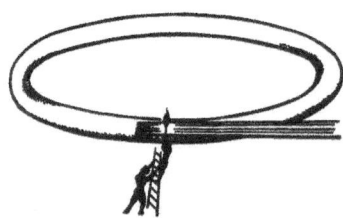

Organized by:
International College of Applied Physics (ICAP)
Comitato Nazionale Energia Nucleare (CNEN)

Sponsored by:
NATO, Association Internationale de Photobiologie
(AIP), Consiglio Nazionale delle Ricerche (CNR),
IBM-Italia, Gruppo Italiano Riparazione DNA (GIRD)

Directed by:
A. Castellani, Comitato Nazionale Energia Nucleare
I. F. Quercia, Istituto Nazionale Fisica Nucleare

Lectures presented at the NATO Advanced Study Institute on
Synchrotron Radiation Applied to Biophysical and Biochemical Research,
held at Frascati, Rome, Italy, August 20–September 1, 1978.

© 1979 Plenum Press, New York
Softcover reprint of the hardcover 1st edition 1979
A Division of Plenum Publishing Corporation
227 West 17th Street, New York, N.Y. 10011

Scientific Committee

B. Alpert   Departement de Physiologie Humaine, Faculté de Mé-
                 decine, Paris, France

C. Biagini   Istituto di Radiologia, Università di Roma, Italy

J.B. Birks[*]   The Schuster Laboratory, University of Manchester,
                 U.K.

A. Castellani   Divisione di Radioprotezione, CNEN-CSN Casaccia,
                 Rome, Italy

P. Fasella   Istituto di Chimica Biologica, Università di Roma,
                 Italy

K.O. Hodgson   Department of Chemistry, Stanford University,
                 Stanford, U.S.A.

K. Holmes   Max Plank Institut für Medizinische Forschung,
                 Heidelberg, B.D.R.

I.F. Quercia   Istituto Nazionale di Fisica Nucleare, Frascati,
                 Rome, Italy

R.B. Setlow   Brookhaven National Laboratory, Upton, New York,
                 U.S.A.

A. Vaciago   Istituto di Fisica, Università di Roma, Italy

[*]The Scientific Committee deeply regrets the untimely death of Dr.
J.B. Birks, who actively participated in the planning of this
Course from the very beginning and was largely responsible for th
section on photochemistry and photophysics.

v

Preface

The study of the interaction between light and matter has
played a fundamental role in the development of natural sciences.
Synchrotron radiation has characteristics of intensity, width and
continuity of wave length range, time structure, tunability and
polarization which are far superior to those of most other sources.
It is possible with synchrotron radiation to perform experiments
which could previously be only thought about and to routinely
carry out measurements which were once made only with great difficulties. The study of the enormously complicated but immensely
interesting biological structures seems to be particularly suited
to this new approach.

The above considerations lead us to consider the opportunity
of critically discussing the achievements and perspectives of the
use of synchrotron radiation in biology and presenting them to a
selected audience within the framework of an advanced school. From
the very beginning we were encouraged in our initiative by many
discussions with colleagues in the Rome area, who were later to
become members of the Scientific Committee of this Course. We were
fully aware that many of the results obtained so far were of a preliminary nature; we felt, however, that they were sufficiently
promising to justify this venture, which seemed particularly important in consideration of the increasing number of centers,
where synchrotron radiation is becoming available.

Frascati seemed a quite suitable site for this school not only
because of its tasty wine and ancient villas but also because of
the new synchrotron radiation facilities at the ADONE storage ring.
The papers presented in this book and the thorough discussions
which followed their presentation during the Course make it evident that the biological use of synchrotron radiation has indeed
come of age and will have a revolutionary impact on several areas
of biomedical research.

We trust that the Authorities responsible for European and national scientific policies will take this into proper consideration.

As it appears from the main groups of lectures in which the Course was structured, at least three areas have emerged: Photobiology, Photochemistry and Structural Research. Particular emphasis is laid on the use of non-ionizing radiation.

We wish to thank the Sponsors, the Participants, the Co-director of the ICAP, N.A. Mancini, and the Staff for their generous and competent help.

We also thank the Scientific Committee and particularly A. Vaciago who actively participated in the running of the Course. The visit of the Head of the Research Group Programme of NATO, M. Di Lullo, was greatly appreciated.

A. Castellani
Division of Radiation Protection
CNEN - CSN Casaccia, Rome, Italy

I.F. Quercia
National Institute of Nuclear Physics
INDN - Frascati, Rome, Italy

Contents

BASIC PROPERTIES OF SYNCHROTRON RADIATION

E. BURATTINI

LABORATORI NAZIONALI DI FRASCATI

FRASCATI , ITALY

1. -Some historical notes.

The emission of the light by relativistic electrons in circular orbits was foreseen before 1940 by A.Liénard (1)and co-workers. Blewett (2),in about 1945, was the first to consider the effect of synchrotron radiation on electron accelerators. At that time the development of circular electron accelerators led to theoretical and experimental investigations of the radiation emitted by relativistic electrons. Schwinger (3) published an extensive and complete discussion of this so called "synchrotron radiation".

The emission of the light was first observed on the General Electric 70-MeV synchrotron in 1947. In 1953, Tomboulian and co-workers (4), made investigations on synchrotron radiation emitted from the 300-MeV synchrotron at Cornell University and showed that it was applicable to spectroscopy experiments. At same time, the properties and the characteristics of synchrotron light were investigated at 660-MeV synchrotron of the Lebedev-Institute in Moscow.

In 1960 Codling and Madden (5) began to use the radiation emitted from the 180-MeV synchrotron at NBS in Washington as a light source for atomic spectroscopy experiments.

At present, more than twenty electro-synchrotrons
and electron-positron storage-rings are utilised over
all the world. In table 1 are listed all the electro-
synchrotrons and all the storage-rings which are used
also as photon sources.

For the unique features of synchrotron radiation, it

Table 1. - Electro-synchrotrons and Storage-rings
          which are used also as photon sources.

Name	Type	Location	E	R	I		
Tantalus I	SR	Stoughton	0.24	0.64	100	0.048	Ded.
Surf II	SR	Washington	0.24	0.83	30	0.037	Ded.
INS-SOR	SR	Tokio	0.30	1.10	50	0.054	Ded.
ACO	SR	Orsay	0.55	1.70	30	0.163	Ded.
Bonn I	ES	Bonn	0.50	1.11	150	0.333	SRL
Vepp-2M	SR	Novosibirsk	0.67	2.00	100	0.350	SRL
Lusy	ES	Lund	1.20	3.65	10	1.050	SRL
Sirius	ES	Tomsk	1.30	4.20	20	1.160	SRL
Pachra	SR	Moscow	1.30	4.00	100	1.220	Ded.
INS-SOR I	ES	Tokio	1.30	4.00	50	1.220	SRL
Adone	SR	Frascati	1.50	5.00	100	1.550	SRL
DCI	SR	Orsay	1.80	3.82	400	3.390	SRL
Vepp-3M	SR	Novosibirsk	2.20	6.15	500	3.800	SRL
Bonn II	ES	Bonn	2.50	7.65	30	4.530	SRL
Doris	SR	Hamburg	3.50	12.12	500	7.850	SRL
Spear	SR	Stanford	4.00	12.70	60	11.200	SRL
Arus	ES	Erevan	6.00	20.65	20	19.500	SRL
Vepp-4M	SR	Novosibirsk	6.00	33.00	100	14.500	SRL
Desy	ES	Hamburg	7.50	31.70	30	2.550	SRL
Cornell IV	SR	Ithaca	8.00	32.00	100	35.000	SRL

E= Electron beam energy,in GeV
R= Bending sections radius,in meters
I= Max. current injected,in mA
ε_c= Critical photon energy,in keV
SR= Storage-Ring
ES= Electro-Synchrotron
Ded.= Dedicated to synchrotron radiation
SRL= Synchrotron Radiation Laboratory

is necessary to create a source which is optimised for the use as photon source. In table 2 are listed all the storage-rings who have been proposed or are in construction,all over the world, "dedicated" for synchrotron radiation pourposes.

Table 2. - Storage-rings proposed or in construction dedicated for synchrotron radiation.

Name	Location	E	R	I		
Brookhaven I	Upton	0.70	2.00	1	0.40	UC
Aladdin	Stoughton	0.75	2.08	1	0.45	UC
Pampus	Amsterdam	1.50	4.17	0.5	1.80	UC
Nina II	Daresbury	2.00	5.55	1	3.20	UC
Brookhaven II	Upton	2.00	8.00	1	2.20	P
IPP	Moscow	2.50	5.00	1	3.50	P
Photon Factory	Tokio		8.00	0.5	4.30	UC

E= Electron beam energy, in GeV
R= Bending sections radius, in meters
I= Max. current injected, in A
ε_c= Critical photon energy, in keV
UC= Under construction
P= Proposed

In the Western Europe first experiments with synchrotron radiation were carried-out in a French-Italian collaboration in 1962 (6-7) at the 1.1-GeV Frascati electro-synchrotron shown in figure 1.
This activity was continued,for many years, mainly by Italian groups directed by Prof. G. Chiarotti (8-12).

In 1975 the C.N.R. (Italian National Research Council) and the I.N.F.N. (Italian National Institute for Nuclear Physic) have agreed to jointly support the construction and operation of a new synchrotron radiation facility at the storage-ring ADONE showed in the figure 2.

This new project, directed by Prof. F. Bassani, has been called P.U.L.S. (Project for the Use of Synchrotron Light) (13).

Figure 1. - Picture of the 1.1-GeV electro-synchrotron
at Frascati.

Figure 2. - Picture of ADONE storage-ring.

2. - Properties of Synchrotron Radiation.

The unique features of synchrotron radiation may be understood by considering an oscillating dipole, that is the accelerated electron in orbit (14). At low energy the radiation is emitted in rather non-directional dipole pattern, such as shown in figure 3a; at high energy, when the velocity of the electron is close to the velocity of the light, the dipole pattern is greatly distorted into the forward direction of the electron, such as shown in figure 3b, because of the relativistic transformation.

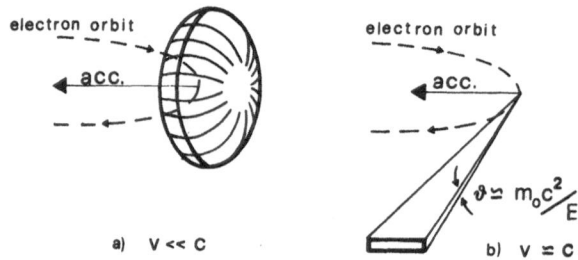

Figure 3. - Angular intensity distribution.

The radiation emitted by a single electron, with energy E, travelling on a circular orbit with radius R, per sec., per rad.,at wavelength λ ,at the azimut angle ϑ relative to the plane of the orbit, per unit wavelength, is given by:

$$P(\lambda,\vartheta) = a_1\, e^2 c\, R^{-3}\, \left(\frac{\lambda_c}{\lambda}\right)^4 \gamma^8 (1 + x^2)^2 \cdot$$

$$\left(k_{2/3}^2 (\xi) + x^2(1+x^2)^{-2} k_{1/3}^2(\xi)\right) \qquad (1-1)$$

where
$$\gamma = E/m_o c^2 \;\; ; \;\; x = \gamma \vartheta \;\; ; \;\; \xi = \frac{\cdot \varsigma}{2\lambda} \; (1+x^2)^{3/2}$$

a_1 = numerical parameter

(1-2)

$K_{2/3}, \; K_{1/3}$ modified Bessel function of second kind

·   The very important parameter λ_c is the so called "critical wavelength" which depends on the radius of curvature R of the orbit, and on the electron beam energy E; λ_c is given by:

$$\lambda_c = \frac{4}{3} \, \pi \, R \, \gamma^{-3} \qquad\qquad (1\text{-}3)$$

The elevation angle ϑ is slightly dependent on the particular photon energy; ϑ decrease slowly with the photon wavelength such as shown in figure 4. In any case, with good approximation is of order of $\gamma^{-1} = m_o c^2 E^{-1}$.

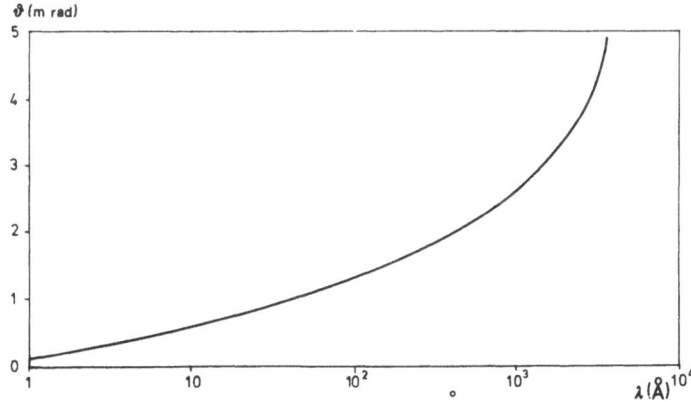

Figure 4. - Dependence of the elevation angle ϑ from the photon energy´.

The integration of eq. (1-1), over all the vertical angles, give the spectral distribution of the total power irradiated by a monoenergetic electron, per rad. of observed orbit, at wavelength λ , into a wavelength band $\Delta \lambda$:

$$P(\lambda) = a_2 e^2 cR^3 \gamma^7 \, G(Y) \, \Delta \lambda \qquad\qquad (1\text{-}4)$$

The " universal function" G(Y), showed in figure 5, is
given by:

$$G(Y) = Y^{-3} \int_Y^\infty K_{5/3} (u) \ du \qquad (1-5)$$

where the $K_{5/3}(u)$ is a modified Bessel function of the
second kind and Y is given by:

$$Y = \frac{\lambda}{\lambda_c} \qquad (1-6)$$

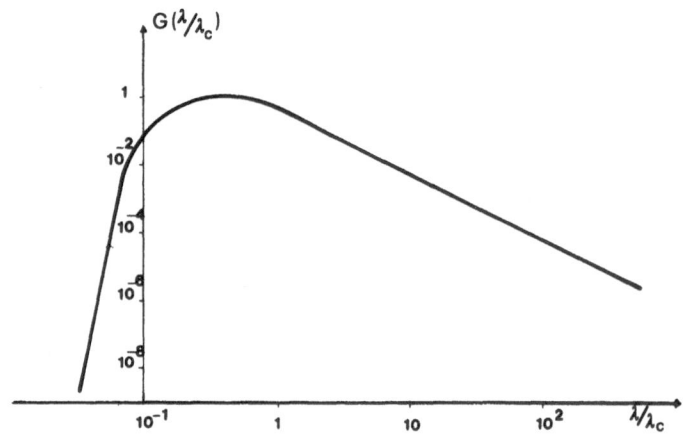

Figure 5. - Universal function $G(\lambda/\lambda_c)$.

    The critical wavelength λ_c characterizes the extent
and the width of the spectral distribution of the synchrotron radiation. The corresponding critical energy
is given by:

$$\varepsilon_c = a_3 E^3 R^{-1} \qquad (1-7)$$

where the a_3 is a numerical parameter. Normally a storage-
ring provides good photon flux up to photon energies:
$\varepsilon = 5 \varepsilon_c$
    Sometimes it is more useful, for a storage-ring,

to describe the spectrum by the number of photons emitted, at wavelength λ , into a relative wavelength band
$\Delta\lambda/\lambda$. Transformation of eq. (1-4) gives:

$$N(\lambda) = a_4 \, Y^2 \, E \, G(Y) \qquad\qquad (1-8)$$

where the numerical parameter a_4 is equal to:

$$a_4 = 2.45 \cdot 10^{10} \qquad\qquad if \ \frac{\Delta\lambda}{\lambda} = 10^{-3}$$

The eq. (1-8) gives the number of photons per sec.,
per mrad., per mA of circulating electron beam current.
Figure 6 shows the spectrum emitted by ADONE at different
energies.

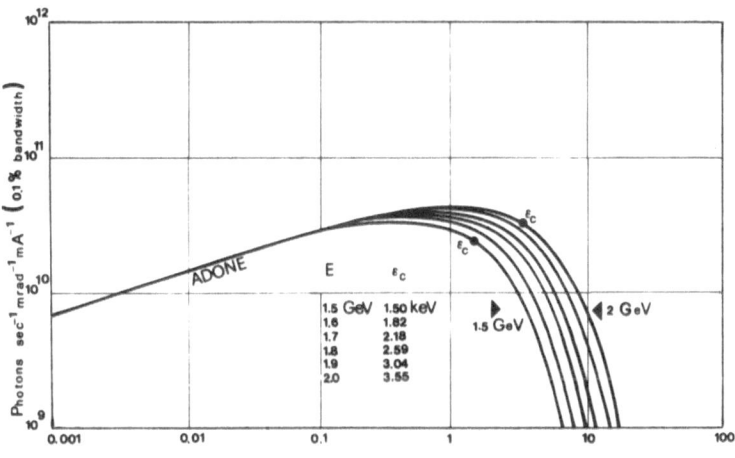

Figure 6. - Spectral distribution of radiation emitted
from ADONE at different energies.

    The spectrum shows a maximum at wavelength $\lambda = 0.42\,\lambda_c$;
at short wavelength the photon flux decrease exponen-
tially: for example at $\lambda = 0.1 \ \lambda_c$, for many experiments,
the photon flux is substantially not useful. At long
wavelength, or at $\mathcal{E} \ll \mathcal{E}_c$ the photon distribution depends
slightly on the electron energy E. In this case, a good
approximation to the number of photons emitted by 1
mrad of electron orbit is given by:

$$N(\lambda) = a_5 \, R^{1/3} \, \lambda^{-1/3} \qquad\qquad (1\text{-}9)$$

where the numerical parameter a_5 is equal to:

$$a_5 = 9.33 \cdot 10^{10} \qquad\qquad \text{if} \quad \frac{\Delta\lambda}{\lambda} = 10^{-3}$$

If all the parameters are given in practical units, in other words:

E, the energy of the electron beam, in GeV
R, the bending radius, in meters
e, the electronic charge, in coulombs
$m_0 c^2$, the electron rest energy, in GeV
c, the velocity of the light, in meters/sec
λ_c , the critical wavelength, in A
I, the current stored, in amperes

therefore it is possible to write same basic synchrotron radiation relationships:

1. Δ E/Turn-electron   =   88.5 E^4/ R   (keV)
2. P   =   88.5 E^4 x I /R   (kwatts)
3. ϑ $\approx$ 1/γ = $m_0 c^2$/E   (radians)
4. ε_c =   2.22 E^3/R (keV)

The first relationship gives the energy lost by electron going in a circular orbit of radius R, per turn.
The second relationship gives the total power radiated by a current I.
The last relationship gives the critical energy.

The synchrotron radiation light is higly polarised with the E vector parallel to the plane of the electron orbit. In figure 7 is shown, in arbitrary units, the distribution of the polarization components as respect to the elevation angle ϑ .

Finally the accelerating radio-frequency apparatus causes the electron beam to be bunched and then the radiation is emitted in sub-nanosecond pulses separated by hundreds of nanoseconds. This time structure of the beam light permits the study of time dependent processes such as fluorescence.

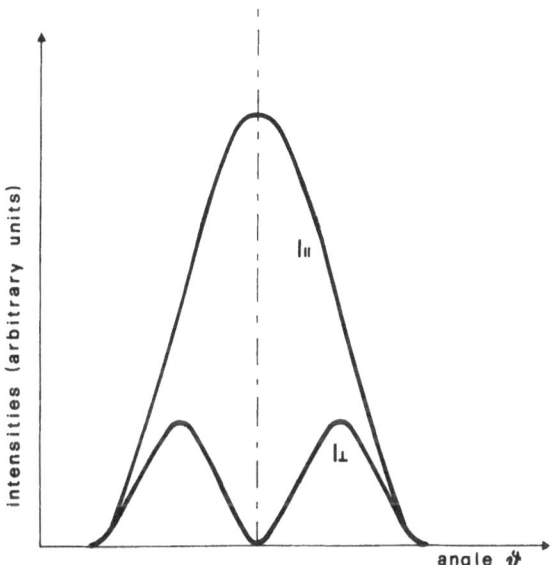

Figure 7. - Relative intensities of the polarization components with respect to the elevation angle ϑ.

Table 3. - Properties of Synchrotron Radiation.

---

1. Spectral distribution extending from the infrared
   to the X-ray region.

2. High photon flux.

3. Very small divergence of the beam: the elevation angle
   ϑ for X-ray, is of the order of 10^{-1} mrad.

4. The position of the beam is stable and exactly reproducible.

5. High polarization of the radiation in the plane of
   the electron orbit.

6. Source of very small sizes:from several 0.1 mm^2 to
   several mm^2.

7. The time structure is very well defined.

8. In contrast to the gas discharge lamps, the synchrotron radiation source exists in an high vacuum
   environment.

The characteristic properties of synchrotron radiation that have made it an interesting light source,
are summarized in table 3.

3. - Comparison of Synchrotron Radiation with other
     conventional sources of light.

In the vacuum ultraviolet region below 2.000 Å,
synchrotron radiation is superior to any conventional
source because of the high intensity,polarization and
tunability.

In the X-ray region the radiation emitted from a
storage-ring must be compared with that emitted from
strong X-ray tubes. In figure 8 is shown a comparison, in arbitrary units, between the synchrotron radiation and the radiation emitted from an X-ray tube with

respect to their characteristics in angle, energy and time.

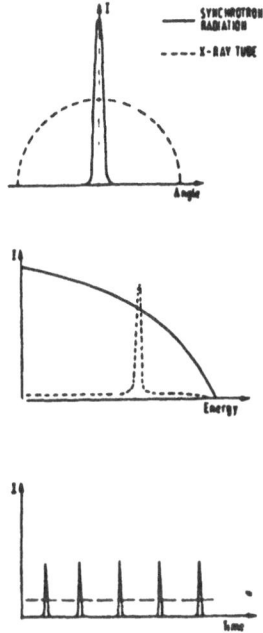

Figure 8. - Comparison in arbitrary units, between the synchrotron radiation and the radiation emitted from an X-ray tube.

4. - Synchrotron Radiation activities at National
     Laboratory of Frascati

     Frascati has been one the first laboratories, in Western Europe, to start research in the field of synchrotron radiation at 1.1-GeV synchrotron. Pionee-ring work on soft X-ray absorption in thin metal film, has been made in sixties by Jaglé, Missoni and co-workers (6). The electro-synchrotron in Frascati was closed down at the end of 1975.

However, as mentioned earlier, a new synchrotron radiation facility is in construction at the storage-ring ADONE. The most important parameters of ADONE, as a synchrotron radiation source, are shown in **table** 4. The pulsed time structure of the photon beam is shown in figure 9.

Table 4. - The most important parameters of ADONE as
            a **synchrotron** radiation source.

---

1. - Maximum energy of the electron beam	E_{max}	= 1.5GeV
2. - Critical photon energy	$\mathcal{E}_c$	= 1.5keV
3. - Bending radius	R	= 5 m
4. - Maximum injected current	I_{max}	= 100mA
5. - Number of the bunches	k	= 3
6. - Number of bending magnets	n	= 12
7. - Emission angle (in the X-ray region)	ϑ	= 0.3mrad
8. - Radio-frequency	f	= 8.54MHz
9. - Beam life-time	τ'	= 10 h
10.- Time structure of the beam	σ_γ	= 0.6nsec
(see figure 9)	τ	= 117nsec

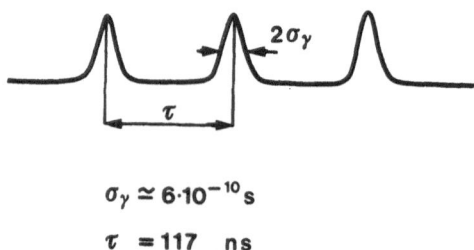

$\sigma_\gamma \simeq 6\cdot10^{-10}s$

$\tau = 117$ ns

Figure 9. - Time structure of the beam.

Figure 10 shows the spectral distribution of synchrotron radiation emitted by ADONE; one can see that with
the machine operating at 1.5 GeV and with a current
stored of 100mA, the critical energy ε_c is 1.5 keV, and
the photon flux, at ε_c , is more than 10^{12} photons x sec^{-1}
x mrad^{-1} into a bandwidth of 0.1 %. This allows to ob-
tain from ADONE a good photon flux up to 10 keV.

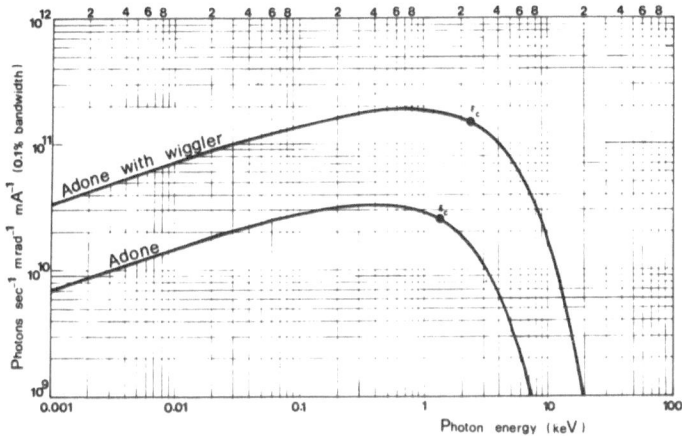

Figure 10. - Spectral distribution emitted from ADONE;
the critical energy is 1.5 keV.

Figure 11 shows the photon spectra emitted from
some storage-rings, working around the world, taking
into account the actual current circulating in the ring.
Figure 12 shows a general plan of ADONE: the shaded
area corresponds to the new PULS laboratory for synchrotron radiation experiments, and figure 13 shows a schematic view of the PULS layout: five experimental stations
will be available to the users.

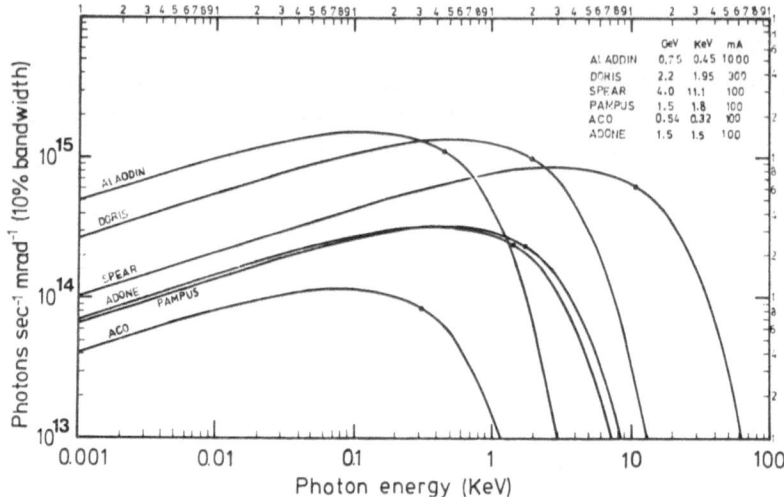

Figure 11. - Photon spectrum emitted from some storage
rings (Aladdin and Pampus under construction).

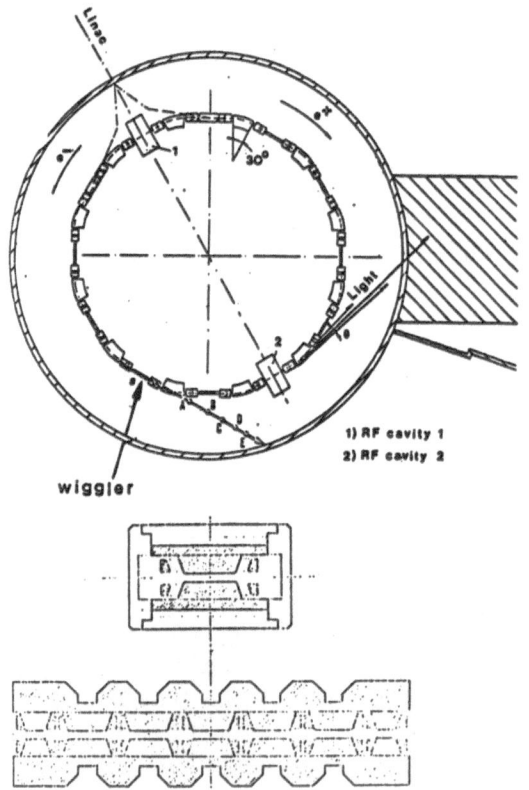

Figure 12. - General plan of ADONE.

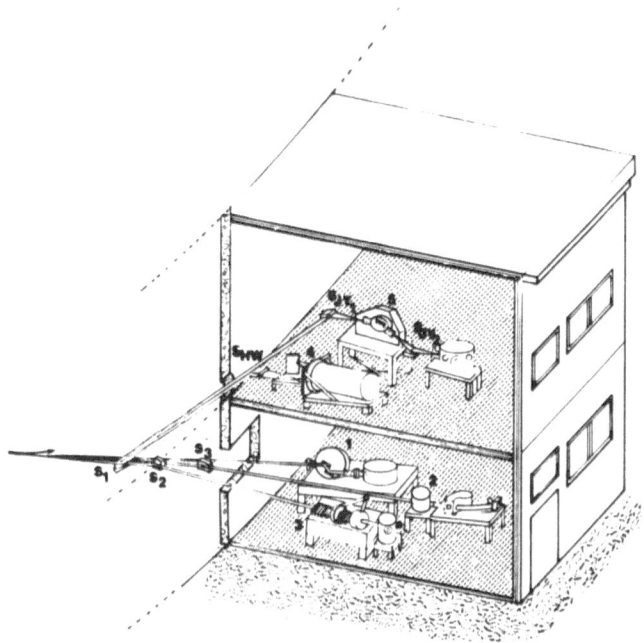

Figure 13. - Schematic view of the PULS
layout.

Finally figure 14 shows the modified bending section
utilised to open three light ports, after connection to
ADONE: each port collects the photons emitted from 10
mrad of the electron path so that roughly 4mrad will
be available for low energy experiments (10-50eV)
and 6 mrad will be available for high energy experiments
(50eV-10keV).The first beam line will be equipped
with the monochromators listed in table 5.
      The vacuum system is equipped with several ion
pumps and with two turbomolecular pumps for fast pre-
vacuum. The ion pumps are preferred because of their
ability to withstand power failures without adverse
effects on the system. The measured working pressure into pipe light is of the order of 10^{-10} torr.
      The pipe light is also equipped with a fast valve,
closing time about 14 msec., to insure the safety of the
main ring in case of accident in the beam lines and

Figure 14. - Picture of the modified bending section.

Table 5. - Monochromators available at the first
          PULS beam line.

Monochromator	Optical system	Radius of curvature	Wavelength range (Å)
Hilger-Watts mod.E766	Normal incidence	1 m	300-1600
Jobin-Yvon mod.LHT30	Seya-Namioka	1 m	100-1600
Mc Pherson mod. 247	Grazing incidence	2.217 m	5-500
Grasshopper Phys. Sc. Lab.	Grazing incidence	2 m	10-500
X-ray monoch.	Channel-cut	Si (220)	1.2-4

with an electro-pneumatic valve. Thus in the event of a
catastrophic vacuum accident, the fast valve interrupt
the shock wave in 14 msec. and the electro-pneumatic
valve provide the complete vacuum isolation.
The complete vacuum system is sketched in figure 15.

At present the first beam line for synchrotron
radiation experiments is practically in operation as
can be seen in figure 16 showing the complete apparatus.
Finally, the figure 17 shows the picture of the synchrotron radiation emitted from ADONE.

The main goal of the project PULS is, at present,
to start with the experiments in the end of 1978 or in
first part of 1979. Scientific programs have been ela-
borated to include optical spectroscopy in the UV and
X-ray region, modulation spectroscopy, EXAFS and photoemission experiments.

Moreover an important aspect of the PULS project
will be the availability of the all facility instrumentation to external users: more than thirty external

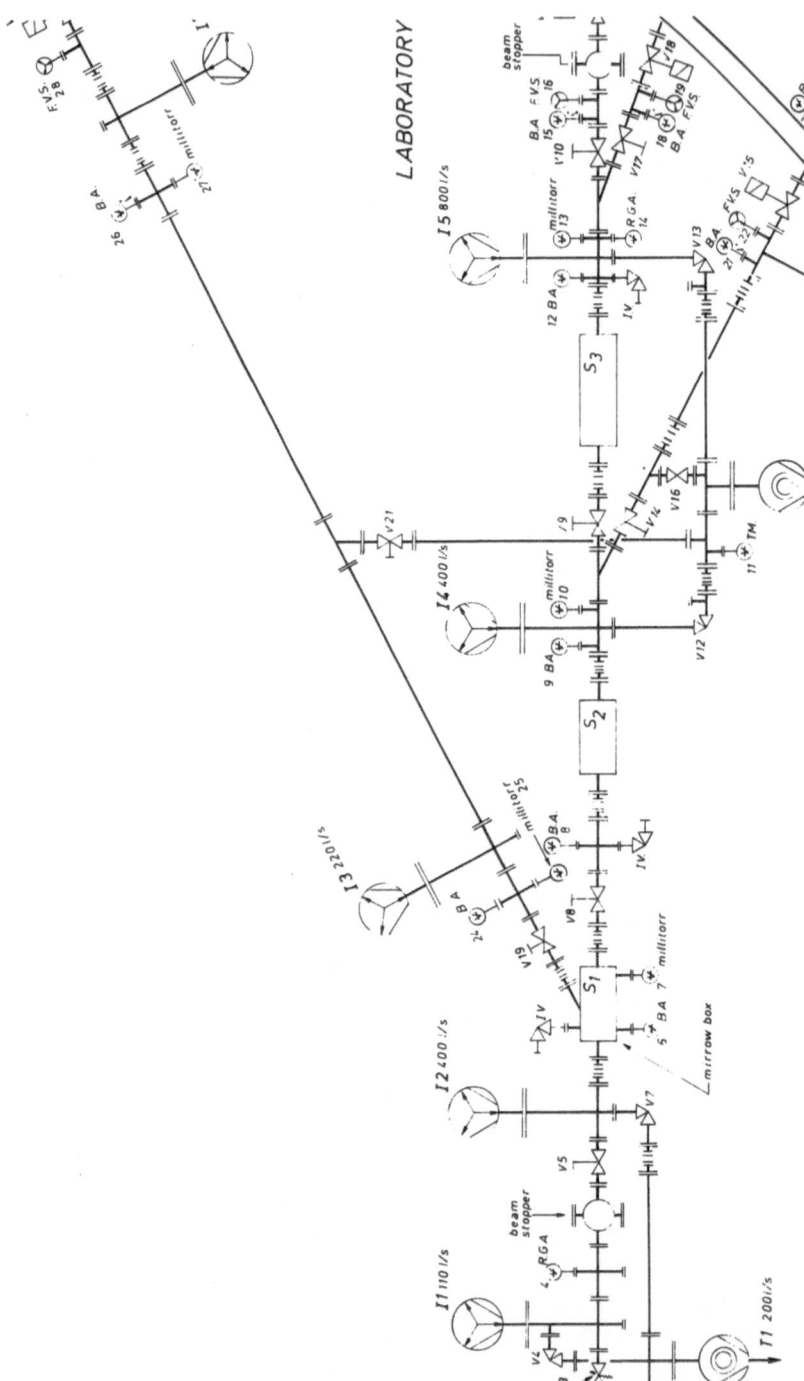

Figure 15. – The vacuum system of the PULS layout.

Figure 16. - First beam line for synchrotron radiation.

Figure 17. - Synchrotron radiation beam
emitted from ADONE.

groups have already been applied for using the facility
to carry out research programs. In this way, the Italian
scientific community will be able to use new experimental techniques based on synchrotron radiation, whose
importance is growing rapidily worldwide. Moreover the
facility will be available to foreign users and this
will help satisfy the increasing need of synchrotron
radiation sources in Europe.

5. - Future developments of Synchrotron Radiation
      activities at Frascati.

     The energy range of synchrotron radiation emitted
by ADONE storage-ring is not suitable for many EXAFS
and biological experiments which require photon energy
more than 10 keV. In order to obtain spectral distribution harder respect to the spectrum emitted from a bending section of ADONE and to produce a larger photon
flux is in construction at Frascati a "six equivalent-
full poles transverse magnet wiggler"(15).
     A wiggler magnet is a particular array of three or

more magnets, with the field B_w showing an alternating polarity inserted in a straight section of the ring. It manifests two advantages:

1. If the magnetic field is higher than that of the bending magnets, the wiggler produce a synchrotron radiation spectral distribution harder than obtained from the normal bendings.

2. The radiation emitted from all poles is substantiallysuperimposed giving in this way a significant increase of the photon spectrum.

From the spectral distribution of the radiation emitted by ADONE equipped with a six equivalent-full poles, shown in figure 10, it is clear that the critical energy is increased from to 1.5 keV to 2.7 keV, and the photon flux, at ε_c , is increased by almost one order of magnitude.

The main parameters of the "transverse wiggler magnet" in construction at National Laboratory of Frascati are shown in table 6.

Table 6. - Wiggler magnet parameters.

---

1. -	N° of poles	5 full + 2 half-poles
2. -	Minimum magnetic gap	40 mm
3. -	Total lenght	2100 mm
4. -	Pole-to-pole distance($\lambda/2$)	327 mm
5. -	Maximum field	1.85 T
6. -	Ampéreturns per pole	31,500
7. -	Current	4500 A
8. -	Electrical power	189 kW
9. -	Iron weight	5000 kg
10.-	Copper weight	270 kg

The wiggler, presently in construction, will be installed in the straight section n° 9 of ADONE storage-ring, such as shown in figure 12. The synchrotron radiation emitted from the wiggler magnet will be extracted through a window in the near bending section.

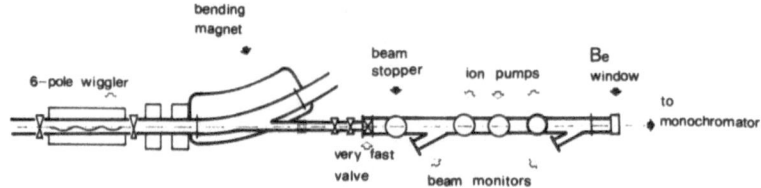

Figure 18. - Sketch of the wiggler beam line.

Figure 19. - Picture of the first section
            of the wiggler optical beam
            line.

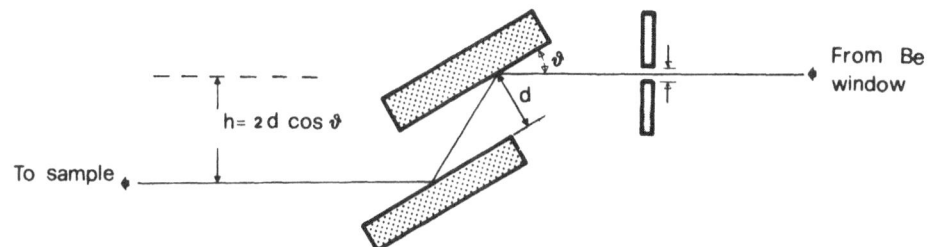

Figure 20. - Sketch of single channel cut silicon crystal monochromator.

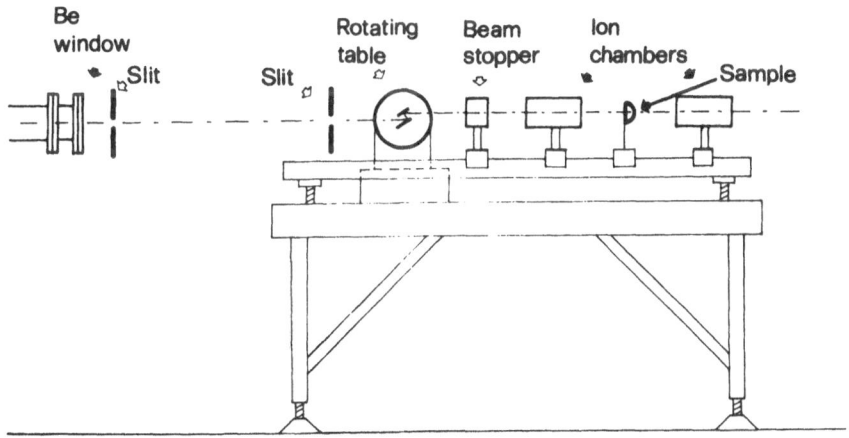

Figure 21. - Sketch of the X-ray monochromator.

Table 7 shows some basic relationships for ADONE with and without magnet wiggler.

Table 7. - Basic synchrotron radiation relationships.

	ADONE		WIGGLER	
1.- Δ E/turn x electron	90	keV	8.3	keV
2.- Total power radiated	8.96	kW	0.83	kW
3.- Energy lost per mrad.	1.43	W	2.64	W
4.- Energy lost per cm of orbit	2.86	W	9.76	W
5.- Critical energy	1.50	keV	2.70	keV
6.- Magnetic field	1	T	1.85	T
7.- Radius of electron orbits	5	m	2.7	m

A first beam-line has been projected and it is in construction at Frascati; figure 18, shows the sketch of the beam-line and figure 19 shows the first section connected to ADONE. The second section will be equipped with a Be window (thickness = 75 μ), and at the end of this section will be installed a very sensitive vacuum sensor connected to a fast valve electronic system.

The first wiggler beam-line will be equipped with a single channel cut silicon crystal monochromator using the Bragg-reflection, such as is sketched in figure 20 and in figure 21.

At present the wiggler magnet has been commissioned and the second section of the optical beam-line, down to the Be window, could be completed in May of 1979; therefore is not optimistic to think that some accurate experiments can start in the summer of 1979.

References

1.- A. Liénard, l'Eclaraige electrique, $\underline{16}$, 5 (1898)
2.- J.P. Blewett, Phys. Rev., $\underline{69}$, 87 (1946)
3.- Schwinger, J. Phys. Rev., $\underline{75}$, 1912 (1949)
4.- D. H. Tomboulian, P.L. Hartman, Phys. Rev.,$\underline{102}$,
    1423 (1956)
5.- R.P.Madden, K.Codling, J. Appl.Phys.,$\underline{36}$,830(1965)
6.- P.Jaeglé,G.Missoni,Compt.Rend.,$\underline{262B}$,71(1966)
7.- P.Jaeglé,F.Combet Farnoux,P.Dhez,M.Cremonese,
    G.Onori,Phys. Letters,$\underline{26A}$,364(1968)
8.-A.Balzarotti,L.Bartolini,A.Bianconi,E.Burattini,
    M.Grandolfo,R.Habel,M.Piacentini,LNF 72/98,Frascati
    (1972)
9.- A.Balzarotti,A.Bianconi,E.Burattini,Phys.Rev.B,
    $\underline{9}$,12(1974)
10.-A.Balzarotti,A.Bianconi,E.Burattini,M.Piacentini,
    LNF74/32(R),Frascati,(1974)
11.-A.Balzarotti,A.Bianconi,E.Burattini,M.Grandolfo,
    R.Habel,M.Piacentini,Phys.Stat:Sol.(b)$\underline{73}$,77(1974)
12.-A.Balzarotti,A.Bianconi,E.Burattini,G.Strinati,
    Solid State Comm,$\underline{15}$,1431(1974)
13.-E.Burattini,International College on Applied Phys.,
    Course on Synchrotron Radiation Research,volII(1976)
14.-R.P.Godwin,Springer Tracts,$\underline{51}$,(1969)
15.-M.Bassetti,A.Cattoni,A.Luccio,M.Preger,S.Tazzari,
    LNF77/26(R),Frascati(1977)

SOME CONSIDERATIONS ON

UV OPTICS FOR SYNCHROTRON RADIATION*

W. R. McKinney

Brookhaven National Laboratory

Upton, New York 11973

Introduction

While synchrotron radiation (SR) has a broad spectrum extending from the hard x-ray region (tens of keV) through the visible and infrared regions (a few eV) the main interest is in those wavelengths which are not readily available from other sources; that is those wavelengths shorter than say 2000Å. This region poses several problems for the optical designer that do not occur in the visible region of the spectrum.

1. Between about 2Å and 2000Å wavelength, air (oxygen) is not transparent, necessitating the evacuation of the equipment.

2. As wavelength decreases an optical surface generally redirects an increasing fraction of the incident light randomly into 2π solid angle rather than the direction desired. This "scattering" of the light is due to the wavelength of the SR becoming more comparable to the size of irregularities of the optical surface.

3. The reflectivity of coatings available for surfaces begins to decrease below 2000Å. Below 1200Å, where aluminum overcoated with MgF_2 to prevent oxidation

---

*This lecture was supported by the U. S. Department of Energy: Contract No. EY-76-C-02-0016.

and enhance reflectivity begins to deteriorate,
there are no good normal incidence reflectors.
(R ≥ 80%)

4. Below ≃ 1100Å, the cutoff wavelength of LiF,
    there are no materials suitable for refractive
    surfaces or rigid windows.

Problem 1 can be overcome with effort and money by various
means of construction and vacuum pumps. I will not cover this
technology except to say that they all introduce hydrocarbons onto
the optical surfaces by one route or another. (O-rings, mechanical
pump fluid, diffusion pump fluid, turbine pump fluid, and finger-
prints.) These hydrocarbons interact photochemically with the high
intensity SR forming a layer of contamination which is often assumed
to be polymerized hydrocarbons but which may not be.[1] To my knowledge, definitive work on the nature of contamination has not been
done and would be a fruitful line of research. There is the possibility that fluorinated pump fluids, such as Fomblin, may be
broken into gaseous products by the SR beam which would then be
pumped away and not deposited on the optical surfaces.[2]

Problem 2 needs investigation in the areas of surface characterization, scattering mechanisms and angular and energy dependence.
Scattering of the SR by optical surfaces may be the most significant
limit to available fluxes at high photon energies.[3]

Items 3 and 4 determine the nature of optics in the UV. Mirrors are, in general, necessary below 2000Å. Although potential
lens materials exist down to ~ 1200Å (BaF_2, CaF_2, LiF_2), their indices of refraction vary significantly near their cutoff wavelengths, introducing chromatic aberration. Polishing these materials for high transmission and low scattered light is often difficult; although for special applications where the wavelength
region of primary interest is not too great, these materials should
not be overlooked.[4]

Below 300Å, where normal incidence reflectivity is very low,
the mirror must be used in grazing incidence geometry. At high
incidence angles (low grazing angles), an electromagnetic effect
occurs which can permit total reflection.

Although there are complications to the following simple picture, a basic understanding of its application to SR should ensue.
There exists an angle called the critical angle (measured from the
tangent to the mirror, i.e., a grazing angle) defined (approximately) by the following equation:

$$\sin \theta c = \lambda \left(\frac{e^2}{mc^2} \frac{N}{\pi} \right)^{\frac{1}{2}} \quad N = electrons/unit\ vol.$$

For a given wavelength, total reflection occurs at all angles smaller than the critical angle. On the other hand, for a fixed angle, all wavelengths longer than a certain one will be reflected with very high reflectance. This forms the basis for what is called the "order sorting" mirror in a synchrotron radiation beam line. The purpose of a beam line, unless all of the SR spectrum is desired, is to throw away all of the SR except the wavelength of interest. The first mirror can be set to absorb heavily all of the wavelengths below a given one, accomplishing about one half of the required job of the beam line in the first reflection. The explanation of the term "order sorting" requires a digression into diffraction gratings. Figure 1 shows schematically the fundamental relation for a diffraction grating.

$$d\sin\theta = \pm\, n\lambda \qquad\qquad n = 0,1,2, \ldots$$

Notice that for a given θ there are an infinite number of λ's which can satisfy the relation. For example, if we were at an angle into which 100Å was diffracted with n (order) = 1, then 50Å with order = 2 would also take the same direction, and 33.33Å with n = 3 would also be there. In practice, we are helped by the fact that the efficiency of the grating often goes down with increasing order. Nevertheless, for some experiments, e.g., photochemical ones where higher energy photons can cause reactions not inducible by the first order wavelength, even 5% higher order contamination can be intolerable. Now that we know something about orders, we can see that if the first mirror is a poor reflector below 50Å, for example, second order interference will not become a problem until 100Å. Given enough money, one could have a beam line for each octave of the spectrum and suffer little second order contamination.

This feature of the first mirror has a disadvantage because most of the power in the SR spectrum is in the short wavelengths. Thus, the order sorting mirror which passes only wavelengths on the long λ side of the critical $\lambda = \lambda_c$ must absorb the major fraction of the unwanted SR. In the case of the storage rings under construction at Brookhaven, this can be many watts of energy, particularly for the 2.5 GeV x-ray ring. With I in amperes, E in GeV and B in kilogauss, the power of the photon beam in kilowatts is

$$P = 2.65\ E^3 IB \ \text{kw (for a circular ring of}$$
$$\text{constant bending field strength)}^5 .$$

Let us assume 10KG, 1 ampere and 0.7 GeV for the UV ring and 2.5 GeV for the large ring. This gives about 9100 watts for the UV ring and 414,000 watts for the x-ray ring. Let's specialize on the UV ring. Assume we will collect 100 milliradians (mr). This could be done by a mirror of 20 cm projected length at 2 meters from the source.

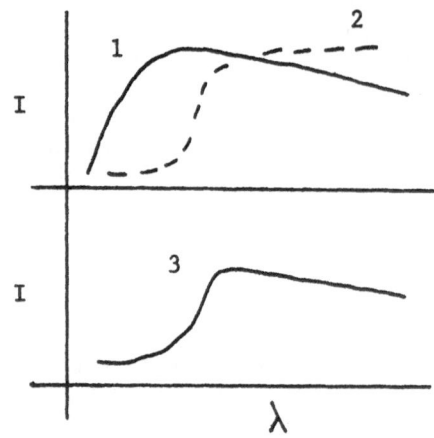

1. SR
2. Mirror reflectivity
3. Transmitted radiation

Fig. 1a Order Sorting Mirror

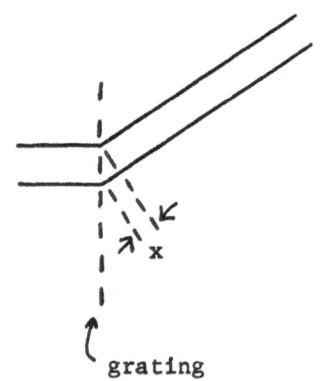

geometry $\Rightarrow$ $x = d\sin\theta$

$x = m\lambda \Rightarrow$ rays in phase
                   constructive
                   interference

grating

Fig. 1b Diffraction Grating

$$\frac{.001}{2\pi} \times 9100 \text{ watts} = 1.44 \frac{\text{watts}}{\text{mr}}$$

$$0.144 \frac{\text{watts}}{\text{mr}} \times 100 \text{ mr} = 144 \text{ watts} \quad .$$

This is a non-negligible power load which must be considered as a source of changes in mirror figure caused by temperature differences. In addition, the narrow vertical divergence of the hardest photons, ≤ 10 mr, will place for some geometries most of the radiation on a narrow strip (~ 1 mr) down the center of the mirror. The cooling problem becomes severe for the x-ray ring as you can see from the numbers above.

So far, we have realized that we are limited to mirrors and often high incidence (low grazing) angles in order to achieve good reflectivity. Because a major goal in SR beam lines is to place as much radiation at the experimenter's disposal as possible, the radiation must be focused or rendered parallel. If we take the previous case of the collection of 100 mr in the horizontal direction, the divergence of the beam must be altered close to the SR source or the beam will become so wide that one can not afford a mirror big enough to reflect it. Remember, also, that at grazing angles mirrors must be much longer than at normal incidence.

The spherical concave mirror is the first and obvious choice for focusing. It has major advantages in being a simple surface to grind, and is routinely made for many other applications. Its major disadvantage is that it is stigmatic only near normal incidence. That is, it images a point source into a point image only near 0° incidence angle. Figure 2 shows this effect. In part (a), if we place the object at the center of curvature of the sphere, it is easy to see that the image will coincide with the object because rays from the object will be perpendicular to the mirror by definition and the ray will reflect back on itself. If we move away from the center of curvature along the optic axis, the image remains a point and remains on the axis of symmetry. Keep in mind that we are talking about a point and not a finite source.

Part (b) shows the effect of moving the object point off the axis. The primary aberration of spherical mirrors has arisen, astigmatism. (Literally, astigmatism means not spotlike, but its use is reserved to only one class of aberration.) Let's pick a spherical mirror of radius 2000 cm and square shape 5 cm on a side with incidence angle = 10°. If we trace the path of four rays from the object point to the points 1 through 4, and see where they reflect, we find not a point focus but the following. The rays first come together in a vertical line at the horizontal focus; and then, farther from the mirror, they converge into a horizontal line at the vertical focus. Between these two foci, we find what is called

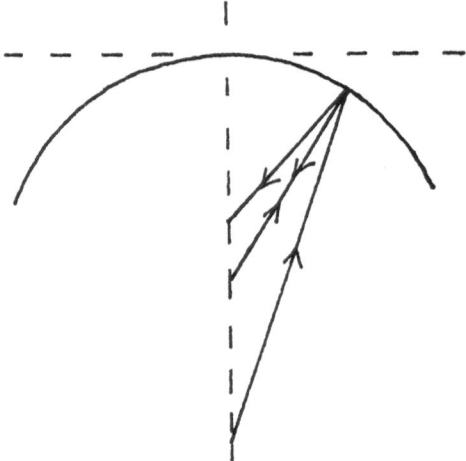

Fig. 2a   Spherical Mirror On Axis

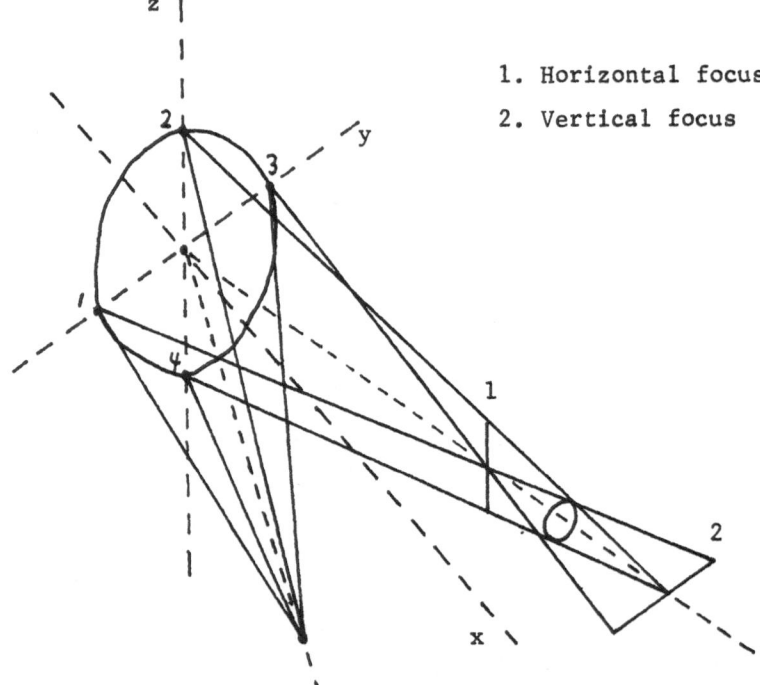

1. Horizontal focus

2. Vertical focus

Fig 2b   Astigmatism of Spherical Mirror

the "circle of least confusion". For our case, the line of horizontal focus will be 2.97 mm in height, the line of vertical focus
3.11 mm in width, and the diameter of the "spot" in between about
1.50 mm. As the incidence angle grows, the situation gets worse.
For our case, holding the object distance of 2000 cm and increasing
the incidence angle moves the horizontal and vertical foci farther
apart. Indeed, at 60° incidence angle the vertical focus disappears,
and the rays do not converge vertically at all. The table in Fig. 2
gives the horizontal and vertical focal lengths and the height of
the image at the horizontal focus for several incidence angles.
Negative focal lengths indicate that the rays diverge and would
cross if their direction were extended behind the mirror. Realize
that positive vertical foci could have been retained by moving the
object point closer to the mirror.

These focal distances have been obtained by a geometrical ray
trace calculation which will be described below. They can also be
calculated by the following relations.

$$\frac{1}{O} + \frac{1}{i_H} = \frac{2}{R \cos\phi} \qquad \frac{1}{O} + \frac{1}{i_V} = \frac{2 \cos\phi}{R}$$

ϕ = incidence angle O = object distance

i_H = distance to horizontal focus i_V = distance to vertical
                                                                                    focus.

This dismal picture of high angle focusing is addressed in the
following manner. Let us rewrite the above equations in the form
of the on axis equation where $i_H = i_V$, and $f \equiv$ focal length.

$$\frac{1}{O} + \frac{1}{i} = \frac{1}{f}$$

We can then identify

$$\frac{R \cos\phi}{2} = f_H \quad \text{and} \quad \frac{R}{2 \cos\phi} = f_V$$

as the horizontal and vertical foci. If we now allow ourselves to
think of two different mirror radii in the two planes, we can solve
these for $f_H = f_V$ —— or point focus.

$$\frac{R_H \cos\phi}{2} = \frac{R_V}{2 \cos\phi}$$

$$\Longrightarrow \cos^2\phi = \frac{R_V}{R_H}$$

This mirror is the toroidal mirror. Geometrically, the surface may be formed by rotating a circular arc of radius R_V about a second point which is farther away. See Fig. 3.

We can now solve for the R_V which will cause the horizontal and vertical foci to coincide. This is a tremendous increase in focal properties at $10°$ incidence, from about $1500\ \mu$ to $1\ \mu$. As we approach a demagnification of 1 to 2 at around $45°$ incidence, however, the size of the image spot goes up to around $30\ \mu$. This behavior supports the rule of thumb for toroidal mirrors which says that they should be used at no higher magnifications than 2 to 1. If we move the object closer to the mirror at the higher incidence angles, and hence move the magnification back to 1 to 1, the point focus returns. This is shown in the table in Fig. 3. In this case, the image and object distances were set equal: $i = 0$; and obtained from the condition for a horizontal focus.

$$\frac{1}{0} + \frac{1}{i} = \frac{1}{f_H} = \frac{2}{R_H\ \cos\phi} = \frac{2}{0}$$

$$\Longrightarrow 0 = R_H\ \cos\phi = i\ \text{(by definition)}.$$

This we can recognize as the representation of a circle expressed in polar form. This is an important fact for grating monochromators; spherical and toroidal reflection gratings form their spectra along this "Rowland" circle, named after the inventor of the concave reflection grating.

We have seen the limits to which our simple formulae may be pushed in describing focal properties of mirrors at very high angles. A hint for the explanation of the defocusing which the 5 cm x 5 cm toroid gave above $70°$ incidence may be seen by looking at the last entry in the table in Fig. 3. The size of the toroid was decreased in this case in the direction perpendicular to the plane of incidence. The defocusing drops from $120\ \mu$ to $6\ \mu$. Remember the mirror's minor radius is now very short, ~ 15.2 cm; and the 5 cm x 5 cm mirror subtends a large angle at this distance. In shortening the mirror to 5 x 1/cm, we have eliminated that portion of the toroid which deviates most from the ideal mirror shape which will focus our point object to an exact point. This shape is the ellipsoid. It has not been greatly used because of difficulty in manufacture. If you will recall the definition of the ellipse and ellipsoid, you will see that it can provide point to point focusing at magnifications other than 1 to 1. The sphere and the toroid perform well in those regions where they are a good approximation to it. At unity magnification and $0°$ incidence angle, the sphere gives an exact focus because it is a degenerate ellipsoid. Near unity magnification, the toroid provides a simple, effective substitute even at relatively

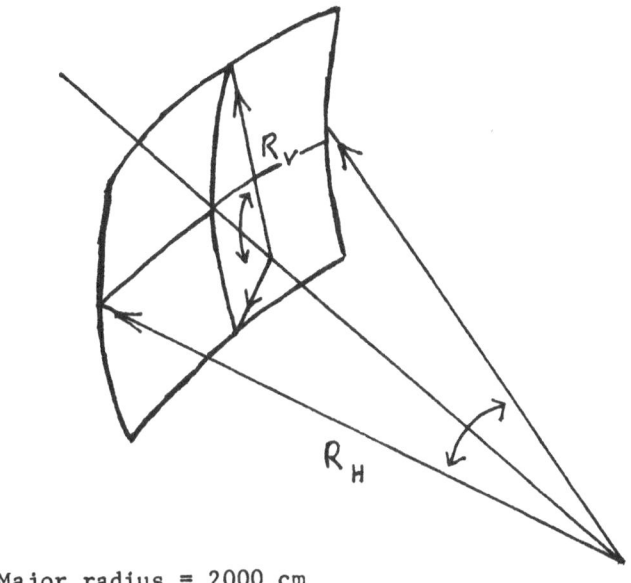

Major radius = 2000 cm
All distances in cm unless noted
* indicates 5 cm x 1 cm mirror, all other cases 5 cm x 5 cm

incidence angle	height of hor. focus in spherical case	distance to horizontal focus in spherical case	distance to vertical focus in spherical case	vertical radius for toroidal case	toroidal case spot dimension	magnification	toroidal case spot dimension m = 1	m = 1 image and object distance
10	0.149	1940	2062.7	1939.7	1 μ	0.97	<1 μ	1970
20	0.552	1772	2274.3	1766.0	6	0.89	<1	1879
30	1.103	1527	2732.1	1500.0	8	0.76	<1	1732
40	1.675	1242	3758.8	1173.6	22	0.62	<1	1532
50	2.163	947.2	7003.4	826.4	35	0.47	<1	1286
60	2.502	666.7	∞	500.0	71	0.33	1	1000
70	2.665	412.6	-6329.9	234.0	150	0.21	2	684
80	2.658	190.2	-3064.2	60.31			20	347.3
85	2.597	91.13	-2422.2	15.19			120	174.3
*85				15.19			6	174.3

Fig. 3   The Toroidal Surface

high incidence angles.

Ray Tracing Mirrors and Gratings

Up to now, we have been investigating the limits of optical formulae using an exact ray trace. We now will describe it in some detail. Let us choose cartesian coordinates with the x axis perpendicular to the center of the mirror and the y axis to be in the horizontal or principle plane. The angle of incidence, α, will be chosen positive implying the angle of reflection, β, to be negative. Directions in space will be specified by direction numbers which are defined by $c\ell$, cm, and cn; where c is any constant and ℓ, m, and n are the cosines of the angles between the direction and the axes. For example, the x direction has direction numbers 1,0,0, or 2,0,0, or 3,0,0, etc.

Our object point will be given as (X_a,Y_a,Z_a) and mirror point will be given as (X_p,Y_p,Z_p). The direction cosine set for the incoming ray is:

$$\frac{X_a-X_p}{D}, \quad \frac{Y_a-Y_p}{D} \quad \frac{Z_a-Z_p}{D} \qquad D = \left[(X_a-X_p)^2 + (Y_a-Y_p)^2 + (Z_a-Z_p)^2\right]^{\frac{1}{2}}.$$

A direction number set is then (X_a-X_p), (Y_a-Y_p), (Z_a-Z_p). (X_a,Y_a,Z_a) is given by the problem at hand, as is α. Y_p and Z_p are known by the assumed size of the mirror. In the case of the sphere, one easily solvable quadratic equation gives X as a function of Y and Z.

$$(X-R)^2 + Y^2 + Z^2 = R^2$$

In the case of the toroid, however, the defining equation is not readily solvable for X.

$$0 = -X^2 -Y^2 -Z^2 + 2RX - 2R(R-\rho) + 2(R-\rho)\left[(R-X)^2+Y^2\right]^{\frac{1}{2}}.$$

One may find X numerically to the degree of precision of a digital computer by Newton's method. This method iterates rapidly to an approximation for X_p good to eight decimal places, and uses the following formula:

$$X_{n+1} = X_n - \frac{f(X_n)}{f'(X_n)}$$

where $f(X) = 0$ is the defining equation for the toroid and

$$f'(X) = \frac{\delta F}{\delta X} = 2(R-X)\left[1 - \frac{R-\rho}{\left((R-X)^2+Y^2\right)^{\frac{1}{2}}}\right].$$

Zero is used as the initial X_n.

We need now the normal to the surface at (X_p, Y_p, Z_p) in order to apply the law of reflection. The direction numbers of the normal to the optical surface are readily available from our defining functions:[6]

sphere $F_s(X,Y,Z) = 0$ $\ell_n = \dfrac{\delta F}{\delta X} = -(X_p - R)$

$m_n = \dfrac{\delta F}{\delta Y} = -Y_p$ $n_n = \dfrac{\delta F}{\delta Z} = -Z_p$

toroid $F_T(X,Y,Z) = 0$ $\ell_N = \dfrac{\delta F}{\delta X} =$ see above

$m_N = \dfrac{\delta F}{\delta Y} = 2Y_p\left[\dfrac{R-\rho}{\left((R-X_p)^2+Y_p^2\right)^{\frac{1}{2}}} - 1\right]$

$N_N = \dfrac{\delta F}{\delta Z} = -2Z_p$

We have enough information to find the direction of the outgoing reflected ray from point P. Let α' be the angle between in incoming ray and the normal to the mirror at P, and let β' be the angle between the outgoing ray and the normal.

We may write down from solid geometry,

$$\cos\alpha' = \frac{\ell_{ap}\,\ell_n + m_{ap}\,m_n + n_{ap}\,n_n}{D_{ap} \cdot D_n}$$

where $D_x \equiv \left[\ell_x^2 + m_x^2 + n_x^2\right]^{\frac{1}{2}}$ and ℓ, m, and n are direction numbers,

and $\cos\beta = \dfrac{\ell_{pf}\,\ell_n + m_{pf}\,m_n + n_{pf}\,n_n}{D_{pf} \cdot D_n}$ (1)

where pf denotes the outgoing ray. By the law of reflection these three lines specified by the two rays and the normal all lie in a plane and $|a'| = |\beta'|$. We can therefore write

$$\cos\left[|\alpha'| + |\beta'|\right] = \cos\left|2\beta'\right|$$

$$= \frac{\ell_{ap}\,\ell_{pf} + m_{ap}\,m_{pf} + n_{ap}\,n_{pf}}{D_{ap} \cdot D_{pf}} \quad . \tag{2}$$

Equations (1) and (2) provide us with two equations in the three
unknown direction numbers ℓ_{pf}, m_{pf} and n_{pf}. We only need these two
because the three direction numbers are not all independent. The
cosine of the sum of two angles is expanded by the identity.

$$\cos (\alpha + \beta) = \cos\alpha \cos\beta - \sin\alpha \sin\beta$$

This can now be repeated for a grid of points on the mirror surface.
In addition, since the object is usually not a point but a circular
or rectangular aperture through which light is passing, a grid of
object points must also be calculated. In general, then, we have
k x ℓ sets of direction numbers where k and ℓ are the number of
mirror and entrance aperture (slit) points, respectively.

In order to display this information in a usable form, we now
take the outgoing principal ray from the center of the mirror, and
choose the point along it which is of interest. This may be anywhere we wish, but is generally the point of horizontal or vertical
focus. For example, we may find (X_f, Y_f, Z_f) the coordinates of this
point from the reflection law and the condition for horizontal focus.
The focal plane is then the plane perpendicular to the outgoing
central ray at this point.

$$X_f(X-X_f) + Y_f(Y-Y_f) = 0$$

A spot diagram may be produced by finding the points of intersection of all of the lines defined by the k x ℓ direction number
sets. Remembering that P denotes the mirror points, any point on
any of the outgoing rays may be described by a single parameter λ
where the coordinates are:

$$(X_p + \lambda\ell, \; Y_p + \lambda m, \; Z_p + \lambda n)$$

If we put these values for X and Y into the equation for the focal
plane, we may solve for λ and find the coordinates of intersection.

$$\lambda = \frac{X_f(X_f-X_p) + Y_f(Y_f-Y_p)}{\ell X_f + m Y_f}$$

We are now one step from plotting our spot diagram. We now must
rotate our original coordinates around the Z axis so that the new
X$'$ axis is along the central outgoing ray. This will allow us to
then plot the values Y$'$ and Z$'$ for our focal spot. X$'$ will now be
the same for all of the intersection points and will be equal to
the distance along the central ray which we chose to observe the
focal properties. The rotation is accomplished with the following
matrix multiplication.

$$\begin{pmatrix} X' \\ Y' \\ Z' \end{pmatrix} = \begin{pmatrix} \cos\phi & \sin\phi & 0 \\ -\sin\phi & \cos\phi & 0 \\ 0 & 0 & 1 \end{pmatrix} \begin{pmatrix} X \\ Y \\ Z \end{pmatrix}$$

where ϕ is the angle of reflection (taken positive).

We have now covered the simpler case of ray tracing a mirror. For the case of the grating we will present the concepts that are necessary, but not treat them in such analytical detail. Let's look at the ruled grating case where grooves are equidistant along the chord of the circular arc. This is more precisely stated that the grooves are located by the intersection of a set of equally spaced parallel planes and the spherical or toroidal surface. The places are spaced a distance D_0 apart and have equations:

$$Y = \pm nD_0 \qquad n = 0,1,2 \dots\dots$$

The law of reflection is replaced by the grating equation which allows $\alpha \neq \beta$;

$$\pm \frac{n\lambda}{D_0} = \sin\alpha + \sin\beta \qquad \text{(a sign convention must be adhered to for } \alpha, \beta)\qquad.$$

The condition for horizontal focus (if we use it) must be generalized to:

$$\frac{\cos^2\alpha}{r} - \frac{\cos\alpha}{R} + \frac{\cos^2\beta}{r'} - \frac{\cos\beta}{R} = 0 \qquad,$$

where r and r' are the entrance and exit slit distances. This reduces to the previous for when $\alpha = \beta$.

The application of the grating equation becomes complicated at all points, except the center of the grating. Note that the grating grooves are <u>not</u> equally spaced on the mirror surface. This requires that D_0 be adjusted by the following equation, producing an effective grating constant as a function of position on the grating.[7]

$$D = D_0 \bigg/ \left[1 - (Y_p/R)^2 \right]^{\frac{1}{2}}$$

This expression may be confirmed somewhat by visualizing what happens at $Y_p = R$ (a hemisphere for the spherical case). In addition, one must be quite careful about defining the α and β of the grating equation. The grating equation applies to cases where α and β are measured in a plane perpendicular to the grooves and containing the surface normal. This is, in general, not the same plane as the plane of reflection. To ray trace properly, the grating equation becomes[7]

$$\pm \frac{n\lambda}{D} = (\sin\alpha' + \sin\beta')\cos\delta$$

where primes indicate that α' and β' are not necessarily the incidence angle and diffraction angle, either at the mirror center or at the point in question. δ is the angle between the plane of reflection and the plane defined above in which α' and β'' are measured.

Application of the Ray Trace Program to a
Hypothetical Toroidal Grating Monochromator

The usual method of concave monochromator design follows the application of Fermat's Principle and the diffraction condition to the optical path function. The optical path is merely the geometrical path in this case because in a vacuum the index of refraction is unity. If F is the optical path from object point to grating point to the intersection with the gaussian image plane, the requirement that F be stationary with respect to displacements along the grooves requires (1), and the diffraction condition requires (2).[*]

$$(1) \quad \frac{\partial F}{\partial Z} = 0 \qquad (2) \quad \frac{\partial F}{\partial Y} = \frac{m\lambda}{d} \quad m = 0,1,2 \ldots$$

F is expanded in a power series, and the different terms are associated with the various aberrations in the image. My colleague, Dr. Malcolm Howells of the National Synchrotron Light Source Staff, has provided me with the parameters of a hypothetical monochromator designed in this fashion.

Figure 4 shows the toroidal grazing incidence monochromator design. It is intended for the 5Å to 50Å region of the spectrum. The entrance and exit directions are fixed. The incidence and diffraction angles are varied by rotation of the grating around the Z axis. The $\Delta\lambda$'s were computed by adding the image blur from the various aberration terms in the following manner:

$$\Delta\lambda = \Delta\lambda_1 + \Delta\lambda_2 \ldots \Delta\lambda_n \qquad ,$$

where n is the number of terms considered to be significant. In this case up to and including terms of 4th power in Y and Z. $\alpha + \beta = 174°$ was chosen to achieve high reflectivity at these short wavelengths. The toroidal shape and short height of the rulings conform to our previously covered ideas about grazing incidence focusing.

[*]This may equal zero, also depending on the definition of F.

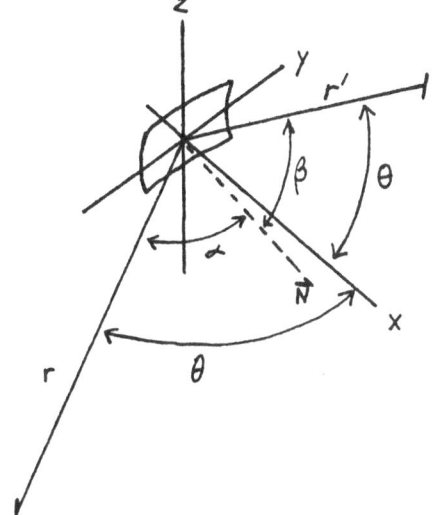

R = 20m
rho = 0.002716R
r   = 0.05187R
r' = 0.04683R
2θ = 174°
grating height 1cm
grating width 5cm
300 grooves/mm

Fig. 4   Toroidal Grating Monochromator

| | | I | II | |
λ (Å)	Δλ from optical path theory Å	width of exit spot from ray tr. μ	dispersion in Å/μ	I x II Å
5	0.158	81.3	0.00180	0.147
10	0.154	80.8	0.00175	0.142
20	0.188	111.5	0.00165	0.185
30	0.233	141.5	0.00155	0.220
40	0.356	231.2	0.00145	0.336
50	0.531	376.5	0.00135	0.509

The values in the table are for a point object.

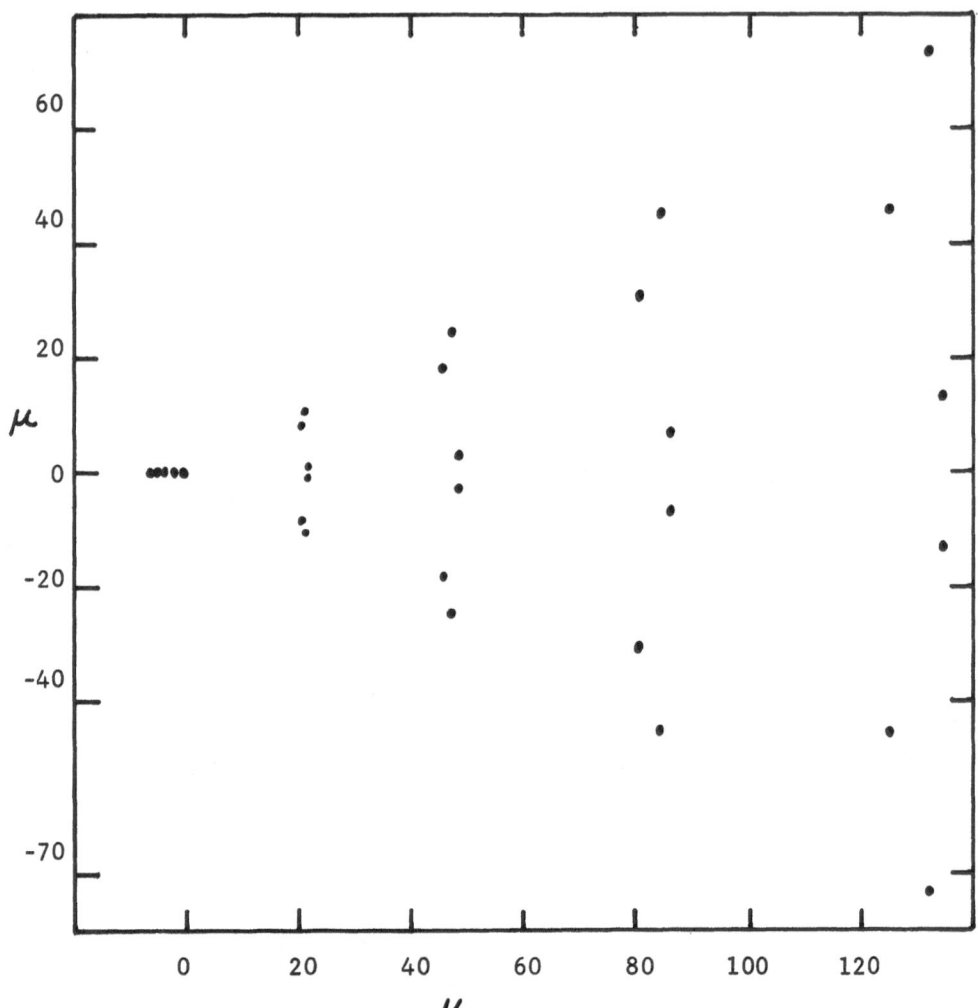

Fig. 5 <u>Focal Plane Ray Trace Diagram for Single Object Point</u>

λ = 30 Å

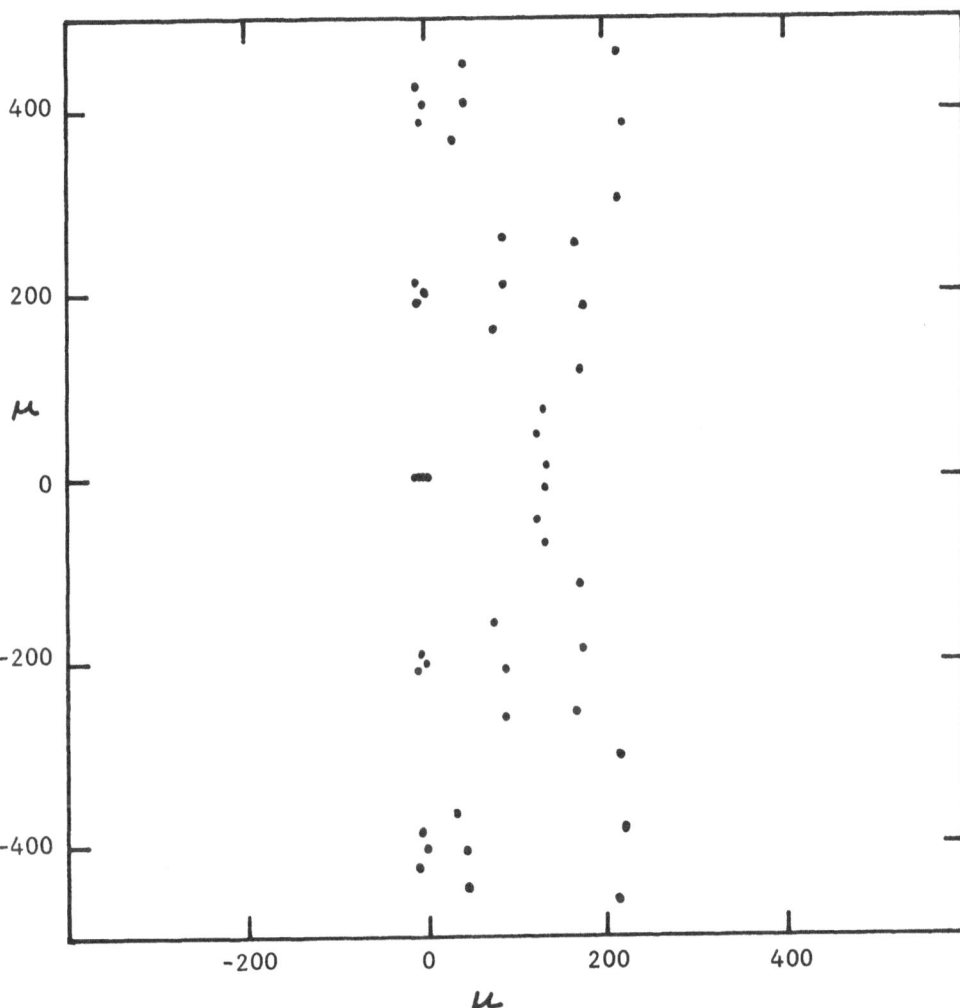

Fig. 6 Focal Plane Ray Trace Diagram for a Grid of Object Points

$\lambda = 30 \overset{o}{A}$

points at z = 0, ±250, ±500 μ

In order to compare the $\Delta\lambda$'s from the aberration theory with the ray trace diagrams, we must calculate the dispersion of the monochromator.

For the central ray, the simpler grating equation holds:

$$\frac{m\lambda}{d} = \sin\alpha + \sin\beta \quad .$$

Holding α constant and differentiating with respect to β, we obtain:

$$\frac{d\lambda}{d\beta} = \frac{d\cos\beta}{n} \quad .$$

Now, in the usual physicist's manner of playing fast and loose with the calculus, we write where ℓ is the horizontal coordinate in the focal (exit slit) plane,

$$\frac{d\lambda}{d\ell} = \frac{d\lambda}{d\beta}\frac{d\beta}{d\ell} \text{ but } r'd\beta = d\ell \Rightarrow \frac{d\lambda}{d\ell} = \frac{d\lambda}{r'd\beta}$$

$$\frac{d\lambda}{d\ell} = \frac{d\cos\beta}{r'n} \quad .$$

These values are tabulated in the figure and allow us to convert between $\Delta\lambda$ and distance in the focal plane.

Figure 5 shows the results of the ray trace for the 30Å case. The points are the intersections with the focal plane of the rays from a grid of points spaced around the edge and inside of the grating surface. As a first approximation we can take the total distance between extreme rays of 141.5 μ as the defocus. Multiplying this by 0.00155Å/μ, reciprocal linear dispersion gives 0.220Å of blur. At first glance, this compares well with the 0.233Å from the optical path theory. A better ray trace would be obtained by using a denser grid of grating points, and considering the density of dots in the image spot diagram. Most of the energy is likely in a smaller width than 141.5 μ. Figure 6 demonstrates the power of the ray trace method. Four object points were added at Z = ± 250 μ, ± 500 μ to evaluate the effects of a finite entrance image size. It is seen that the spot distribution changes as the object point moves out of the principal plane.

We plan to generalize our ray trace program in the future to include deviations from the ideal figure of the mirror surface. We also plan to include other cases, such as holographically recorded gratings, to supplement the straightforward approximation used here of treating the grating as a set of small plane gratings located at each mirror point.

Acknowledgements

I am indebted to Mrs. Joan Weisenbloom for assistance in programming and analytic geometry, and to Barbara Gaer for careful typing of this manuscript. Dr. Malcolm Howells kindly loaned his results for the aberration theory and ray trace comparison.

References

1. Shirley, D. A., "Beam Line Chemistry", Workshop on X-ray Instrumentation for Synchrotron Radiation Research, SSRL Report No. 78/04, p. VII - 80, 1978.

2. Laurenson, L., "Perfluoropolyethers as Vacuum Pump Fluids", Research/Development, p. 61, November, 1977.

3. Rehn, V. and Jones, V. O., "VUV and Soft X-ray Mirrors for Synchrotron Radiation", SSRL Report No. 77/13, 1977.

4. Giles, J. W., McKinney, W. R., Freer, C. S., and Moos, H. W., "An Image Stabilized Telescope-Ten Channel Ultraviolet Spectrometer for Sounding Rocket Observations", Space Science Instrumentation, Vol. 1, No. 1, p. 51, 1975.

5. Blewett, J. P., ed., "Proposal for a National Synchrotron Light Source", BNL 50595, Vol. 1, 1977.

6. Widder, D. V., Advanced Calculus, 2nd Ed., Prentice Hall, Englewood Cliffs, N. J., p. 107, 1961.

7. Kastner, S. O. and Neupert, W. M., "Image Construction for Concave Gratings at Grazing Incidence by Ray Tracing", JOSA, Vol. 53, No. 10, pp. 1180-1184, 1963.

General References

Techniques of Vacuum Ultraviolet Spectroscopy, James A. R. Sampson, John Wiley & Sons, Inc., New York, 1967.

Vacuum Ultraviolet Spectroscopy, A. N. Zaidel' and E. Ya Schreider, Halsten Press, New York, 1970.

VUV MONOCHROMATORS AT SYNCHROTRON RADIATION SOURCES

Volker Saile

Sektion Physik, Universität München
D-8000 München 40, Germany

1. INTRODUCTION

Synchrotron Radiation (SR) with its intense continuum from the infrared up to photon energies of some ten keV is an extremely versatile tool for various kinds of spectroscopy, structure analysis and a number of other applications in physics, chemistry, biology and technology[1,2]. Interest in synchrotron radiation sources has led to a rapid increase in the number of radiation laboratories and to a worldwide demand for storage rings dedicated exclusively to the generation of synchrotron radiation[3,4].

In spite of all unquestionable advantages of SR, access to it is not a trivial task. Experimentalists intending to work at a synchrotron radiation source have to consider some inconveniences and unusual factors which range from the immobility of the source to special and often expensive experimental equipment, e.g. for obtaining monochromatized light. In the past this might have been a barrier for many scientists to use this source – mostly those who are not specialists in monochromator design and operation and who refrain from handling rather heavy and complicated set ups. It is the aim of this paper to introduce into the principles and to describe examples for monochromator instrumentation at SR sources for the spectral range extending from 2000 Å to 20 Å (photon energies between 6 eV and 1000 eV). This is quite a natural short wavelength limit: For higher photon energies the instrumentation is quite different changing from grating monochromators to crystal monochromators and from ultra high vacuum conditions to operation with Be-windows and set ups under He atmosphere or even air. While in some cases the vacuum ultraviolet region was defined[5] for photon energies up to 6000 eV the limit lies at about 1000 eV from an experimental point of view. Above this energy typical X-ray experimental equipment is used.

In the VUV the dominant scientific interest is to investigate the _electronic_ properties[1,2] of atoms, molecules and solids. Experimental methods are absorption and reflection spectroscopy and the study of secondary processes like photoelectron spectroscopy, luminescence and photofragmentation. More recently great efforts have been made to exploit short wavelengths in the order of some ten Å for microscopy[6] and lithography[7]. Calibration of light sources and detectors[8] are examples for applied research. Others are the determination of optical constants[9] and development and tests of high efficiency mirrors and gratings[10].

In a number of articles SR instrumentation is discussed. The basic ideas can be found in Ref. 5 while more special articles appeared recently[1,2,11,12]. The aim of the present review is to give an introductory survey for the existing monochromators and to outline some new developments.

2. SOURCE CHARACTERISTICS AND ITS IMPLICATIONS

Synchrotron Radiation is emitted by electrons (or positrons) radially accelerated in a synchrotron or storage ring. The highly relativistic velocities of the particles result in a characteristic **radiation pattern sketched in Fig. 1.** At each point of the orbit light is emitted in a very narrow cone. While horizontally the angular extension is 2π as a result of a superposition of these cones, the angular spread is vertically (ψ in Fig. 1) confined to about 1 mrad[13].

The often noted outstanding features of SR are
- a continous spectrum from the infrared to x-rays (see Fig. 2)
- linear polarization of the light in the orbit plane; outside elliptical or circular polarization (see Fig. 3)
- time structure of the radiation with pulses as short as 100 psec (see Fig. 4)
- the already mentioned azimuthal collimation of about 1 mrad
- the small size of the source with several 0.1 mm^2 to several mm^2
- stable operation with respect to beam position and intensity for many hours
- the clean environment - SR is emitted under high or ultrahigh vacuum conditions
- the exact calculability of its properties (spectral and angular distribution as well as polarization) (as examples see Figs. 2,3).

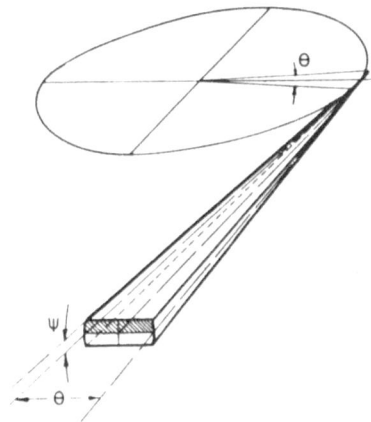

<u>Fig. 1:</u> Geometry at a SR-source (from Ref. 13)

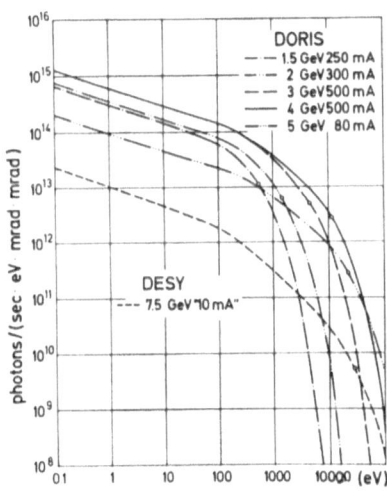

<u>Fig. 2:</u>

Spectral distribution
of SR into a solid angle
for the DORIS storage
ring at different
operating conditions
and for the DESY syn-
chrotron (from Ref. 14)

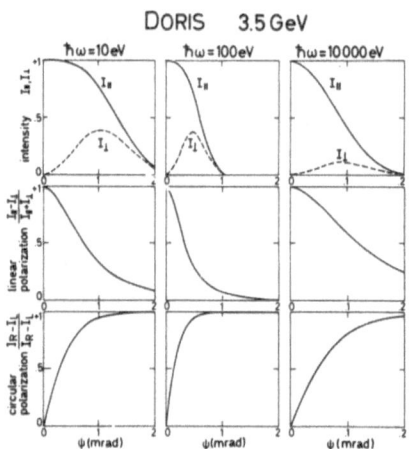

<u>Fig. 3:</u> Angular distribution of intensity components with
E-vector parallel ($I_{||}$) and normal ($I_\perp$) to the plane
of the DORIS storage ring. (I_L) and (I_R) denote left
and right hand circularly polarized components (from
Ref. 15). ψ is defined in Fig. 1

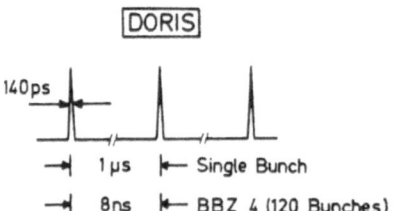

<u>Fig. 4:</u> Time structure at the DORIS storage ring for two
operating conditions with one electron bunch and
with 120 bunches in the ring

For photon energies $\hbar\omega \ll \varepsilon_c$, the characteristic energy
of a SR source (these energies are marked in Fig. 2 by circles), the
intensity in a slit with infinite extension vertical to the orbit
plane (this is a good approximation because of the collimation of
SR) is given[16] by:

$$N(\text{photons}/(\text{sec eV})) \approx 4.5 \times 10^{12} \times \Theta(\text{mrad}) \times R^{1/3}(\text{m})$$

$$\times j(\text{mA})/\varepsilon^{2/3}(\text{eV})$$

where Θ is the angle accepted by the slit (see Fig. 1), R the radius
of the electron orbit, j the beam current and ε the photon energy.

SR sources can be roughly divided into two groups

- low energy machines (E $\lesssim$ 2 GeV) and
- large machines (E > 2 GeV)

as illustrated in Table 1 [17]:

Table 1:   Some characteristics of small and large
           storage rings

type of machine	small	large
magnet radius	< 5 m	> 5 m
particle energy	< 2 GeV	> 2 GeV
distance experiment - source	< 10 m	> 10 m
charact. energy ε_c	< 2000 eV	> 2000 eV
advantages	no radiation hazard and damage, much cheaper operation	x-rays available, better time structure

In addition to the "good" characteristics of SR one should keep in
mind several restrictions and disadvantages if compared to conventional laboratory sources. Some of them are summarized in Table 2.

Table 2: Properties of SR, some applications and constraints

Properties	Useful applications	Constraints and Difficulties
collimation	simple monochromator design	
intense continuum	absorption, reflection, secondary effects	monochromators necessary, problems with higher orders, radiation damage by x-rays
polarization	optical anisotropy, ellipsometry, photoemission	optical components should only be used in s-polarizing geometry
time structure	time resolved fluorescence, coincidence experiments, time of flight analysers	simple counting techniques are limited to rates below the repetition frequency of the light pulses
calculability	radiation standards	
clean source	surface physics	at storage rings ultra-high vacuum requirements for beam lines and monochromators
big machines		immobility of the source, large SR sources are often optimized for high energy physics and not for SR, beam times dictate sometimes the progress of the experiments instead of physical arguments, expensive light sources

3. MATCHING THE EXPERIMENT TO THE SOURCE

Let us assume an experiment at a SR source aimed at the investigation of processes in atoms, molecules or solids by means of VUV photons with variable wavelength λ and spectral width $\Delta\lambda$. Such a typical experiment can be sketched by the block diagram in Fig. 5.

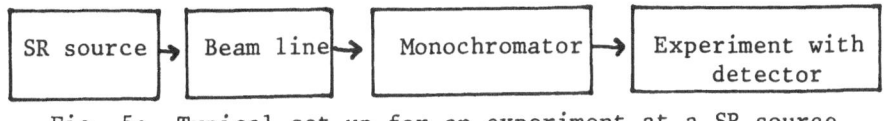

Typical set up for an experiment at a SR source

Unlike in conventional spectroscopy at SR facilities the source, beam line with optical components, **monochromator** and experiment have to be treated as a whole. What are the constraints for such a set up?

The basic requirement is to maximize the photon flux $N_E(\lambda)$ at the sample within the bandwidth $\Delta\lambda$. For many experiments, e.g. photoelectron spectroscopy not only the number of photons per second but the illumination of the sample $N_E(\lambda)$ per target area is the relevant quantity. In that case the sample has to be very close to the exit slit, or the monochromatic light has to be focussed by an appropriate mirror.

A word of caution regarding detectors should be mentioned here: The particles detected (photons, electrons, ions) arrive at the detector with the time structure of the SR source (see Sec. 2). The short pulses are beyond the time resolution of most detectors available. That implies that counting techniques are limited to count rates lower than the repetition frequency of the SR pulses. Otherwise it is possible that one pulse contains more than one particle to be detected. In most cases this is no serious restriction since the repetition frequencies are so high (e.g. for DORIS 1 MHz) that analog measurements can be performed easily; but it can become crucial in coincidence   experiments where only a few pulses of those available are detected. For such experiments charge sensitive detectors and adequate electronics have to be used.

Considering the geometry of the set up[18] we may characterize the SR source by the horizontal and vertical dimensions x_o and y_o of the beam, the photon flux $N_o(\hbar\omega)$ for a photon energy $\hbar\omega$ with a bandwidth $\Delta\hbar\omega$, and the angular spread ψ in azimuthal direction which is depended on photon energy but a very small quantity in the order of 1 mrad for VUV-energies. Therefore we can integrate over the y-direction as explained in Section 2 and get $N(\hbar\omega)$ in photons/ (sec·eV·mrad). Radiation from the source is accepted by the beamline which may contain optical elements like mirrors with an aperture Θ (see Fig. 1). The beamline transfers the emittances of the source into those of the beamline at the entrance slit of the monochromator: $\Theta \cdot x_o \rightarrow \Theta_1 \cdot x_1$; $\psi y_o \rightarrow \psi_1 \cdot y_1$ with image dimensions x_1, y_1 and a transmission coefficient $T_B \leq 1$. As for any optical **device**, the brightness (measured in photons/(sec·mrad·mrad·area) of the image cannot exceed that of the source. A monochromator with slits of dimensions x_M, y_M, apertures Θ_M, ψ_M and a transmission coefficient T_M

disperses the incoming radiation. The photon flux at the exit slit
is given[18] by:

$$N_E(\hbar\omega) = N(\hbar\omega) \cdot T_B \cdot T_M \cdot \Theta \cdot (\Theta_M \cdot x_M / \Theta_1 \cdot x_1) \cdot (\psi_M \cdot y_M / \psi_1 \cdot y_1)$$

with $(\Theta_M \cdot x_M / \Theta_1 \cdot x_1) \leq 1$ and $(\psi_M \cdot y_M / \psi_1 \cdot y_1) \leq 1$. If these two factors exceed 1 they have to be set identically to 1 in that equation.

For a given monochromator the photon flux is mainly determined
by the angle Θ accepted horizontally and beam emittances have to be
carefully matched to monochromator acceptances.

4. MIRRORS AND COATINGS

Due to the lack of transmitting materials, mirrors are essential
components for focussing and deflecting VUV beams. In the design of
VUV monochromators mirrors are incorporated

- as beam splitters which deflect the incoming beam or part of it
  into a given experiment
- as a device to focus the SR, e.g. onto the entrance slit of a
  monochromator or onto the sample

Apart from these obvious applications mirrors offer further advantages:

- The polarization of the light can be enhanced as a result of the
  different reflectivities for the E-vector of the light being
  parallel (R_p) or perpendicular (R_s) to the plane of incidence (see
  Fig. 6). As can be seen s-polarizing reflection is the preferable
  geometry for mirrors and gratings.
- The reflectivity of mirrors in the VUV is strongly dependent on
  photon energy and angle of incidence as shown in Fig. 7. By an
  appropriate choice of the coating material and angle of incidence one is able to suppress short wavelengths in the reflected
  beam. This filter-like behaviour is extremely important at large
  storage rings where hard x-rays would otherwise destroy gratings
  and expensive mirrors with complicated shapes.

Figures 6 and 7 demonstrate that reflectivities of 30 % or more
can only be obtained under grazing incidence and the angle of incidence has to be increased with photon energies. This fact implicates
serious difficulties for the fabrication of mirrors. The grazing angle
result in large reflecting areas, the importance of surface roughness
increases dramatically for shorter wavelengths[21]. Focussing mirrors
under grazing incidence have complicated shapes - in the ideal case
ellipsoids which are approximated by paraboloids, toroids or two
cylindrical mirrors for separate horizontal and vertical imaging.

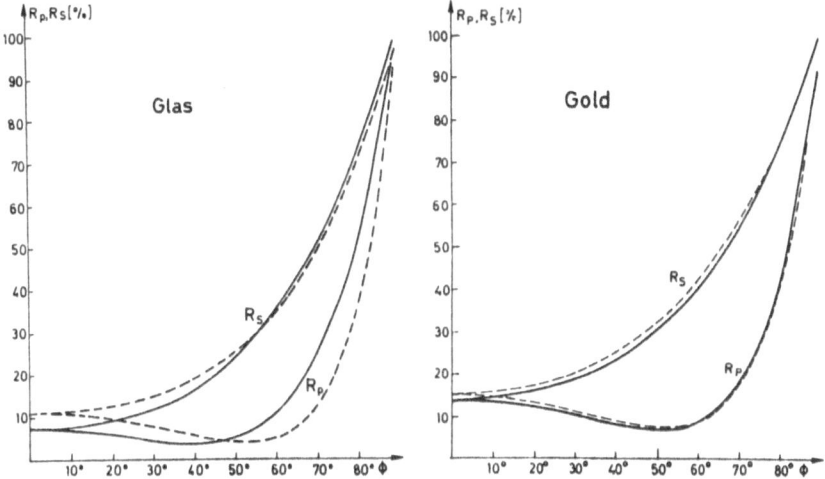

Fig. 6:   Reflectivities of glass and gold vs. angle of incidence Θ. R_s and R_p denote the reflectivities for the
E-vector of the light being perpendicular and parallel
to the plane of incidence. The curves are calculated
for two wavelengths $\lambda = 600$ Å (full line) and $\lambda = 900$ Å
(dashed line). From Ref. 19.

With two mirrors it becomes possible to collect horizontally large
angles and to demagnify the vertical extension of the source simultaneously[22]. With a horizontal entrance slit for the monochromator
(vertical dispersion plane) this geometry can lead to high photon
fluxes.

For large storage rings like SPEAR and DORIS the choice of the
material for the first mirror hit by SR is of great importance. The
collimation and flux of X-rays at these machines results in power
densities of some 100 W/cm² at the mirror in a narrow line along
the surface in the orbital plane of the machine. Heat dissipation,
thermal conductivity and expansion become crucial. Several attempts
have been made to solve this problem. In Stanford massive copper [23]
mirrors are successfully used for some years. At DORIS glass-cera-
mics material (Cerodur) with negligible thermal expansion coefficient
up to 300° C has proven to work up to about 2 GeV. At higher particle
energies these mirrors show irreversible deformations of the surface[24].
For steel mirrors similar distortions have been found but they have
beeen reversible i.e. could be observed only during illumination
with synchrotron radiation[24].

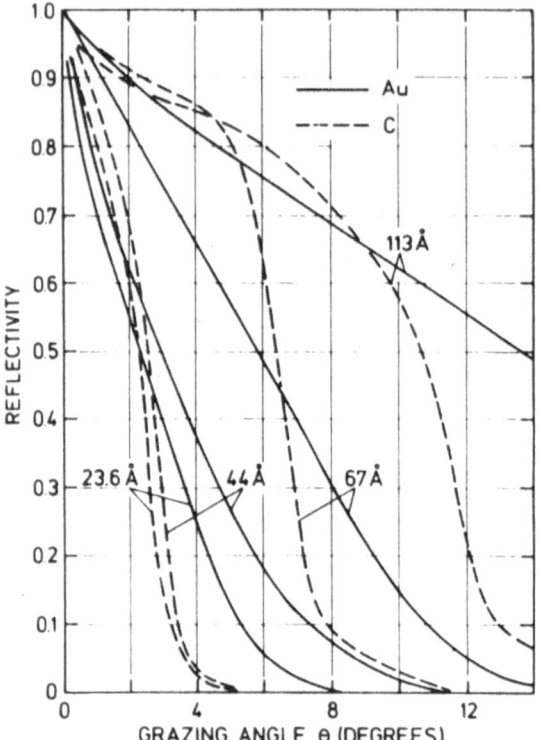

<u>Fig. 7:</u> Reflectivity of Au and C as a function of the grazing
           angle for various wavelengths. (From Ref. 20).

A promising new material – chemical vapor deposited SiC – has
been carefully investigated by Rehn et al.[25]. It fulfills several
conditions: extremely high reflectivity even at normal incidence
(see Fig. 8), good thermal conductivity and perfectly smooth surface with about 7 Å rms roughness.

Coatings are evaporated on mirrors and gratings for increased
reflectivity. In principle Fresnel's formula can be applied with
known optical constants to calculate this reflectivity. In practice
one has to take into account technical problems like increase of
roughness with coating, sticking coefficients etc. Commonly used
coating materials for photon energies below 17 eV are Al coated

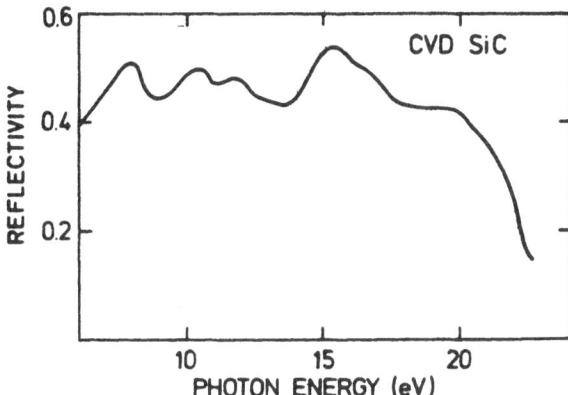

Fig. 8: Measured reflectivity of chemical vapor deposited SiC at normal incidence. (From Ref. 25).

with MgF_2 and noble metals like Au and Pt for higher energies. Recently Haelbïch and Kunz[26] made considerable progress in developing multilayer coatings for the VUV (see Fig. 9). The maximum reflectivities obtained are much higher than those for pure Au or C. Furthermore these multilayer coatings with their narrow maxima can be used as interference filters (see Fig. 9).

5. MONOCHROMATORS

Powerful monochromators and spectrographs for conventional VUV sources have been developed since almost 100 years. A comprehensive description of design principles and realization is available in the literature (see Ref. 27). Basic requirements for monochromators are optimum wavelength resolution and high transmission. Due to the lack of mirrors with sufficient reflectivities in the VUV at small angles of incidence, as few reflections as possible should be applied in monochromator designs. Instruments used at SR facilities can be roughly divided into two categories.

- Normal incidence monochromators with gratings illuminated at near normal incidence. As a consequence the wavelength range is limited to about 300 Å (40 eV).

- grazing incidence instruments with gratings illuminated at grazing angles. The instruments cover the wavelengths between 600 Å (20 eV) and 10 Å (1000 eV).

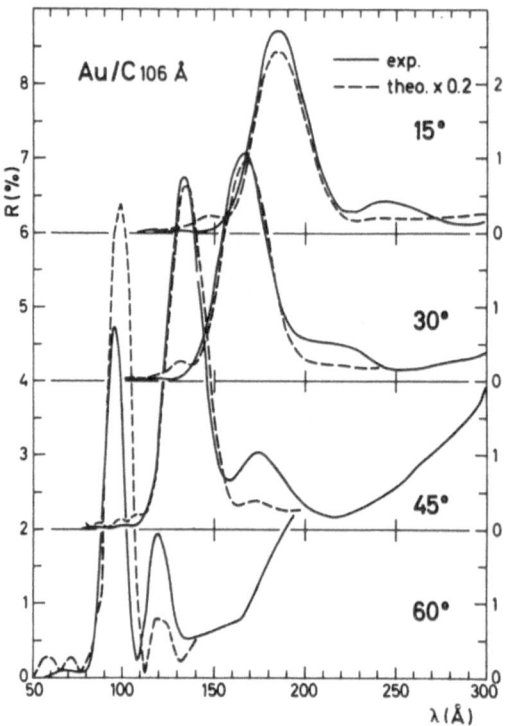

Fig. 9: Reflectivities of Au/C multilayer coatings on a
glass substrate for various angles of incidence
as a function of wavelengths. (From Ref. 26).

     This separation is at present somewhat artificial as monochro-
mators with widely ranging angles of incidence have been constructed
(see Sec.5.3,4).Many instruments are based on the Rowland concept to
combine dispersion and focussing in one optical component — the
spherical grating[28] thus minimizing the number of reflecting surfaces. The optical properties of such a grating with radius R can
be described by a circle tangentially to the surface of the grating.

If the entrance slit is located on that circle, the spectrum extends along the same circle. For real monochromators the spectrum is recorded either by a photographic plate bent along the circle or an exit slit where a detector is moved along the circle for photoelectric detection. Some instruments based on this concept will be discussed in Sections 5.2 and 5.3.

5.1 Gratings

Classical gratings for the VUV are reflection gratings either plane or spherically concave with mechanically ruled, equidistant grooves up to 3600 1/mm[27]. For the dispersion relation one obtains:

$$m\lambda = d \ (\sin\alpha + \sin\beta)$$

for order m of diffraction, wavelength λ, groove separation d, and angles α and β for incoming and diffracted beam. With a sawtooth profile for the grooves (blaze) it is possible to enhance the photon flux for special wavelengths in the various orders. An example for the performance of such gratings is shown in Fig. 10[29].

For practical use the efficiency i.e. the intensity ratio between light of wavelength λ falling onto the grating and diffracted by it, is of great importance. While the total intensity reflected is given by the coating material, the efficiency is determined by the blaze. Recently various experimental results dealing with these efficiencies of quite different types of gratings have been reported[1,2,11].

In the last few years grating fabrication was revolutionized by the application of coherent laser beams. These so called "holographic gratings" are produced by fixing interference fringes in a photoresist on a substrate[31]. Compared to mechanically ruled gratings holographic ones offer the following advantages[32]: They are free from ghosts, low level of stray light, very large diffracting areas and high line densities (6000 1/mm) can be produced, and the blanks for the grating may have complicated shapes, e.g. toroids or ellipsoids. The most remarkable feature of holographic gratings is the possibility to introduce focussing properties by adequate interference fringes in the production process[32]. This can be achieved by a variable spacing of the grooves or curved grooves[33]. With that method it is possible to correct aberrations like astigmatism in conventional monochromators and even to produce gratings which are totally stigmatic for three wavelengths.

For VUV wavelengths two new promising components for dispersion have been introduced recently. The first one is the free standing transmission grating[34] with 1000 1/mm. It has been tested successfully with line sources[35] and at the SPEAR facility[36] and a monochro-

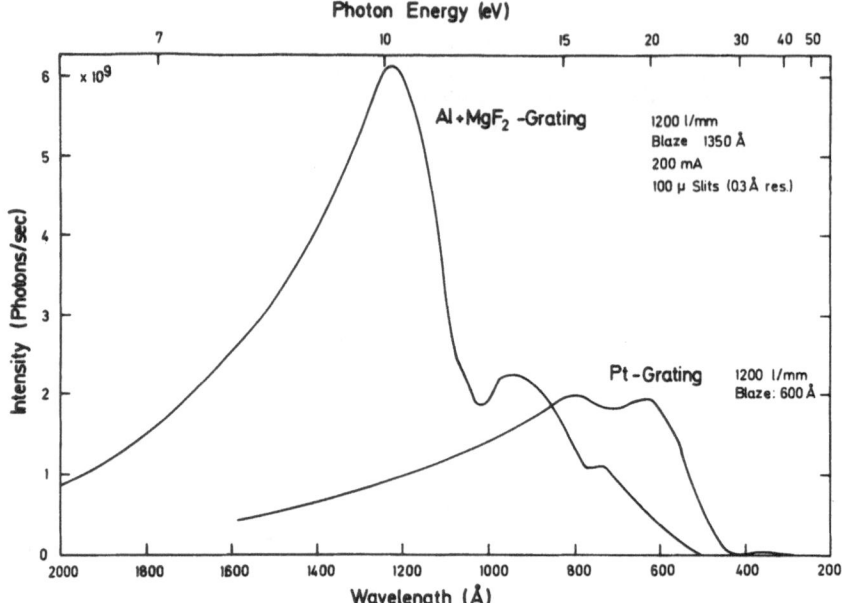

Fig. 10:   Intensity available in first order at the exit slit
of the 3m normal incidence monochromator at DORIS for
an electron beam current of 200 mA (from Ref. 29). The
distribution entering the instrument is totally flat.
The absolute photon flux is measured by a double ionization chamber[30]. It should be noted that the spectra
are convolutions of blaze characteristics with spectral
dependence of reflection for the coating material.

mator based on it is planned for a new beam line[37]. The second is
the Fresnel zone plate[31]. Holographically produced zone plates are
already utilized for monochromators[31] and as lenses for x-rays at
DESY[6].

5.2 Normal Incidence Monochromators

For the normal incidence range, that is the part of the spectrum
extending from 2000 Å to about 300 Å conventional sources though
limited in their usefulness are available. Among these alternative
light sources are: Excimer lasers[38] and methods like generation of

VUV frequencies by nonlinear mixing[38] cover more and more wavelengths between 1000 Å and 2000 Å. For some photon energies up to 40 eV harmonic generation of laser wavelengths has been reported[40]. Rare gas line sources[41] have been used successfully for many years but they cover only a very few wavelengths. Above 800 Å carefully developed lamps producing the He-continuum yield a photon flux comparable to that from a SR source[42]. A detailed comparison of the properties of these sources with SR appears e.g. in Ref. 12.

The availability of these sources has led to the development of a great variety of Normal Incidence Monochromators[5] and most of the designs currently in use at the various SR laboratories are obvious extensions of the existing mountings. Some examples realized at SR facilities are illustrated in Fig. 11. The first three mountings are found at most SR-facilities. Scanning of the wavelengths is performed by a rotation of a concave grating. With fixed entrance and exit slits this results in a serious defocussing which is overcome by different techniques.

For the McPherson type monochromator (Fig. 11a) the scanning mechanism performs a simultaneous rotation and translation of the grating along the bisector of the angle subtended by the slits at the grating center[5]. With such an instrument high resolution given by the focal length and angular dispersion of the grating can be obtained. At the DORIS storage ring a 3m – instrument with vertical dispersion plane provides a bandwidth of 0.03 Å with 10 μm slits and grating with 1200 1/mm. The wavelength range covered extends from 3000 Å to 300 Å (4 eV to 40 eV)[30,44]. In Fig. 10 the spectral distribution of the radiation behind the exit slit is given in absolute values for 100 μm slits (0.3 Å resolution). A toroidal mirror with a reflecting surface of 500 x 60 mm^2 focusses under a deflection angle of 15° the SR onto the entrance slit. A cone of 1 mrad × 1 mrad is accepted in a s-polarizing geometry. An example for the capabilities of this instrument is shown in Fig. 12 where the absorption of Ar gas is recorded photoelectrically up to very high quantum numbers. In the last 3 years a number of experiments in solid state physics[45], molecular physics[46] and atomic physics[44] have been performed with this instrument with great success.

The second mounting shown in Fig. 11b is the Seya-Namioka monochromator[47] with the simplest scanning mechanism of all types, i.e. a rotation of the grating about an axis through the center of the grating: To minimize the deviation from the Rowland circle, the angle between incident and diffracted beam is chosen as 70°30'. The most severe aberration of a Seya-Namioka monochromator is the astigmatism. This drawback can be corrected by appropriate mirrors[48] or specially corrected holographic gratings[49]. For the instrument at the Stanford SR facility the aberrations are reduced by astigmatic source optics and long focal length. A resolution of 0.18 Å is obtained with 20 μm slits and a 1 m grating with 1190 1/mm[22].

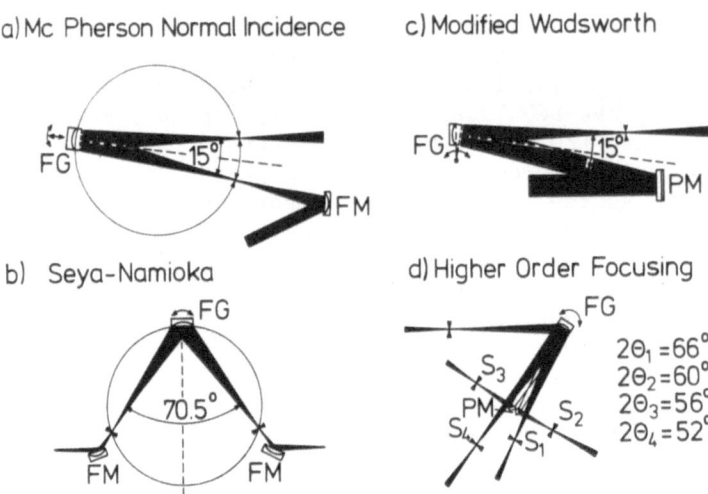

Fig. 11: Some types of normal incidence monochromators (from
Ref. 43).
FM: focussing mirror, FG: spherical concave grating,
PM: plane mirror, S_i: **exit slits. For d) the**
angles $2\Theta_i$ subtended by the fixed slits are given.
For references see text.

The third type of normal incidence monochromator (Fig. 11c) is the
modified Wadsworth mount first realized by Skibowki and Steinmann[50].
This instrument works without an entrance slit thus exploiting the
small divergence of the SR beam. To reduce defocussing at the exit
slit an eccentric pivot for rotation of the grating is chosen. The
finite size of the source which acts as a virtual entrance slit in
combination with defocussing limits the resolution to 1 - 2 Å for
a monochromator with a grating of 2m radius and 1200 1/mm and a
source like DESY/DORIS. In a mounting with a vertical dispersion
plane[51] the resolution is increased slightly. Recently a similar
mount with a holographic grating corrected for astigmatism was

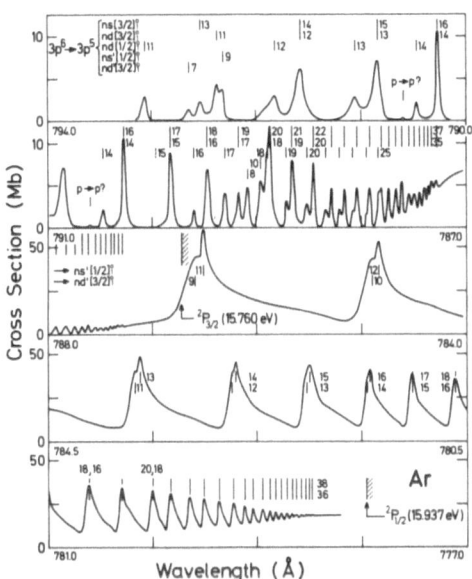

Fig. 12: Absorption spectrum for Ar near the first two ioniza-
          tion limits. Between the ionization limits Fano-Beutler
          profiles are observed. Note that the whole spectrum
          covers only 450 meV.

brought into operation at LURE[52]. The high transmission (only one
slit, no focussing optics besides the grating) and the simplicity
favor this type of instrument for experiments requiring a high photon flux. As an example a set up for time resolved luminescence
measurements[53,54] is shown in Fig. 13.

The radiàtion monochromatized by a modified Wadsworth monochromator is focussed onto a sample. A Seya-Namioka monochromator disperses the light emitted by the sample. The lifetimes of excited
states are determined by monitoring the exponential decay of emitted
intensity after excitation with a light pulse from DORIS (see Fig.
14).

Based on various new theoretical treatments of imaging and
diffraction with gratings[33,56-58] a number of new mountings have
been suggested. One of them with exit slits is shown in Fig. 11d.[59]
The idea of such asymmetrical mounts is to introduce a defect in
focus to balance higher order aberration terms and to maximize the
product of resolution times luminosity.

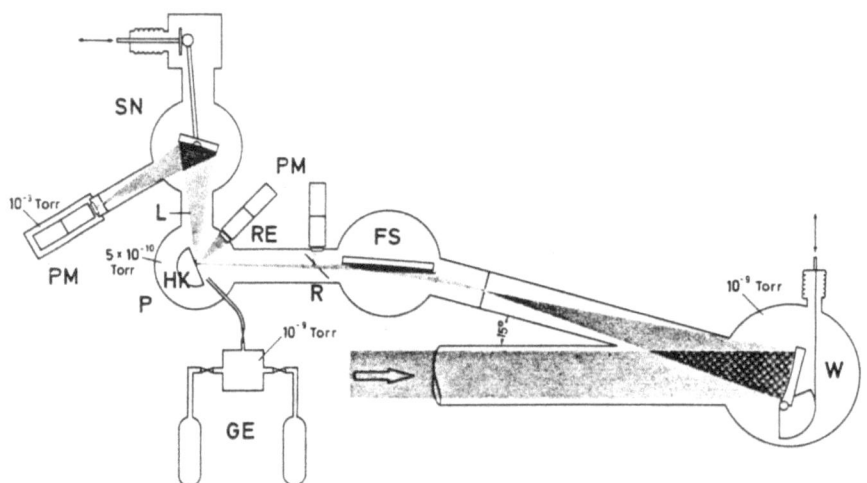

Fig. 13: Experimental set up including a modified Wadsworth
and a Seya-Namioka monochromator for luminescence
measurements (from Ref. 53).
W: modified Wadsworth monochromator, FS: focussing
mirror, P: experimental chamber, HK: Helium flow cryostat,
SN: Seya-Namioka monochromator, PM: photomultiplier,
R: reference signal, RE: reflected light, L: emitted
light, GE: gas handling system.

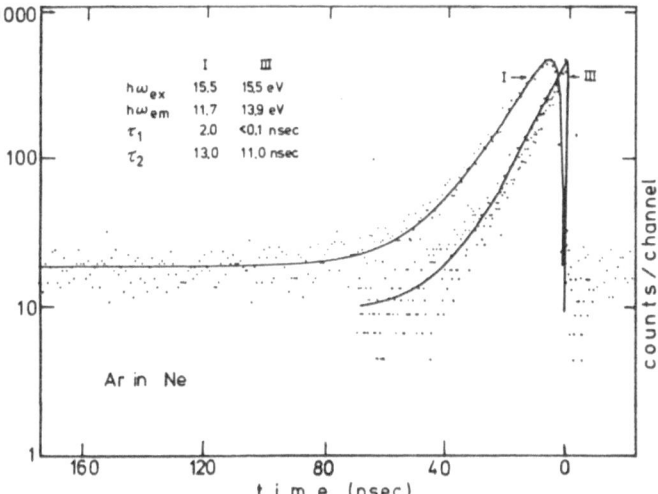

Fig. 14:   Time resolved luminescence of Ar atoms in a Ne matrix
(from Ref. 55).
$\hbar\omega_{ex}$ is the excitation energy, $\hbar\omega_{em}$ the emitted photon
energy. τ_1 are population and τ_2 decay times extracted
from these curves.

5.3 Grazing Incidence Monochromators

The grazing incidence range presents more difficult problems
than the normal incidence region, e.g. superposition of higher
orders and complicated imaging. As there exists no unique solution
which meets all requirements simultaneously (that is high resolution,
high efficiency (aperture), good order sorting capabilities, a fixed
exit slit and simple scanning mechanism,) quite a lot of different
instruments have been designed, each of which aiming at a particular compromise concerning these requirements. A comprehensive survey
on grazing incidence monochromators used at SR facilities and the
problems involved for this wavelength region is given in Ref. 43,
60, 61. Some of the instruments tested and in use at SR facilities
are sketched in Figs. 15 and 16.

The classical concept for a high resolution grazing incidence
monochromator is that developed by Rowland[27] as already discussed
in Sec. 5.1. In the simplest version[63] the SR is focussed by a

Rowland Monochromators

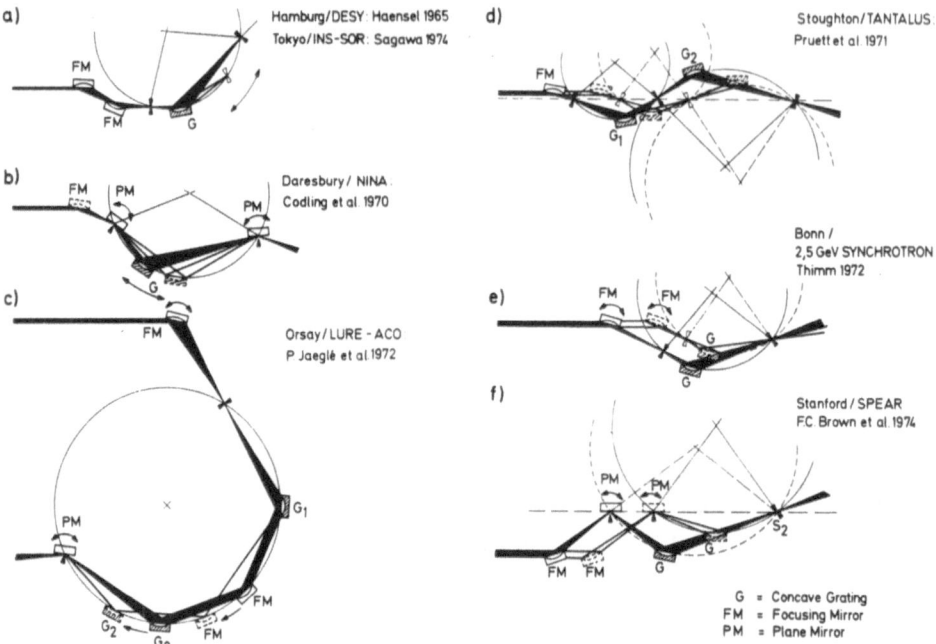

Fig. 15: Some monochromators based on the concept of a Rowland circle (from Ref. 62).

mirror onto the entrance slit. A moving exit slit allows experiments with massive equipment to be located only in front of the monochromator or spectrograph. This restricts the number of possible experiments to transmission measurements. With bent photographic plates for a simultaneous registration of the whole spectrum a Rowland monochroamtor is extremely powerful especially for very low intensities or exploitation of higher orders for increased wavelength resolution.

To overcome the limitations of a travelling exit slit, the design in Fig. 15b[64] introduces rotating mirror-slit combinations. One of the major drawbacks, the contribution of higher orders at longer wavelengths is minimized with double grating monochromators. Examples are the instruments in use at ACO[65] and at the Tantalus I storage ring[66]. Due to the number of reflecting surfaces the transmission of these systems is rather limited. This led Thimm et al.[67] to construct a monochromator (see Fig. 15e) with only two reflecting surfaces by taking into account a small variation of the exit beam direction. The so called "Grasshopper" monochromator[18,68] in use at

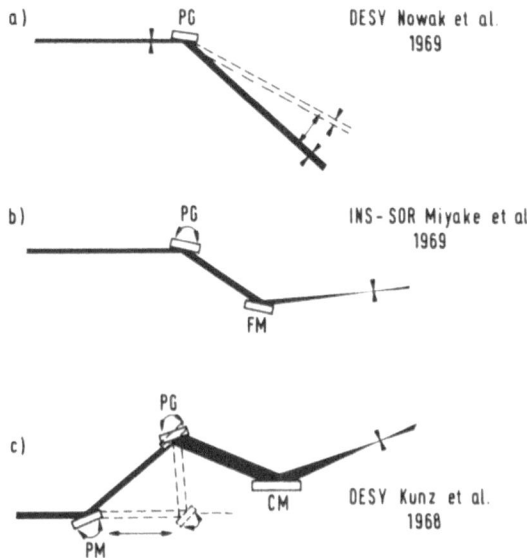

Plane Grating Monochromators

Fig. 16: Plane grating monochromators at SR facilities exploiting the small divergence of the SR beam (from Ref. 60). PG: plane grating, PM: plane mirror, FM: spherical mirror, CM: parabolic mirror

SPEAR is shown in Fig. 15f. A mounting with a fixed exit slit is achieved by rotating a 1m Rowland circle around the exit slit S_2[69,70] Great efforts have been made to connect this instrument in an optimum way to the high power storage ring in Stanford: An ultrasmooth platinum coated copper mirror 6.5 m from the orbit and 9.7 m from the monochromator is illuminated under a grazing angle of 2° and focusses horizontally 2 mrad of SR onto the entrance slit. This mirror not shown in Fig. 15f works as a beam splitter and a filter with a cut-off energy of about 2.5 keV. As a consequence the heat load on the following mirror is minimized. Thus this spherical mirror FM can be a quartz mirror with Pt coating. It moves parallel to the beam under a grazing angle of 2°. The entrance slit - mirror combination is similar to that developed by Codling et al.[64]. With a grazing angle between 3° (Zeroth order) and 8° (400 Å) this mirror

determines the short wavelength limit to about 1 keV and rejects
higher orders at longer wavelengths. With a 2m grating, 600 1/mm
and 15 µm slits a bandwidth of 0.15 Å has been achieved; for
2400 1/mm a band pass of 0.25 eV at 280 eV has been reported[18].
The "Grasshopper" works up to very high photon energies of about
1 keV. Its capability even for a very critical energy region – the
carbon K edge where cracked hydrocarbons reduce the reflectivity
of mirrors and gratings drastically[21] is shown in Fig. 17.

Plane grating monochromators as illustrated in Fig. 16 exploit
the collimation of SR to a nearly parallel beam. The simplest instrument possible[72] (Fig. 16a) is limited in resolution and order sorting
capability. A design with a spatially fixed exit beam is realized at
the Tokyo SR facility[73] (Fig. 16b). Modified versions have been developed at the Daresbury SR source[74]. For the third type (Fig. 16c)[75]
wavelength scanning is achieved by rotating a plane mirror and grating
in such a way, that they remain parallel. Additionally the plane
mirror moves along the incoming beam. The dispersed light is focussed
by a parabolic mirror onto a fixed exit slit. A great advantage of

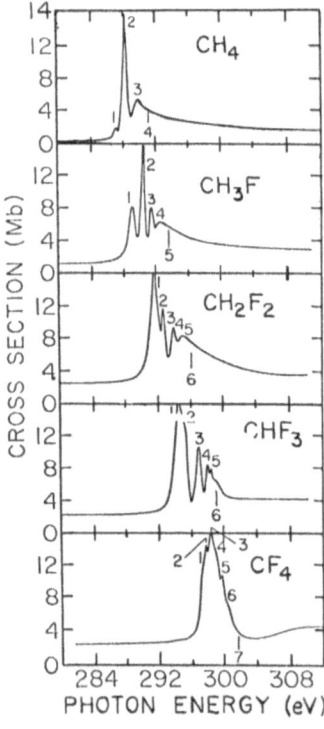

Fig. 17:
Rydberg transitions in methane
vapor and fluoromethanes at
the carbon K-edge (from Ref. 71)

this instrument is the rejection of higher orders due to the fact
that for longer wavelengths the angles of incidence are more normal incidence like. The finite size of the source defines as a virtual entrance slit the wavelength resolution. For the UHV requirements at the DORIS storage ring the scanning mechanism had to be
simplified. The result is the "Flipper" monochromator[76] developed
for photoelectron spectroscopy experiments (see Fig. 18). Instead
of a moving premirror it is equipped with 6 mirrors which can be
slided into the beam with high precision. Each mirror corresponds
to a certain wavelength range. The movements are reduced to a
simple rotation of the grating. The instrument accepts horizontally
0.75 mrad of SR and provides with a grating with 1200 1/mm, a parabolic mirror of 1m focal length and the DORIS source a resolution
$E/\Delta E$ between 5600 and 1250 from 13 eV to 250 eV. By changing the
angle of incidence over a wide range such instruments bridge the
gap from 25 eV to 50 eV where neither normal incidence nor grazing
incidence monochromators of the Rowland type work statisfactorily.

5.4 New Developments

The remarkable progress made in the development of new disper-
sive optical components has been discussed already in Sec. 5.1.
Studies of secondary effects like angular resolved photoemission
require monochromators with a very high transmission over a wide
photon energy range from approximately 10 eV to 300 eV. For such an
application instruments with holographically produced and corrected
gratings[32,33] on toroidal blanks[77] are very promising (see Fig. 19).
A small instrument with 0.3 m focal length at ACO[78] has proven to
be extremely useful: It accepts 5 mrad of SR under a grazing angle
of 19° and provides a photon flux up to 4×10^{11} photons/(sec·Å) with

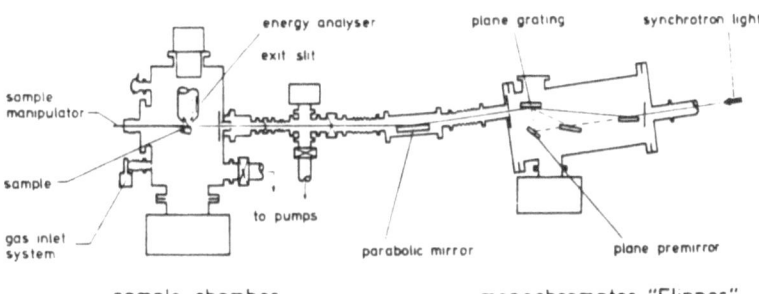

energy analyser   exit slit   plane grating   synchrotron light

sample
manipulator

sample

gas inlet
system   to pumps

parabolic mirror   plane premirror

sample chamber   monochromator "Flipper"

<u>Fig. 18</u>: Photoemission experiments at the DORIS storage
         ring with the "Flipper" monochromator (from Ref. 76)

Toroidal Holographic Grating Monochromator

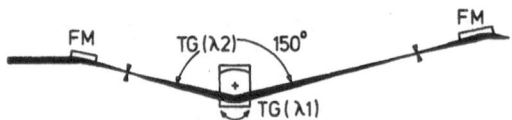

<u>Fig. 19:</u>   Proposed monochromator with two switchable toroidal
            gratings TG(λ1) and TG(λ2). FM: focussing mirrors
            (from Ref. 43)

1 Å resolution (100 μm slits) for photon energies up to 110 eV.
Instruments with larger focal lengths   higher resolution and
for shorter **wavelengths are under** construction[78].

   The photon energy range between 300 eV and 1000 eV gains more
and more interest for example for EXAFS applications. Unfortunately
it is inaccessible for most of the monochromators discussed. The
reasons for this are hydrocarbon contamination and roughness of
optical surfaces[21] as well as the extreme grazing angles necessary.
As an alternative approach to conventional instruments with reflecting components a monochromator with a free standing transmission
grating has been proposed with a resolution of 1:1000 with 1000 1/mm
for 10 eV to 1000 eV[37]. A second approach for a new instrument in
this region is a monochromator based on zone plates[79] fabricated
by X-ray lithography or holographically[31].

                              6. Conclusion

   Selecting a monochromatized beam of photons of a given energy
from the continuum of synchrotron radiation is not an easy task.
While a number of mountings based on conventional concepts have been
developed to a high degree of sophistication much remains to be done.
We observe today very interesting developments in the area of grating technology, the manufacturing of mirrors and optical elements
which may eventually lead to completely new and powerful concepts
in the monochromator design for the VUV-range.

Acknowledgement

The author wishes to thank Dr. E.E. Koch for critically reading the manuscript and making many valuable suggestions. He is indebted to Drs. C. Kunz and W. Gudat for providing figures and their manuscript Ref. 43 prior to publication. Careful and skillful typing of Mrs. E. Thumann is gratefully acknowledged.

References

1. Vacuum Ultraviolet Radiation Physics, ed by E.E. Koch, R. Haensel and C. Kunz (Pergamon-Vieweg, Braunschweig, 1974)
2. Vth Int.Conf. on Vacuum Ultraviolet Radiation Physics, Extended Abstracts, ed. by M. Castex, M. Pouey and N. Pouey (Montpellier, Sept. 5-9, 1977)
3. An Assessment of the National Need for Facilities Dedicated to the Production of Synchrotron Radiation (Solid State Sciences Committee, Assembly of Mathematical and Physical Sciences, National Research Council, Washington, 1976)
4. Synchrotron Radiation a Perspective View for Europe, prepared by ESF (European Science Foundation, Strasbourg, France 1978)
5. a) J.A.R. Samson, Techniques of Vacuum Ultraviolet Spectroscopy (Wiley and Sons, New York 1976)
   b) A.N. Zaidel' and E.Ya. Shreider, Vacuum Ultraviolet Spectroscopy, (Ann Arbor-Humphrey Science Publishers, Ann Arbor, London, 1970)
6. G. Schmahl, D. Rudolph and B. Niemann in Ref. 2, Vol. III, p. 40
7. E. Spiller and R. Feder, X-Ray Lithography, in Topics in Appl. Physics. Vol. 22, ed. H.J. Queisser (Springer Verlag, Berlin, Heidelberg, New York, 1977); W. Gudat in Ref. 11, p. 279
8. D. Einfeld, D. Stuck and B. Wende in Ref. 2, Vol. III, p. 114; R.P. Madden in Ref. 2, Vol. III, p. 120; E. Pitz and A. Schulz in Ref. 11, p. 243
9. H.J. Hagemann, W. Gudat and C. Kunz, J.Opt.Soc.Am. 65, 742 (1975); R.P. Haelbich, M. Iwan and E.E. Koch, Optical Properties of Some Insulators in the Vacuum Ultraviolet Region Physik Daten, Physics Data ZAED, Karlsruhe, Germany, Vol. 8-1 (1977)
10. Several contributions in Ref. 1, 2, 11
11. Proc. Intern. Conf. on Synchrotron Radiation Instrumentation and New Developments, ed. by F. Wuilleumier and Y. Farge, Special issue of Nuclear Instruments and Methods 152 (1978)
12. Synchrotron Radiation, ed. by C. Kunz, Topics in Current Physics (Springer-Verlag, Berlin, Heidelberg, New York, in press)
13. P.M. Guyon, C. Depautex and G. Morel, Rev.Sci.Instr. 47, 1347 (1976)
14. E.E. Koch, C. Kunz and E.W. Weiner, Optik 45, 395 (1976)
15. C. Kunz, Phys. Bl. 32, 9 (1976)
16. C. Kunz in Ref. 1, p. 753
17. A more detailed dicussion is found in Chap. I of Ref. 12

18. F.C. Brown, R.Z. Bachrach and N. Lien in Ref. 11, p. 73

19. U. Backhaus, Diplomarbeit Universität Hamburg, 1973;
    calculated with the optical constants given by K. Platzöder,
    Diplomarbeit Universität München, 1967

20. A.P. Lukirskii, E.P. Savinov, O.A. Ershov, V.A. Fomichev and
    I.I. Zhukova, Opt. Spectrosc. 19, 237 (1965)

21. R.Z. Bachrach, S.A. Flodstrom, R.S. Bauer, V. Rehn and
    V.O. Jones in Ref. 11, p. 135

22. V. Rehn, A.D. Baer, J.L. Stanford, D.S. Kyser and V.O. Jones
    in Ref. 1, p. 780

23. J.L. Stanford, V. Rehn, D.S. Kyser and V.O. Jones in Ref. 1,
    p. 783

24. B. Niemann, private communication

25. V. Rehn, J.L. Stanford, V.O. Jones and W.J. Chyoke, Proc. Int.
    Conf. on Physics of Semiconductors, Rome, August 1976, p. 985;
    V. Rehn, J.L. Stanford, A.D. Baer, V.O. Jones and W.J. Chyoke,
    Appl.Opt. 16, 1111 (1977); feasibility of SiC for gratings is
    discussed in W.J. Chyoke, W.D. Partlow, E.P. Supertzi, F.J.
    Venskytis and G.B. Brandt, Appl.Opt. 16, 2013 (1977)

26. R.P. Haelbich and C. Kunz, Optics Commun. 17, 287 (1976)

27. G.W. Stroke, Encyclopedia of Physics, Vol. XXIX, ed. by S. Flügge
    (Springer-Verlag, Berlin, Heidelberg, New York, 1967), p. 426;
    Diffraction Grating Handbook, ed. Bausch and Lomb Inc.
    (Rochester, New York, II. edition 1972)

28. H.A. Rowland, Phil.Mag. 16, 197 and 210 (1883)

29. V. Saile in Ref. 11, p. 59

30. V. Saile, P. Gürtler, E.E. Koch, A. Kozevnikov, M. Skibowski
    and W. Steinmann, Appl.Opt. 15, 2559 (1976)

31. G. Schmahl in Ref. 1, p. 667

32. Diffraction Gratings Ruled and Holographic - Handbook, ed.
    Jobin-Yvon Company (Longjumeau, France, 1976)

33. H. Noda, T. Namioka and M. Seya, J.Opt.Soc.Am. 64, 1031 (1974)

34. J.A. Dijkstra and L.J. Lantwaard, Opt.Commun. 15, 300 (1975)

35. S.A. Flodstrom and R.Z. Bachrach, Rev.Sci.Instr. 47, 1464 (1976)

36. E. Källne, H.W. Schopper, J.P. Delvaille, L.P. van Speybroeck
    and R.Z. Bachrach in Ref. 11, p. 103

37. J. Stöhr, V. Rehn, I. Lindau and R.Z. Bachrach in Ref. 11, p. 44

38. D.J. Bradley, M.H.R. Hutchinson and C.C. Ling, Tunable Lasers
    and Applications, ed. by A. Mooradian, T. Jaeger and P. Stokseth,
    Proc. of the Loen Conf. Norway, 1976 (Springer-Verlag, Berlin,
    Heidelberg, New York, 1976) p. 40

39. P.P. Sorokin, J.A. Amstrong, R.W. Dreyfus, R.T. Hodgson, J.R.
    Lankard, L.H. Manganaro and J.J. Wynne, Laser Spectroscopy,
    ed. by S. Haroche, J.C. Pebay-Peyroula, T.W. Hänsch and S.E.
    Harris, Proc. II. Int.Conf., Megève, 1975 (Springer-Verlag,
    Berlin, Heidelberg, New York (1975))

40. Physics Today, Dec. 1976, p. 17

41. E.E. Koch, Interaction of radiation with condensed matter, Vol. II,
    L.A. Self (editor), publication of the Trieste Center for
    Theoretical Physics Int.Atomic Energy Agency, Wien 1976, p. 225

42. K. Radler, private communication
43. W. Gudat and C. Kunz, Chapter 3 in Ref. 12
44. V. Saile, Thesis, Universität München, 1976
45. V. Saile, M. Skibowski, W. Steinmann, P. Gürtler, E.E. Koch
    and A. Kozevnikov, Phys.Rev.Lett. 37, 305 (1976)
46. W.B. Peatman, B. Gotchev, P. Gürtler, E.E. Koch and V. Saile,
    J.Chem.Phys. (in press)
47. M. Seya, Sci. Light 2, 8 (1952); T. Namioka, Sci. Light 3, 15
    (1954), J.Opt.Soc.Am. 49, 951 (1959)
48. N. Rehfeld, U. Gerhardt and E. Dietz, Appl.Phys. 1, 229 (1973)
49. H. Noda, T. Namioka and M. Seya, J.Opt.Soc.Am. 64, 1043 (1974)
50. M. Skibowski and W. Steinmann, J.Opt.Soc.Am. 57, 112 (1967)
51. E.E. Koch, Thesis, Universität München, 1972
52. C. Depautex, M. Lavollee, G. Jezequel, J.-C. Lemonnier and
    J. Thomas in Ref. 11, p. 69
53. U. Hahn, N. Schwentner and G. Zimmerer in Ref. 11, p. 261
54. U. Hahn, Thesis, Universität Hamburg, 1978
55. U. Hahn and N. Schwentner, in preparation
56. M. Lavollee and S. Robin, J.Opt.Soc.Am. 64, 319 (1974)
57. C.H.F. Velzel, J.Opt.Soc.Am. 68, 38 (1978)
58. M. Pouey, Some Aspects of Vacuum Ultraviolet Radiation Physics,
    ed. by N. Damany, J. Romand and B. Vodar (Pergamon Press,
    Oxford, 1974) Chapter 9
59. M. Pouey in Ref. 1, p. 728
60. C. Kunz, Proc. Intern. Symposium for Synchrotron Radiation Users,
    ed. by G.V. Marr and I.H. Munro (Daresbury Nucl.Phys.Lab.
    Report DNPL:R26, 1973) p. 68
61. E.E. Koch, Problems of Elementary Particle Physics, Proc. of the
    8th All Union School of High Energy Particle Physics (Yerevan,
    1975), p. 502
62. K. Thimm, J. Electr. Spectr. Rel. Phenom. 5, 755 (1974)
63. R. Haensel and C. Kunz, Z. Angew. Physik 23, 276 (1967)
64. K. Codling and P. Mitchell, J.Phys. E3, 685 (1970)
65. P. Jaeglé, P. Dhez and F. Wuilleumier, Rev.Sci.Instrum. 48,
    978 (1977); P. Dhez, P. Jaeglé, F.J. Wuilleumier, E. Källne,
    V. Schmidt, M. Berland and A. Carillon in Ref. 11, p. 85
66. C.H. Pruett, N.C. Lien and S.D. Steben, III. Int. Conf. on
    Vacuum Ultraviolet Radiation Physics, Tokyo 1971, 31a A2-5
67. G. Puester and K. Thimm in Ref. 11, p. 95; K. Thimm see Ref. 62
68. F.C. Brown, R.Z. Bachrach, S.B.M. Hagström, N. Lien and
    C.H. Pruett in Ref. 1, p. 785
69. M. Salle and B. Vodar, C.R. Acad.Sci. Paris 230, 380 (1950)
70. A design with the Rowland circle rotating around the grating
    has been realized by H. Sugawara and T. Sagawa in Ref. 1, p. 790
71. F.C. Brown, R.Z. Bachrach and A. Bianconi, Chem.Phys.Lett. 54,
    425 (1978)
72. J. Römer, Diplomarbeit, Universität Hamburg, 1970
73. K.P. Miyake, R. Kato and H. Yamashita, Sci. Light 18, 39 (1969)

74. J.B. West, K. Codling and G.V. Marr, J.Phys. E7, 137 (1974);
    M.R. Howells, D. Norman, G.P. Williams and J.B. West,
    J.Phys. E11, 199 (1978)

75. C. Kunz, R. Haensel and B. Sonntag, J.Opt.Soc.Am. 58, 1415
    (1968); H. Dietrich and C. Kunz, Rev.Sci.Instrum. 43, 434 (1972)

76. W. Eberhardt, G. Kalkoffen and C. Kunz in Ref. 11, p. 81;
    W. Eberhardt, Thesis, Universität Hamburg, 1978

77. A monochromator based on a toroidal grating ruled mechanically
    has been realized by R.P. Madden and D.L. Ederer, J.Opt.Soc.Am.
    62, 722 (1972)

78. Y. Petroff, P. Thiry, R. Pinchaux and D. Lepere in Ref. 2,
    Vol. III, p. 70; D. Depautex, P. Thiry, R. Pinchaux, Y. Petroff,
    D. Lepere, G. Passereau and J. Flamand in Ref. 11, p. 101

79. E. Spiller, Workshop on X-Ray Instrumentation for Synchrotron
    Radiation Research, ed. by H. Winick and G. Brown, SSRL Report
    No. 78/04 (May 1978)

BIOPHYSICAL SPECTROSCOPY IN

THE VISIBLE AND ULTRAVIOLET USING SYNCHROTRON RADIATION

John Clark Sutherland

Biology Department, Brookhaven National Laboratory

Upton, New York 11973 USA

ABSTRACT: Spectroscopy in the ultraviolet and visible regions of
the electromagnetic spectrum is extremely important in the study of
biological materials. These lectures consider the applications of
synchrotron radiation (SR) to these regions. They are limited to
measurements which do not make use of the time structure inherent
in SR. Comparisons of SR with conventional sources suggests that
the greatest improvements will be realized in the far and vacuum
ultraviolet regions -- wavelengths less than about 300 nm.
Consideration of the transitions of the valence electrons of most
organic and biologically important materials indicate that
wavelengths less than about 120 nm will not be especially
informative in studies of the structure and function of biological
materials (although they are important for understanding the
interactions between these materials and high energy electrons).
In addition, spectroscopic experiments at wavelengths less than 105
nm become more difficult because of the loss of window materials
and surfaces with high normal incidence reflectance. The
characteristic visible and ultraviolet absorption bands of
proteins, nucleic acids and sugars are reviewed. Important
spectroscopic techniques such as absorption, natural and magnetic
circular dichroism, fluorescence and various fluorescence
polarization spectroscopies are described and their potential use
in the far and vacuum ultraviolet (120-300 nm) using synchrotron
sources is discussed.

I.   INTRODUCTION

A.   Classification of Visible and Ultraviolet Experiments.

Experiments which use ultraviolet and visible synchrotron radiation to probe the structure and function of biological molecules can be classified as shown in Table I. The characteristic feature of the class of experiments described as "Photochemistry and Photobiology" is that one observes some change in the chemical composition or biological function of the irradiated specimen, while in "Spectroscopy" one observes photons which have passed through, been emitted by, scattered by or reflected from the sample. I have divided the spectroscopic experiments into those which make explicit use of the time structure of SR and those to which the time structure is irrelevant. To date (August 1978), the chief use of uv and visible SR to study biological materials has been time resolved fluorescence which is an example of time structure dependent spectroscopy. However, experiments representing the other classes are in the prototype or planning stages. While these lectures will emphasize the potential of SR in several types of time-structure-independent spectroscopies, some of the material -- particularly that related to the design of spectrometers -- will also be applicable to the other classes of experiments.

Table I.
Classification of Visible and Ultraviolet Experiments
of Biological Materials Using Synchrotron Radiation

I. Spectroscopy
a.   time structure independent
b.   time structure dependent
II. Photochemistry/Photobiology

B.   Time Structure Independent Biophysical Experiments Using Synchrotron Radiation.

The experiments I shall discuss are listed in Table II, roughly in order of increasing "photon requirement". That is, as one goes down the list, the experiments tend to require more intense radiation sources. The properties which make SR an attractive source for these experiments are its a) high intensity, b) broad spectral range, c) short term stability, d) polarization and e) pseudo d.c. character (i.e., the pulse frequency is so high (1 to 300 MHz) that SR can be treated as if it were continuous in time rather than pulsed).

The many useful features of SR have lead to it being called "The source for all reasons". Yet the complexity, cost and for most scientists the remoteness of synchrotron sources mean that SR should also be called "The source of last resort". We must carefully define the areas in which the advantages of using SR outweigh the drawbacks.

Table II.

Time Structure Independent Spectroscopic Experiments

Absorption
Circular Dichroism
Magnetic Circular Dichroism
Fluorescence
Fluorescence Polarization (linear)
Fluorescence Detected CD and MCD
Fluorescence Polarization (circular)
Photoacoustic Spectroscopy

C.   Factors Important in Design of Experiments.

In Table III I list broad areas which must be considered in the design of spectroscopic experiments. In evaluating the potential of SR we must consider all five areas, not just the properties of SR alone. For example, the existence of a good light source in a particular spectral region is of no use if the samples we wish to study are "uninteresting" in that region or if the medium in which they are suspended is completely opaque. Nor should SR be considered if a less expensive or more convenient source would be adequate.

Table III.

Components of a Spectroscopic Experiment

1.   Source
2.   Optical Components
3.   Sample
4.   Detector
5.   Signal Processing/Analysis

SOURCES: For the experiments listed in Table II it is either necessary or at least extremely desirable to use a source which can be tuned over a broad spectral range. The emission spectra of the three conventional sources used for these experiments, the xenon arc, hydrogen discharge and tungsten lamp, are shown in Figure 1. The performance of these sources deteriorates with decreasing wavelength. While commercial CD spectrometers using Xe arcs can record spectra down to almost 180 nm, the intensity of their beam is reduced by several orders of magnitude compared to the visible region. Naturally their performance suffers accordingly. The hydrogen discharge has a continum which extends to even shorter

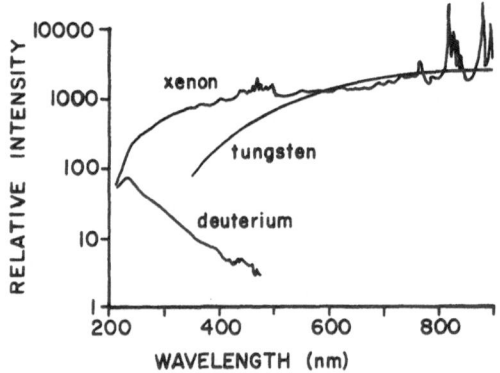

Figure I/1. Relative intensity (power) of radiation from a 150 W
xenon lamp, a 100 W tungsten lamp and a 50 W deuterium lamp.
Adapted from ref. (1).

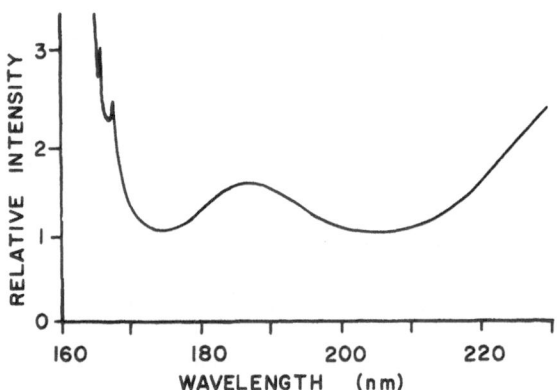

Figure I/2. Relative intensity (photons/sec) from a hydrogen
discharge in the far and vacuum ultraviolet. Adapted from Samson
(2).

wavelengths. Even for equivalent input power, however, the radiant power density in the far uv is never comparable to that achieved by a Xenon arc in the visible. Even worse, the H_2 continuum has relative minima in the biologically important 170-220 nm region -- see Figure 2.

SR spans the entire visible, uv and vacuum uv (vuv). Its intensity is increasing as the wavelength decreases (in all existing or proposed storage ring sources the critical wavelength is less than 100 nm). Thus, the superiority of SR compared to conventional broadband sources will be greatest in the far and vacuum uv -- i.e. wavelengths less than about 250 to 300 nm. While SR may prove superior to conventional sources at longer wavelengths as well, the greatest _relative_ improvement is gained in the shorter wavelength range.

Comparisons between SR and laser sources are more complicated because of the diversity of laser technologies and the speed with which they are evolving. Lasers are clearly superior to any other sources in experiments such as Raman spectroscopy which require high power densities and very narrow spectral band widths. No laser, however, comes close to the broad tunability provided by SR. While a dye laser can, in principle, be tuned from about 330 to 800 nm, many different dyes are required and the resultant intensities vary dramatically with wavelength. Wavelengths below 330 are produced by doubling a lower frequency -- an inherently inefficient process. The various pulsed lasers (except mode-locked ion lasers) have pulse rates and pulse-to-pulse variations in amplitude which are incompatible with experiments such as circular dichroism in which the pulse rate of the source must greatly exceed the polarization modulation frequency. While it is not possible to predict how laser technology will evolve, at the moment there are domains in which SR clearly offers significant advantages.

OPTICAL COMPONENTS: The second area which must be considered in designing a spectrometer for biological research is the optical components through which the radiation passes or from which it is reflected between source and detector. Note that the sample and the detector must be evaluated as optical components. In the previous section I argued that the greatest impact of SR on biophysical spectroscopy will be in the far and vacuum uv. Thus, the focus in this section will be on the availability of components in that region.

As one proceeds from the visible through the uv and into the vuv, the selection of materials suitable for lenses, mirrors, polarizers, windows, etc. decreases. Pyrex, other glasses, and Al + SiO coated mirrors are unusable for wavelengths less than 300 nm. Calcite prisms are good only to about 220 nm. The very best

quality quartz starts absorbing about 180 nm and sapphire at about
145 nm. However, optical experiments can be designed along fairly
usual (albeit more expensive) lines down to about 120 or 130 nm.
Down to these wavelengths windows, lenses (but watch the chromatic
aberration) and photoelastic modulators (more on this device later)
can be fabricated from synthetic CaF_2 crystals-which are
isotropic and polarizers can be made from MgF_2. For a more
detailed description of the optical properties of CaF_2, MgF_2
and LiF see refs. 2 and 3. Some commercial photomultipliers use
MgF_2 or sapphire as windows. Such optically anisotropic
materials must be avoided in experiments such as CD in which the
detector must respond only to the intensity and not to the
polarization of the radiation incident upon it. Special aluminized
mirrors and diffraction gratings overcoated with evaporated MgF_2
retain high normal incidence reflectivity down to 120 nm as shown
in Fig. I/3. The wavelength frequently quoted as the limit for
transmission optics is 105 nm -- the cutoff wavelength for high
quality LiF. Unfortunately the mechanical properties of LiF are
inferior to those of CaF_2. Thus careful attention must be paid to
the trade-off between using less fragile materials versus the
information to be gained from extending the spectral range of an
experiment by 25 nm or so. One limitation of LiF_2 and CaF_2 is
that the X rays present in unfiltered SR may form color centers
which cause these materials to absorb in the uv and vuv. It may be
necessary to remove the "hard" components of SR before the first
window by a nongrazing-angle reflection.

Spectroscopy does not stop at 100 nm, but it gets rather
more difficult since normal incidence reflections should be
avoided and no materials* are available to serve as windows to
isolate the experimental chamber and monochromator from the light
source. The loss of window materials is a serious matter for
those who would use SR to study biological materials in
(more-or-less) their natural state. Sucessful operation of an
electron storage ring requires that it be maintained at a high
vacuum. Many biological samples contain water, hydrocarbons and
other molecules which the scientists charged with the operation of
the storage ring regard as "noxious" materials. Thus extending an
experiment to wavelengths below 100 nm will require removing any
windows and adding differential pumping and fast acting automatic
values. It may also mean that a longer portion of the beam line
must meet the vacuum standards of the storage ring (e.g. the use
of rubber "0" rings may be forbidden).

---

*Very thin films of Al or other metals may be used as windows be-
tween 20 and 80 nm but they will not support atmospheric pressure.

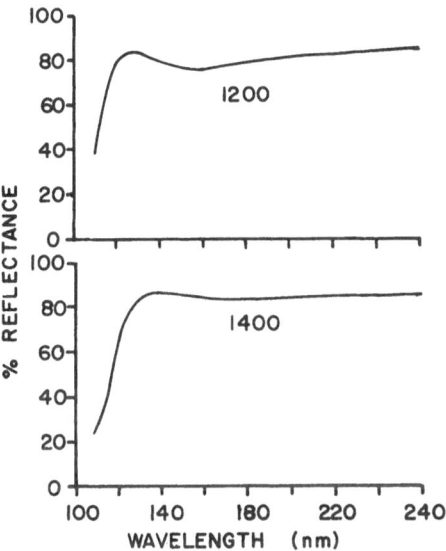

Figure I/3. Normal incidence reflectance of aluminum mirrors
coated with MgF$_2$ from ref. (4). The designations 1200 and 1400
indicate that these mirrors are optimized for 120 and 140 nm
respectively.

The vacuum uv derives its name from the fact that molecular oxygen starts absorbing at about 195 nm. Conventional (i.e. non-vacuum) spectrometers can be operated down to 170 nm by purging with N_2. Instruments designed for operation at shorter wavelengths usually resort to vacuum technology even though gasses such as argon and helium remain transparent. Apparently the thought of flowing several liters per minute of He through the spectrometer is even worse than buying a vacuum tight monochromator and accessories.

Biological materials are generally found in aqueous environments. Thus water is the "optical component" which may limit the range of an experiment. Water is transparent from about 185 nm in the uv through the visible and into the near infrared. Figure 4 shows the onset of absorption in the uv. A variety of techniques have been developed to extend optical measurements of biological materials past the onset of absorption by water at both the short and long end of the "water-window". Some of these techniques will be discussed in the sections on Absorption Spectroscopy and Circular Dichroism which follow.

DETECTORS: The principle detector used in the uv and vuv is the photomultiplier. The major limitation on the spectral response is the material used as a window. Photomultipliers using CaF_2 windows are available on special order from several vendors, while tubes with MgF_2 and LiF windows are catalogue items. Note that MgF_2 is birefringent and will not be suitable for windows of detectors whose response must be independent of the polarization of the incident radiation. For wavelengths less than the CaF_2 or LiF cutoffs, windowless photomultipliers can be used or the face of a windowed tube can be covered with a fluorescent material such as sodium salicilate. The later procedure reduces the quantum efficiency of the detector by about an order of magnitude compared to the same tube with a transparent window (2,6).

SAMPLES: Having the instrument to perform an experiment is of no use unless the wavelengths used provide useful information about the material being studied. For the experiments listed in Table II, the first requirement is that the sample absorb some of the incident radiation. As an example, consider the near infrared wavelengths between 1000 and 2000 nm. Almost the only biological materials which absorb in this region are those proteins which contain transition metal ions such as iron. Thus use of the 1000-2000 nm region has been limited even though instrumentation is not a great problem. Figure I/5 indicates the wavelengths absorbed by a variety of biological materials. All of the absorption indicated in Figure I/5 is electronic in origin i.e. it results in a change of the electronic state of the molecule. Absorptions due solely to changes in vibrational state usually occur at longer wavelengths and will not be considered here.

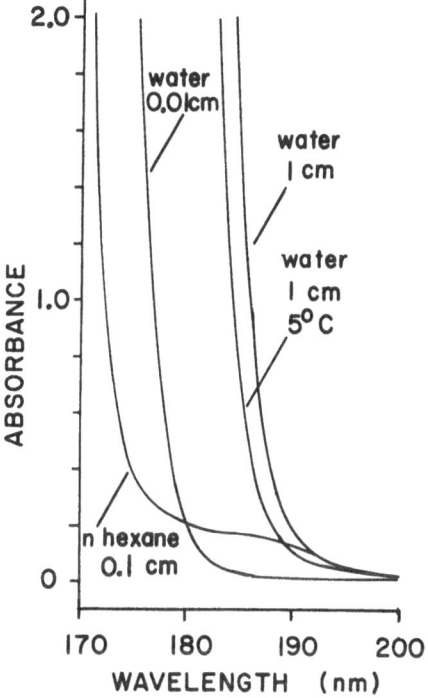

Figure I/4.   Absorbance of 1 cm and 0.01 cm of water and
0.1 cm of n-hexane at room temperature and 1 cm of water
at 5°C.   From ref.(5).

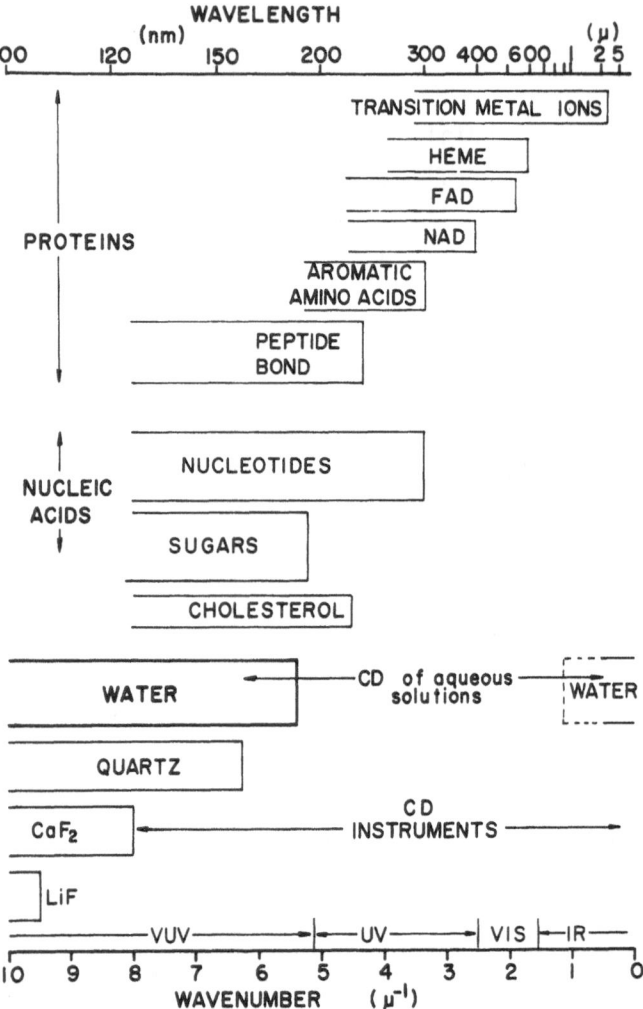

Figure I/5.   Regions of absorption of a variety of biological
molecules as well as water, CaF$_2$ and LiF and the limits for
CD experiments.

Some biological molecules absorb in the visible and near uv. For example, some proteins are associated -- covalently or noncovalently -- with small molecules which absorb in this region. The spectroscopic "handles" provided by their visible absorption bands have made these some of the most extensively studied and best understood of all proteins. A far larger number of proteins while not absorbing at longer wavelengths exhibit absorption below 310 nm due to the aromatic amino acids: tryptophan, tyrosine and phenylalanine. The peptide bonds of all proteins absorb below 240 nm with a maximum around 200 nm. At shorter wavelengths there is additional absorption due to nonaromatic amino acids.

Nucleic acids exhibit little absorption before about 300 nm where the bases start to absorb as shown in Figure I/7. At shorter wavelengths the ribose moieties and even the phosphate groups contribute. As with proteins, absorption tends to become more intense as the wavelength decreases because the oscillator strength of more and more electrons are recruited.

Figure I/5 demonstrates that there is no lack of biologically interesting materials which absorb below 300 nm -- the wavelength range where SR will have the greatest impact. But do optical experiments in this spectral region provide significant information about the structure and function of the materials being studied? The answer to this question depends on the nature of the transition responsible for the absorption. The charge-transfer and $d \leftarrow d$ transitions of transition metal ions and the $\pi^* \leftarrow \pi$ and $\pi^* \leftarrow n$ transitions of aromatic electrons which are responsible for all of the absorption at wavelengths greater than about 180 nm are of proven utility in the study of biological structure and function. While $\pi^* \leftarrow \pi$ transitions are also observed at wavelengths below 180 nm, transitions involving σ and σ orbitals (e.g. $\sigma^* \leftarrow n$ or $\pi^* \leftarrow \sigma$) and Rydberg transitions are also observed. Some of these transitions will undoubtably prove useful. At shorter wavelengths, however, the spectrum will be dominated by $\sigma^* \leftarrow \sigma$ transitions, which are unlikely to be of great use because of the ubiquitousness of carbon-carbon and carbon-nitrogen bonds in biological structures. An example to illustrate this point is given in the section on absorption spectroscopy.

Still shorter wavelengths have sufficient energy to excite electrons from closed shells into the valence shell or produce ionization. The spectra in the vicinity of these absorption edges are of great importance in the study of biological materials, but they are usually treated in discussions of X ray spectroscopy and will not be considered here.

The strength of absorption tends to increase towards shorter wavelengths. Figure I/6 shows the absorption spectrum of a "typical" heme protein and Figure I/7 shows actual data for the

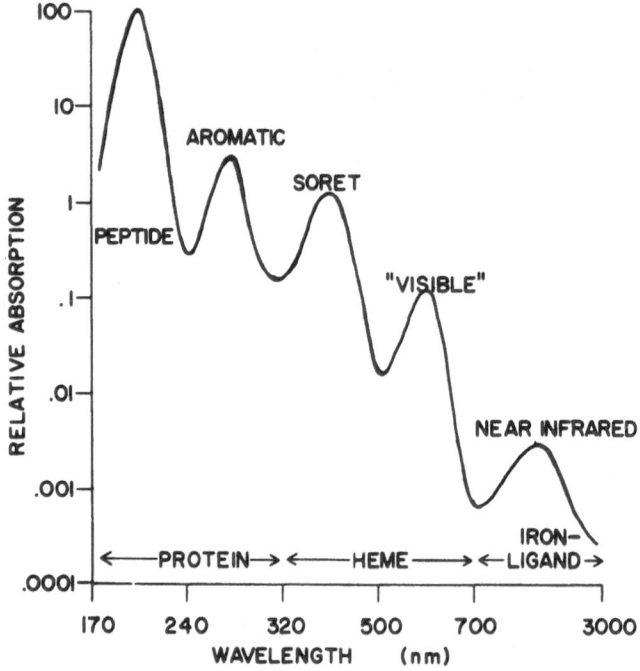

Figure I/6: Absorption spectrum of a "typical" heme protein in the ultraviolet, visible, and near infrared. Absorption increases by several orders of magnitude and thus is plotted on a log scale. The wavelength scale is also nonlinear, and is roughly linear in units of photon energy. Data for two oxidation states of a real heme protein above 350 nm are shown in Figure I/7. The aromatic and peptide bands of BSA, a protein which does not contain components which absorb above 300 nm, are shown in Figure II/3. The ratios of the magnitudes of absorption of the various bands depends on the particular protein being studied.

longer wavelength bands of cytochrome c. Note that absorption is plotted on logarithmic scales in both figures.

SIGNAL PROCESSING AND ANALYSIS: The techniques used to process spectral information after radiation reaches the detector are for too numerous to discuss here. I will mention briefly the system used in my laboratory. Several years ago, we constructed a spectrometer capable of measuring all of the parameters listed in Table II (7,8,9). All of the electronic signal processing in this

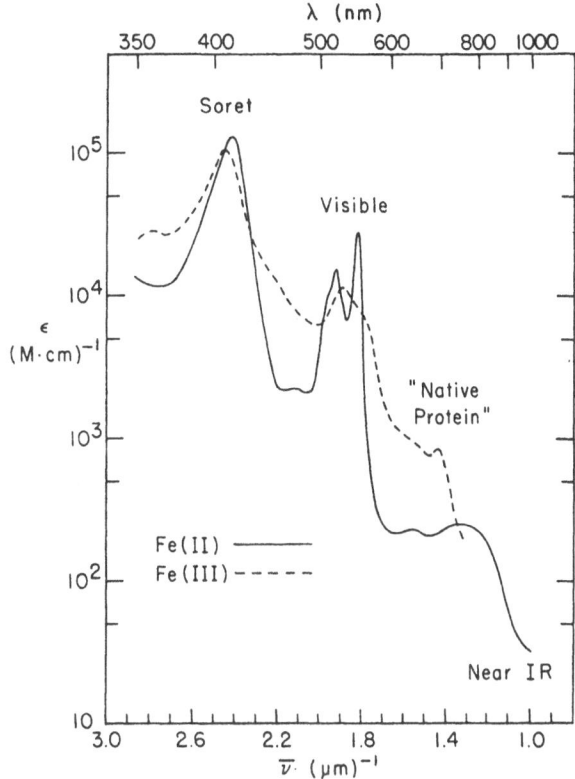

Figure I/7. Molar extinction coefficient of reduced and oxidized cytochrome c versus wavenumber (photon energy) in the visible and near infrared. From ref. (36).

spectrometer is performed by a phase-sensitive amplifier/detector (i.e. a "lock-in" amplifier). The objective of the original design was to provide the capability of performing a wide range of experiments with a limited amount of money. We are presently constructing an instrument which is similar in concept -- although greatly different in detail -- which will use SR as its source. While retaining the advantage of economy, the main virtue of the "multi-mode" approach for storage ring instrumentation is that a variety of experiments can be performed without the down-time which would be incurred if a separate spectrometer or even a separate sample chamber and set of electronics were used for each class of measurement. Our present design anticipates operation down to about 130 nm in the vuv. In the subsequent sections I shall discuss in greater detail the classes of experiments we shall perform.

II. ABSORPTION SPECTROSCOPY

A. Rationale

Absorption spectroscopy is the most fundamental and most widely used of the methods that I shall discuss. Besides the inherent
utility of an absorption measurement whenever possible it is desirable to provide an absorption spectrum to accompany "higher order"
measurements such as circular dichroism or fluorescence excitation
spectra. Most types of absorption spectroscopy do not require as
intense a source of radiation as the other experiments listed in
Table II. Thus, except for photoacoustic spectroscopy, it will
usually not be necessary to use SR at wavelengths above 100 nm if
the only objective is the recording of an absorption spectrum. I
discuss absorption spectroscopy here both becaues an understanding
of the principles involved is necessary for analysis of the more
complicated experiments discussed later and because of the abun-
dance of absorption spectra of biological materials in the vuv compared to other types of measurements.

B. Definition of Units

As with all electromagnetic radiation, the absorption of a
visible or uv photon is a discrete event. The rate of decrease of
the intensity of a collimated beam of photons passing through an
absorbing medium is proportional to the density of absorbing centers. If we imagine a monochromatic beam of intensity $I_o(\lambda)$ and
wavelength λ (nm) incident on a planar slab of homogeneous material
of thickness d(cm) containing a single component which absorbs
wavelength λ then, ignoring scattering and reflections at the front
and rear surfaces of the slab, the intensity of the beam which
emerges from the slab, $I(\lambda)$ will be given by

$$I(\lambda) = I_o(\lambda) \, e^{-\sigma(\lambda)nd}$$

where n is the number of absorbers per cubic centimeter and $\sigma(\lambda)$
is the cross section for absorption of wavelength λ. The units of σ
are cm^2/absorber.

An equivalent set of units is usually used in the chemical
and biological literature to describe the absorption process. In
this system we say

$$I(\lambda) = I_o(\lambda) \, 10^{-A(\lambda)}$$

where A is the absorbance at wavelength λ while I and I_o have the
same meaning as above.

If there is a single absorbing species then we can express the absorbance in terms of the product

$$A(\lambda) = \epsilon(\lambda) C d$$

where C is the concentration (moles/liter) of the absorber, d(cm) is path length and $\epsilon(\lambda)$ is the molar extinction coefficient at wavelength λ. The absorption spectrum of a sample is a plot of A or ϵ versus λ. It is easy to show that $\sigma = 3.82 \times 10^{-21} \epsilon$.

I must mention another set of units frequently used by the physics community since a substantial fraction of vuv spectra are presented in this fashion. In this system

$$I = I_o e^{-(4\pi nkd/\lambda)}$$

where n is the refractive index of the material, d is pathlength (measured in the same units as λ), k is the (wavelength) extinction coefficient and λ is the wavelength of the radiation measured in vacuum -- i.e. the monochromator setting. Note that λ/n is the wavelength measured in the absorbing material. Thus k is related to the rate of decrease in intensity per wavelength as measured in the material. The decrease in intensity can also be expressed in terms of the imaginary part of the complex dielectric constant, ϵ_2 which is equal to 2nk.

In addition to absorptive processes, biological samples frequently scatter light and thereby reduce the fraction of the incident radiation reaching the detector. Thus in general we write

$$I(\lambda) = I_o(\lambda) 10^{-D(\lambda)}$$

where $D(\lambda)$ is the optical density at wavelength λ. This parameter can be expressed as the sum of true absorption plus scattering, i.e.

$$D = A + S.$$

A variety of techniques exist for measuring the true absorption spectrum of a sample in spite of interference from scattering (10, 11, 12).

C. Sample Preparation.

Biological materials are usually associated with aqueous media. Thus samples used in spectroscopic experiments are prepared in aqueous solution whenever possible. The pH and ionic strength of solutions should be controlled by addition of appropriate

buffers and salts. Liquid samples are usually contained in
cuvettes (or cells) with parallel transparent front and rear
surfaces. An arrangement for measuring absorption spectra is shown
in Figure II/1. The sample is in the cuvette marked "sample" while
an identical cuvette marked "reference" contains the same
solvent/buffer system but lacks the component(s) whose absorption
is to be recorded. Suppose I_o^* is the intensity incident on the
front of each cuvette. Then I is the intensity incident on the
detector when the sample cuvette, is in the beam and I_o is the
intensity observed when the reference cuvette is in the beam. The
values of I and I_o are then used in the equation for optical
density defined above.
That is

$$D = \log (I_o/I).$$

This procedure automatically corrects for reflective losses at the
front and rear surfaces of the cuvette and for weak scattering or
absorption due to the buffer. Differences between sample and
reference cuvettes (or in the case of double beam
spectrophotometers between the optical paths transversed by the
sample and reference beams) are accounted for by subtracting the
spectrum obtained with the reference solution in both cuvettes.

The absorption spectrum of aqueous solutions can be measured
for wavelengths down to the absorption edge of water at about 180
nm although care must be exercised below 220 nm because some salts
and buffers start to absorb (13). Even if some radiation is
transmitted, absorption measurements at wavelengths less than 180
nm are extremely difficult because of differential attenuation of
the beam passing through the sample and reference cells due to
slight differences in path length and slight differences in the
temperature of the two cuvettes (c.f. Figure I/4). If the
concentration of the absorber is high, displacement of water in S
will also unbalance the system.

Some biologically interesting materials can be dissolved in
solvents which transmit to shorter wavelengths. For example,
Johnson and Tinoco (14) measured the absorption spectrum of poly-γ-
methyl glutamate dissolved in hexafluoro-2-propanol down to 160 nm,
see Figure II/2.

Absorption spectra can be recorded at any wavelength -- to the
soft X-ray region and beyond -- by forming dry films of the
material on transparent substrates or microgrids. Preiss and
Setlow (15) used this type of preparation to measure the absorption
spectra of amino acids, peptides, proteins, nucleic acids to 145 nm
over twenty years ago. Yamada and Fukutome (16) measured the
absorption spectra of vacuum evaporated films of adenine, guanine,
cytosine, thymine and uracil down to 120 nm while Arakawa, Birkhoff

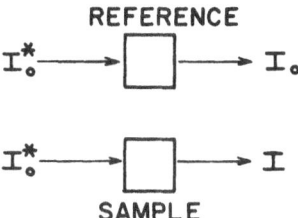

Figure II/1. Arrangement for absorption spectroscopy which corrects for loss of radiation due to reflections at the surfaces of the cell and weak absorption by the buffer solution.

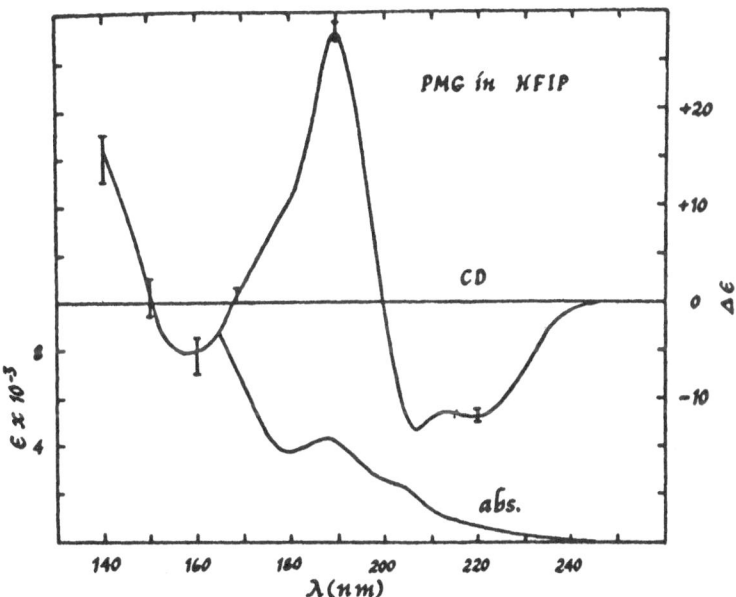

Figure II/2. Absorption and CD spectra of poly-γ-methyl glutamate in hexafluoro-2-propanol from ref. (14).

and their associates have measured the absorption spectra of a wide range of biological materials from the visible deep into the vacuum uv (17,18,19,20). Several of their spectra extend to 82 eV (15 nm!).

D. Absorption Spectra In The Far and Vacuum Ultraviolet.

Figures II/3 and II/4 show the optical constants of bovine
serum albumin (BSA) (19) and calf thymus DNA respectively (17).
The curve of k for BSA shows the relative weak absorption due to
aromatic amino acids at 280 nm (4.4 eV) and the much stronger band
near 200 nm (6.2 eV) due to the peptide bonds. In agreement with
solution spectra (14,21,22,23,24,25) there are hints of additional
structure to about 10 eV (124 nm). At higher photon energies there
is a broad structureless band peaking at 14 eV (88 nm) which is
presumably due to transitions involving σ electrons. In the case
of DNA (Figure II/4), peaks are observed at 264 nm (4.7 eV) and 190
nm (6.5 eV) which are also observed in solution spectra (25).
Although the individual bases exhibit some structure at higher
energies (16,18) the superposition of their spectra and the
contribution of their moieties appear to conspire to render the
spectrum of DNA structureless at higher energies except for a broad
peak near 14 eV similar to that observed for BSA. The greater
relative absorbance of DNA (17) above 10 eV compared to guanine
(18) presumably reflects the contribution of deoxyribose and
phosphate moieties. Above 14 eV, no structure is observed for
either BSA or DNA. The extinction coefficient decreases and, in
effect, the molecules become transparent. These data suggest that
above photon energies of 10 to 12 eV (roughly 125-100 nm) there
will be great similarities among the valance band spectra of a wide
variety of biological molecules, thus limiting the usefulness of
this region for studies of biomoleculal conformation. This region
is useful however for understanding the interaction of a material
with high energy electrons passing through it and may thus be
valuable to the fields of radiation chemistry and
electron microscopy. It is a happy coincidence that the spectra of
many biological materials are becoming less interesting at the same
wavelengths at which spectroscopy using SR becomes more difficult
due to the loss of suitable windows. These observations have
obvious implications concerning the design of spectrometers and
experiments.

E. Photoacoustic Spectroscopy.

An intense beam of monochromatic radiation "chopped" at an
audio frequency can be used to periodically heat a solid sample
which absorbs at that wavelength. The periodic heating generates
sound waves in a transparent gas in contact with the sample. The
amplitude of these waves, detected by a microphone, can be used to
measure the absorption-like "photoacoustic" spectrum of the sample
(26). Similar measurements can be made on absorbing gasses.
Photoacoustic spectroscopy differs from conventional absorption
spectroscopy in two important respects. First it is inherently
insensitive to light scattering due to irregularities in sample
shape. Thus powders of biological materials, which can be prepared

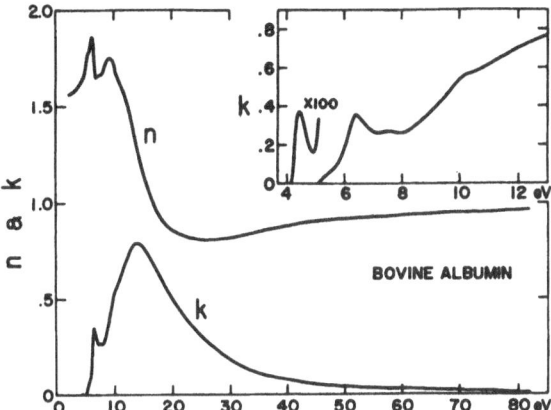

II/3. Optical constants of dry films of bovine serum albumin in
the uv and vuv. For photon energies less than 10 eV there is good
qualitative agreement with solution spectra. Data from ref. (19).

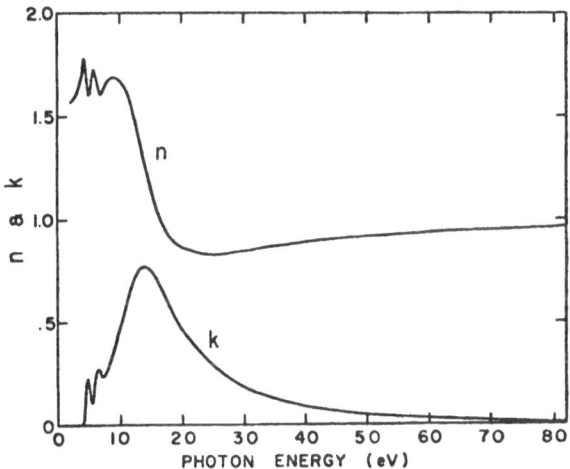

Figure II/4. Optical constants of a dry film of DNA in the uv
and vuv. For photon energies less than 10 eV, k is similar to
absorption spectra measured in solution. Note the similarity to
the spectra for BSA above 10 eV. From ref. (17).

by lyophilization, can be studied. Besides being a simple and easy
procedure, lyophilization frequently preserves the biological
activity of proteins and nucleic acids although their conformations
may be different in the dry state. Second, photoacoustic
spectroscopy, unlike transmission measurements, demands intense
radiation. Photoacoustic spectrometers using Xenon arcs do not
operate below about 240 nm. Thus SR appears to be an ideal source
for extending photoacoustic measurements to shorter wavelengths.

III. CIRCULAR DICHROISM AND MAGNETIC CIRCULAR DICHROISM

Introduction: Circular dichroism (CD) is the difference
between the absorption of left and right circularly polarized light
induced by the structural assymmetry of a chromophore or by the
assymmetry of its environment. CD is an important technique for
studying biological materials over the spectral range extending
from 5000 nm (5 μ) in the infrared to 125 nm in the vacuum
ultraviolet (13,27,28,29,30,31). The higher intensities provided
by SR may permit improvements in the sensitivity of CD measurements
throughout this range of wavelengths. As indicated in section I,
the greatest improvements are likely to be in the far and vacuum
uv. In the far uv and vuv regions improvements in CD measurements
will be particularly welcome because of the superiority of CD
compared to absorption spectroscopy in these regions. The first
advantage of CD is that it frequently resolves structure -- the
indication of distinct transitions -- which are poorly resolved or
completely unresolved in the absorption spectrum. The second
important advantage of CD is that spectra of samples in aqueous
solution can be measured down to 160 nm (perhaps further using SR)
since water introduces no CD of its own.

Magnetic circular dichroism (MCD) is the difference between
the absorption of left and right circularly polarized light induced
in a chromophore by an external magnetic field (parallel to the
direction of the light beam). While the apparatus required for an
MCD experiment is similar to that for CD — with of course the
addition of the magnet -- the information obtained from the two
types of date are quite different and frequently complementary.
For reviews of biological applications of MCD see (32,33,34).

Units: In the section on absorption spectroscopy I noted that
the absorbance of a sample, A, was equal to $\log(I_o/I)$ in the
absence of scatter. Probing the sample with left and right
circularly polarized light we can, using similar expressions,
define A_L and A_R respectively. The CD for wavelength λ, is
given by

$$\Delta A_{CD} (\lambda) = A_L (\lambda) - A_R(\lambda).$$

An expression similar to Beer's law gives $\Delta A_{CD} = \Delta \epsilon_{CD} C d$ and
thus defines the differential extinction coefficient (35). The
ratio $\Delta A_{CD}/A$ (which is equal to $\Delta \epsilon_{CD}/\epsilon$) is rarely greater
than 10^{-2} and typically in the 10^{-3} to 10^{-5} range for
biological materials. Modern CD spectrometers can measure values
of ΔA_T as low as 2 or 3 x 10^{-6} in the visible and near uv. Of
course, performance deteriorates in the far uv because of the
reduced intensity of the xenon arc.

In an MCD experiment we measure the sum of CD and MCD. Since the magnetic component is a linear function of field intensity, H, it is convenient to write

$$\Delta A = \Delta A_{CD} + H \cdot \Delta A_{MCD}.$$

The MCD is separated from the CD by measuring ΔA -- the net CD -- first with the field on and then with it off. Another procedure is to reverse field polarity before the second measurement (34). Using Beer's law we can convert ΔA_{MCD} to Δe_{MCD}. The proper unit for measuring field strength is the Tesla (T) which equals 10^4 Gauss.

The extrinsic parameters ΔA and ΔA_{CD} are, like absorbance, dimensionless quantities. ΔA_{MCD} has the units of T^{-1}. The intrinsic parameters, Δe_{CD} and Δe_{MCD} have dimensions of $M^{-1}cm^{-1}$ and $M^{-1}cm^{-1}T^{-1}$ respectively. Another system of measurement expresses the extrinsic and intrinsic CD parameters in terms of ellipticity, θ, and specific ellipticity, $[\theta]$. These can be converted to ΔA and Δe by the relations

$$\Delta A = \theta/33$$

and $\quad \Delta e = [\theta]/3300.$

Measurement of CD and MCD: A typical CD/MCD spectrometer is shown in Figure III/1. Linearly polarized monochromatic light enters a photoelastic modulator (41,42,43,44,45) the strain axis of which is oriented at 45° with respect to the plane of polarization of the light. The plane polarized light is converted alternately into left and right circularly polarized light at a frequency of 10 to 50 kHz. The amplitudes of the two circularly polarized components are the same when they enter the sample. If the sample exhibits circular dichroism, the beam which reaches the detector will be amplitude modulated by an amount $\Delta I = I_R - I_L$. If I is the time average intensity of the beam reaching the detector (I $= (I_R + I_L)/2$), it is easy to show that ΔA is proportional to $\Delta I/I$. Very small values of ΔI are measured easily with a "lock-in" amplifier while I is determined by amplifying and low - pass filtering the d.c. coupled signal from the detector. Note that the ratio $\Delta I/I$ and thus ΔA is independent of I_0 the intensity entering the sample. (Of course the signal-to-noise ratio is not independent of I_0!) Measurement of CD is thus compensated for variations in source intensity as is measurement of absorbance. There are, however, important differences. In CD, only one cuvette is necessary so there can be no problems with mismatched path lengths or temperatures. Also, both of the parameters, ΔI and I, whose ratio gives the CD are measured for radiation which has passed through the cuvette. Radiation lost from

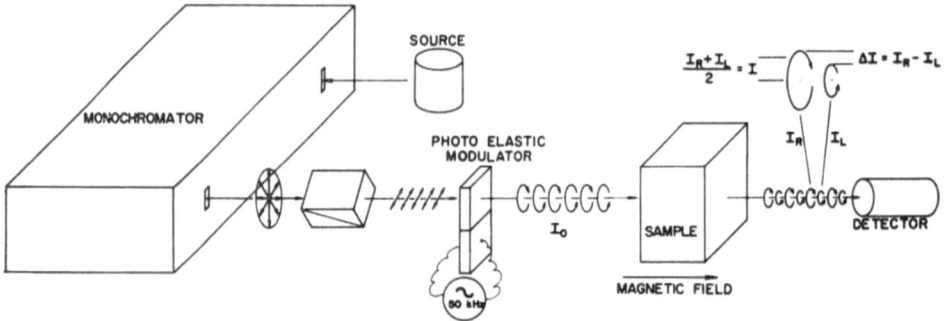

Figure III/1.   Schematic diagram of a typical CD/MCD spectrometer.

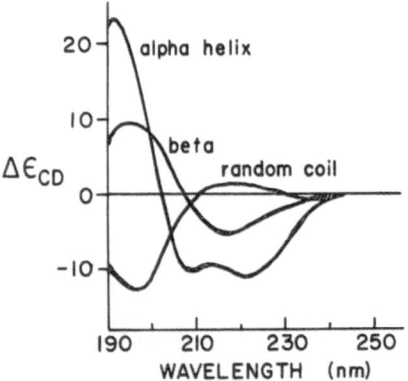

Figure III/2.   Circular dichroism of poly-1-lysine in an
α-helix, β-sheet and "random" coil.   From ref. (41).

the beam to processes which do not distinguish between the two
circular polarizations has no effect on the value of the CD
measured other than reducing the signal-to-noise ratio. The water
molecule has a plane of symmetry and thus absorbs left and right
circularly polarized light to exactly the same extent. Thus
circular dichroism ignores the absorption by water so long as
sufficient light reaches the detector. CD thus permits deeper
penetration into the vuv using aqueous samples. Using vacuum
spectrometers equipped with hydrogen discharge sources, several
groups have measured the CD of biological molecules in water down
to 160 nm -- for reviews of this work see refs. (30, 31, and 37).
The CD of biological molecules can, as in the case of absorption,
be measured when the material is dissolved in certain nonaqueous
solvents or cast as a film. These techniques have permitted CD
measurements to 140 nm (Figure II/2).

Proteins: Circular dichroism is an important tool for
studying several different aspects of protein conformation. From
300 to 2000 nm, information is obtained on cofactors, prosthetic
groups and substrates. Between 240 and 300 nm the spectrum of most
proteins are dominated by the aromatic amino acids -- particularly
tryptophan and tyrosine. Changes in CD in this spectral region may
reflect changes in chromophore environment due to allosteric
interactions. For examples of the uses of CD (and MCD) in the near
infrared, visible and near uv see the papers on hemoglobin by Eaton
et al (38) and Perutz et al. (39).

From 240 to 170 the spectra of proteins are chiefly due to
absorption by the peptide bond. The CD of the transitions of this
spectral region are a unique and powerful tool for studying the
structure of proteins in solution and changes in structure
associated with protein function. Formation of a peptide chain
results in a sequence of strongly interacting transition dipoles.
The spatial arrangement of sucessive dipoles leads to interactions
which result in circular dichroism. The exact nature of these
interactions depends on protein secondary structure. The CD
spectra of α helix being strikingly different from that of
β -sheet, β -turns and disordered (or "random") structures. The CD
of three conformations of poly-1-lysine are shown in Figure III/2.
The CD of an actual protein may be expressed as a linear sum of the
three curves shown in Figure III/2. The resulting coefficients
thus give estimates of the fraction of each type of secondary
structure present in protein (40). In practice, this procedure
works well for proteins which contain significant fractions of
their amino acids in α -helical structures. One of the
difficulties for proteins which do not have a high fraction of
residues in α helical structure is that for most CD spectrometers
the precision of the CD measured for wavelengths less than 200 nm

is significantly degraded. Synchrotron radiation should improve
the quality of spectra obtained in this region.

At wavelengths less than 170 nm proteins have structure in
their CD spectra which is also important in determining the
conformation of the peptide units and selecting from among several
theoretical treatments of the spectra of coupled peptide
transitions. The CD of poly- γ-methyl glutamate, shown in Figure
II/2, extends down to 140 nm (14). The error bars on the CD
spectrum are an indication of the need for more intense sources for
CD in this wavelength region. Note that for wavelengths less than
180 nm there may be contributions by non-aromatic side chains as
well as peptide bonds and aromatic residues (15,24).

Nucleic Acids: Circular dichroism has also been important in
elucidation of the physical properties of nucleic acids. Most work
has centered on the CD of the 260 nm band. The binding to DNA
of dyes which absorb in the visible and near uv has also been
investigated. For a review of this work see Bloomfield et al.
(29).

Recently, Wells and Yang (46,47) and Johnson and his
collaborators (48,49,50) have extended measurements of the CD of
DNA below 200 nm. Johnson's group has gone as far as 164 nm. These
spectra reveal several well-defined bands, the amplitudes of which
are sensitive to the configuration of the DNA. An example is shown
in Figure III/3. Also note the superior resolution of CD (Figure
II/3) compared to absorption spectroscopy (Figure II/4).

Protein-Nucleic Acid Complexes: Numerous processes critical
to cellular function involve interactions between proteins and
nucleic acids. The structure of chromatin and the function of
histones in the formation of nu bodies has been an area of intense
activity. No less important are interactions of DNA and RNA with
polymerases, repair enzymes, binding proteins and ribosomes. CD is
one of the most often used tools in the study of nucleic acid:
protein complexes (for a recent review see ref. 13). The recent
work on the CD of nucleic acids and proteins below 180 nm suggests
that data from this region may be valuable in studying these
complexes.

Sugars and Polysaccharides: Unlike proteins and nucleic
acids, sugars do not absorb at wavelengths less than 200 nm. Thus
CD has been one of the principal optical tools in the study of
these molecules. For recent reviews see refs. 30, 31 and 37.

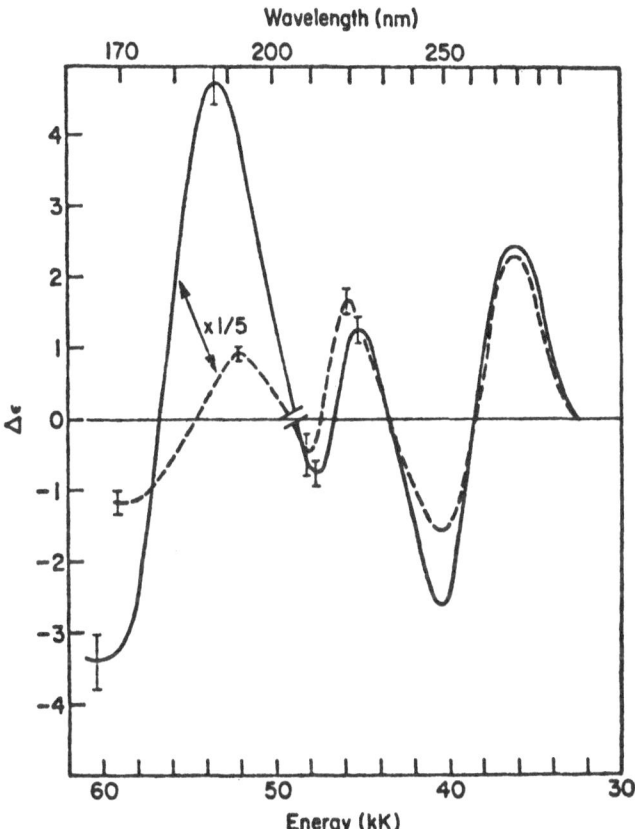

Figure III/3. Circular dichroism of native (——) and heat denatured (----) M. lysodeikticus DNA. From ref.(49).

IV. EMISSION AND EMISSION POLARIZATION SPECTROSCOPIES.

The energy introduced into a molecule by the absorption of a
photon can eventually reappear as another photon usually of a lower
energy. If the emitting state has the same spin multiplicity as
the ground state the transition is spin allowed, the process is
prompt (typically nanoseconds) and is called fluorescence. If the
the multiplicities of the excited and ground state differ, the
process is spin forbidden. This emission is much slower and is
called phosphorescence. For a comprehensive review of emission
spectroscopies of proteins and nucleic acids see ref. 51.

Since Jameson (this conference) is discussing the applications
of SR to fluorescence spectroscopy, my remarks will be limited to a
few special topics which are related to the other areas which I
have discussed. Thus I shall mention the far uv excited
fluorescence of aromatic amino acids, emission polarization
measurements using the photo elastic modulator, and fluorescence
detected CD and MCD .

Far UV Excitation of Fluorescence and Phosphorescence: Figure
IV/1 shows an energy level diagram indicating absorption events
leading to excitation to one of the sequence of excited states.
Usually fluorescence occurs from the lowest singlet state and
phosphorescence from the lowest triplet. Both occur after
vibrational relaxation. The probability of absorption may change
significantly as photon energy is increased. Figure IV/2 shows the
absorption spectra of the three aromatic amino acids. Normally the
fluorescence of phenylalanine in a protein is ignored because of
the low absorption cross section of the longest wavelength band,
the low quantum yield of fluorescence of phenylalanine compared to
tyrosine and tryptophan and energy transfer to these residues. At
shorter wavelengths however the absorption of phe becomes
comparable to the others. Fluorescence from phe excited by $\lambda <$ 200
nm might be an especially useful tool for investigating the
conformation of peptides, peptide hormones and antibiotics which do
not contain trp or tyr.

Teale and Weber (53) showed that the quantum yield for
fluorescence of all three aromatic amino acids was independent of
wavelength down to 210 nm. Still shorter wavelengths may be less
efficient because of competing processes such as ionization. There
are also data which suggest that phosphorescence is favored over
fluorescence when excitation is in the vuv (54). Yet fluorescence
excitation spectra are rarely reported below 240 as a result of the
almost universal use of xenon arcs in modern spectrofluorometers.
SR should make it possible to extend fluorescence excitation
spectra down to the absorption threshold of water and, using the
sample preparation techniques describe above, further into the vuv.

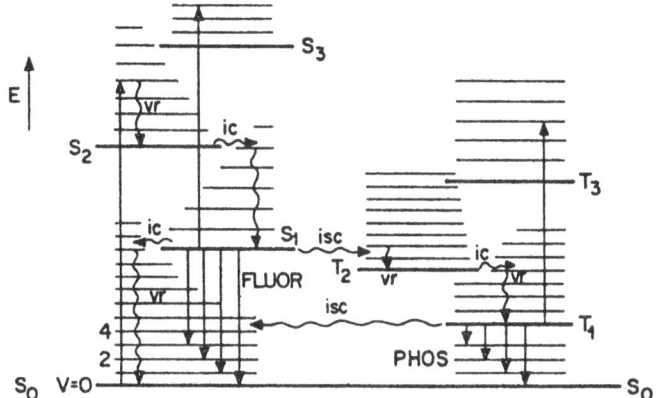

Figure IV/1. ETergy level diagram showing absorption (↑) from both ground and excited states, vibrational relaxation (vr), internal conversion (ic), fluorescence (fluor), intersystem crossing (isc) and phosphorescence (phos.) From ref. (52).

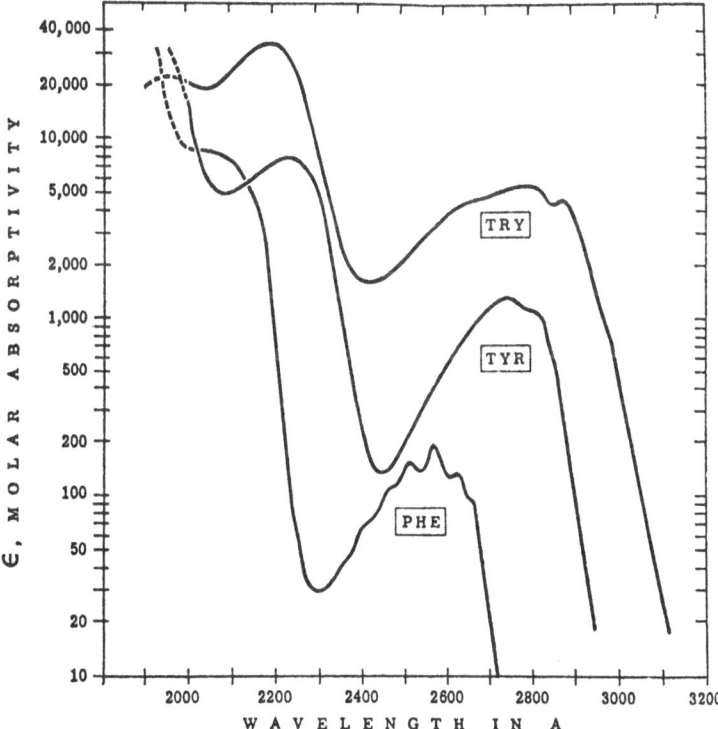

Figure IV/2. Absorption spectra of tryptophan, tyrosine and phenylalanine. The extinction coefficient of phe becomes comparable to those of trp (try) and tyr below 200 nm. From ref. (24).

One of the problems to be investigated is whether photons absorbed by other amino acids can stimulate emission by the aromatic amino acids -- c.f. Yeargers and Augenstein (54).

Fluorescence Polarization Measurements: The impact of SR on measurements involving both polarized light and emission spectroscopy is likely to be particularly strong since these "higher order" measurements have an even greater need for an intense source i.e. a higher quantum requirement. SR will permit measurements to be made with greater precision, in less time -- an important aspect for kinetic measurements -- and to shorter wavelengths.

Three types of polarization measurements are directly compatible with the instrumentation used for CD/MCD and fluorescence. These are measurement of the degree of linear polarization of emitted light, measurement of the circular polarization of emitted light, and measurement of the CD or MCD of the sample detected by monitoring fluorescence emission rather than the transmitted beam.

The linear polarization of emitted light has been a useful means of detecting the existence of multiple transitions which are unresolved components of a single absorption envelope. In the case of fluorescence, linear polarization measurements can be used to determine rotational diffusion times and the viscosity of the microenvironment of the fluorophore. Fluorescence polarization is also an excellent tool for detecting excitation energy transfer between identical chromophores (55).

The circular polarization of emitted light (called circularly polarized luminescence, CPL or circularly polarized emission, CE) provides the same type of information about the excited states of a chromophore that CD provides for the ground state. The circular polarization of emission can be due to the chiral nature of the emitter or of the environment of the emitter or to an external perturbation such as a magnetic field. Thus CE is the emission analog of CD and MCE is the emission analog of MCD. The extensive bodies of theory developed for CD and MCD are applicable to the emission measurements with only minor modifications. Yet CE and MCE measurements are made in only a few laboratories. One of the factors which has limited the application of these measurements particularly in the ultraviolet is the inadequacy of conventional sources; a situation which should be improved by the availability of intense uv radiation from synchrotron storage rings. For a recent review of CE and MCE see Steinberg (56) and Sutherland (34) respectively.

Fluorescence detected circular dichroism (57,58,59) and magnetic circular dichroism (8) (FDCD and FDMCD respectively) monitor the time dependent amplitude modulation of the intensity of the fluorescence from a sample in order to measure the CD or MCD of the sample. This arrangement greatly reduces the signal-to-noise ratio compared to conventional transmission detected CD/MCD. The advantage of this type of measurement is that one can separate the CD/MCD of the fluorescent components of a multichromophoric mixture from the CD/MCD of the nonfluorescent components. Thus it might be possible to determine the CD of the tryptophan residues in a protein in the 240 - 170 nm wavelength region and subtract their contribution from that of the peptide bond. FDCD may also be useful for studying protein-nucleic acid complexes since it may permit separation of the CD of the aromatic amino acids from that of the nucleic acids in the 240 - 310 nm region. The apparatus required for FDCD and FDMCD is rather similar to that required for CD and MCD (8,59) but widespread use has been retarded by the requirement for more intense, continuously tunable uv sources.

V. TIME RESOLVED SPECTROSCOPIES.

Much of the interest to date in the biophysical applications of SR has involved the use of the time structure of SR and is thus limited to the temporal resolution of events in the nanosecond range, since the pulse-to-pules interval is at most about 1 microsecond (at SSRL) and may be much less for storage rings which operate in a multi bunch mode. Certainly some interesting events occur in the time domain provided by SR time structure. In biological work, however, equally interesting events occur on much longer time scales. For example, consider the rearrangement of the peptide chains of the subunits of hemoglobin in response to changes in ligation of one of the heme groups. Various techniques have been developed to trigger conformational changes in a controlled manner. For example, in the millisecond range events can be initiated with a "stopped flow" apparatus while in the microsecond range one can use "temperature jump" perturbations. Other perturbations such as flash photolysis or pressure jumps can also be used. The change in the biological material resulting from these perturbations is followed via some spectral parameter — absorption, CD, fluorescence, etc. Since the ability to make a precise measurement in a short time generally requires an intense source, time resolved measurements are generally limited to wavelengths greater than 250 nm. In some cases it would be desirable to follow spectral changes at shorter wavelengths. For example, allosteric changes in the secondary structure of hemoglobin might be detected by monitoring the CD of the peptide band. Although a number of instruments have been reported which record time resolved CD, none operate in the region of peptide absorption because of the poor signal-to-noise ratio achievable with conventional light sources (60,61,62,63,64,65). SR may make it possible to perform time resolved CD and MCD in this biologically important region of the spectrum. The same potential exists for the other classes of experiments I have described.

ACKNOWLEDGMENTS

    We thank all of the authors and publishers who permitted use
of figures published elsewhere and Dr. Wayne McKinney for a
critical reading of the manuscript. This work was supported by the
U.S. Department of Energy and a Research Career Development Award
(CA 00465) from the National Cancer Institute, DHEW

REFERENCES

1.   Oriel Corporation, 1975 Catalog, Stamford, Connecticut, U.S.
2.   Samson, J. A. R. (1967) Techniques of Vacuum Ultraviolet
     Spectroscopy. John Wiley and Sons, New York, New York, U.S.
3.   Optical Crystals, Harshaw Chemical Company, Solon, Ohio, U.S.
4.   Product Data Bulletin 210, Acton Research Corporation, Acton,
     Massachusetts, U.S.
5.   Bramston-Cook, R. and Erickson, J. O. (1973) Far-Ultraviolet
     Spectroscopy Instrumentation and Applications. Varian
     Instrument Applications, 7(3), 5-7.
6.   Robin, M. B. (1974) Higher Excited States of Polyatomic
     Molecules. Vols. 1 and 2. Academic Press, New York, New
     York, U.S.
7.   Sutherland, J. C., Cimino, G. D. and Lowe, J. T. (1976)
     Emission and polarization spectrometer for biophysical
     spectroscopy. Rev. Sci. Instrum., 47, 358-360.
8.   Sutherland, J. C. and Low, H. (1976) Fluorescence-detected
     magnetic circular dichroism of fluorescent and nonfluorescent
     molecules. Proc. Nat. Acad. Sci. U.S.A., 72, 276-280.
9.   Sutherland, J. C., Cimino, G. D. and Lowe, J. T. (1976)
     Magnetic circularly polarized fluorescence of
     tetraphenylporphyrin. In Excited States of Biological
     Molecules, (J. Birks, ed.), pp. 28-33. John Wiley and Sons,
     New York, New York, U.S.
10. Latimer, P. and Eubanks, C. A. H. (1962) Absorption
     spectrophotometry of turbid suspensions: A method of
     correcting for large systematic distortions. Arch. Biochem.
     Biophys., 98, 274-285.
11. Butler, W. L. (1972) Absorption spectroscopy of biological
     materials. Meth. Enzymol., 27, 3-25.
12. Dorman, B. P., Hearst, J. E. and Maestre, M. F. (1973) UV
     absorption and circular dichroism measurements on light
     scattering biological specimens; fluorescent cell and related
     large-angle light detection technique. Meth. Enzymol., 27D,
     767.
13. Fasman, G. D. (1978) Circular dichroism analysis of chromatin
     and DNA-nuclear protein complexes. Meth. Cell Biol., 18,
     327-349.
14. Johnson, W. C. and Tinoco, I. (1972) Circular dichroism of
     polypeptide solutions in the vacuum ultraviolet. J. Am. Chem.
     Soc., 94, 4389-4390.

15. Preiss, J. W. and Setlow, R. (1955) Spectra of some amino acids, peptides, nucleic acids and protein in the vacuum ultraviolet. J. Chem. Phys., 25, 138-141.

16. Yamada, T. and Fukutome, H. (1968) Vacuum ultraviolet absorption spectra of sublimed films of nucleic acid bases. Biopolymers, 6, 43-54.

17. Inagaki, T., Hamm, R. N., Arakawa, E. T. and Painter, L. R. (1974) Optical and dielectric properties of DNA in the extreme ultraviolet. J. Chem. Phys., 61, 4246-4250.

18. Emerson, L. C., Williams, M. W., Tang, I., Hamm, R. N. and Arakawa, E. T. (1975) Optical properties of guanine from 2 to 82 eV. Radiat. Res., 63, 235-244.

19. Inagaki, T., Hamm, R. N., Arakawa, E. T. and Birkhoff, R. D. (1975) Optical properties of bovine plasma albumin between 2 and 80 eV. Biopolymers, 14, 839-847.

20. Williams, M. W., Arakawa, E. T., Birkhoff, R. D., Hamm, R. N., Schweinler, H. C. and MacRae, R. A. (1975) Optical properties of chloroplasts and red blood cells in the vacuum uv. Radiat. Res., 61, 185-190.

21. Goldford, A., Saidel, L. and Mosovich, E. (1951) The ultraviolet absorption spectra of proteins. J. Biol. Chem. 193, 397-404.

22. Setlow, R. B. and Guild, W. R. (1951) The spectrum of the peptide bond and other substances below 230 nm. Arch. Biochem. Biophys., 34, 223-225.

23. Ham, J. and Platt, J. (1952) Far u.v. spectra of peptides. J. Chem. Phys., 20, 335-336.

24. Wetlaufer, D. B. (1962) Ultraviolet Spectra of Proteins and Amino Acids. In Advances in Protein Chemistry, (C. B. Anfinsen, Jr., M. L. Anson, K. Bailey and J. T. Edsall, eds.), Vol. 17, pp. 303-390. Academic Press, New York, New York, U.S.

25. Setlow, R. B. and Pollard, E. C. (1962) Molecular Biophysics, Addison-Wesley.

26. Rosencwaig, A. (1975) Photoacoustic spectroscopy: A new tool for investigation of solids. Analyt. Chem., 47, 592A.

27. Jirgensons, B. (1973) Optical Activity of Proteins and Other Macromolecules. Springer-Verlag, New York, New York, U.S.

28. VanHolde, K. E. (1971) Physical Biochemistry, Chapt. 10. Prentice-Hall Inc., Englewood Cliffs, New Jersey, U.S.

29. Bloomfield, V. A. Crothers, D. M. and Tinoco, I. (1974) Physical Chemistry of Nucleic Acids. Harper and Row, New York, New York, U.S.

30. Pysh, E. S. (1976) Optical activity in the vacuum ultraviolet. Ann. Rev. Biophys. Bioengr., 5, 63-75.

31. Johnson, W. C. Jr. (1978) Circular dichroism spectroscopy and the vacuum ultraviolet region. Ann. Rev. Phys. Chem. (in press).

32. Djerassi, C., Bunnenberg, E. and Elder, D. L. (1971) Organic chemical applications of magnetic circular dichroism. Pure and Appl. Chem., 25, 57-90.

33. Holmquist, B. and Vallee, B. L. (1978) Magnetic circular dichroism. Meth. Enzymol., 49, 149-178.

34. Sutherland, J. C. (1978) The magnetic optical activity of porphyrins. In The Porphyrins, (D. Dalton, ed.), Vol. 3. Academic Press, New York, New York, U.S. (in press).

35. Velluz, L., Legrand, M. and Grosjean, M. (1965) Optical Circular Dichroism. Academic Press, New York, New York, U.S.

36. Sutherland, J. C. and Klein, M. P. (1972) Magnetic circular dichroism of cytochrome c. J. Chem. Phys., 57, 76-86.

37. Stevens, E. S. (1978) Far (vacuum) ultraviolet circular dichroism. Meth. Enzymol., 49, 214-221.

38. Eaton, W. A., Hanson, L. K., Stephens, P. J., Sutherland, J. C. and Dunn, J. B. R. (1978) The optical spectra of oxy- and deoxyhemoglobin. J. Am. Chem. Soc., 100, 4991-5003.

39. Perutz, M. F., Heidner, E. J., Ladner, J. E., Beetlestone, J. G., Ho, C. and Slade, E. F. (1974) Influence of globin structure on the state of the heme. III. Changes in heme spectra accompanying allosteric transitions in methemoglobin and their implications for heme-heme interaction. Biochemistry, 13, 2187-2200.

40. Greenfield, N. J. and Fasman, G. D. (1969) Computed circular dichroism spectra for evaluation of protein conformation. Biochemistry, 8, 4108-4116.

41. Billardon, M. and Badoz, J. (1966) (a) Modulateur de Birefrigence. C. R. Acad. Sci., 262, 1672-1675. (b) Mesure des variations d'indice de réfraction en region d'absorption par une méthode polarimetrique. C. R. Acad. Sci., 263, 26-29.

42. Kemp, J. C. (1969) Piezo-optical birefringence modulators: New use for a long-known effect. J. Opt. Soc. Am., 59, 950-954.

43. Mollenauer, L. F., Downie, D., Engstrom, H. and Grant, W. B. (1969) Stress plate optical modulator for circular dichroism measurements. Appl. Opt., 8, 661-665.

44. Jasperson, J. N. and Schnatterly, S. E. (1969) An improved method for high reflectivity ellipsometry based on a new polarization modulation technique. Rev. Sci. Instrum., 40, 761-767.

45. Breeze, R. H. and Ke, B. (1972) A circular dichroism spectrophotometer using an elasto-optic modulator. Anal. Biochem., 50, 281-303.

46. Wells, B. D. and Yang, J. T. (1974) A computer probe of the circular dichroic bands of nucleic acids in the ultraviolet region. I. Transfer ribonucleic acid. Biochemistry, 13, 1311-1316.

47. Wells, B. D. and Yang, J. T. (1974) A computer probe of the circular dichroic bands of nucleic acids in the ultraviolet region. II. Double-stranded, ribonucleic acid and deoxyribonucleic acid. Biochemistry, 13, 1317–1321.

48. Li, H. J., Isenberg, I. and Johnson, W. C. (1971) Absorption and circular dichroism studies on nucleohistone IV. Biochemistry, 10, 2587–2593.

49. Lewis, D. G. and Johnson, W. C. (1974) Circular dichroism of DNA in the vacuum ultraviolet. J. Mol. Biol., 86, 91–96.

50. Sprecher, C. A. and Johnson, W. C. (1977) Circular dichroism of the nucleic acid monomers. Biopolymers, 16, 2243–2264.

51. Steiner, R. F., and Weinryb, I. (1971) Excited States of Proteins and Nucleic Acids. Plenum Press, New York, New York, U.S.

52. Lamola, A. A. and Turro, N. J. (1977) Chapter 2. In The Science of Photobiology, (K. C. Smith, ed.), pp. 27–61. Plenum Press, New York, New York, U.S.

53. Teale, F. W. J. and Weber, G. (1957) Ultraviolet fluorescence of the aromatic amino acids. Biochem. J., 65, 476–482.

54. Yeargers, E. and Augenstein, L. (1968) Vacuum ultraviolet studies on the nature of the radiation inactivation of trypsin. Biophys. J., 8, 500–509.

55. For a review of measurements of the linear polarization of fluorescence see J. W. Longworth's article in ref. 51.

56. Steinberg, I. Z. (1978) Circularly polarized luminescence. Meth. Enzymol., 49, 179–199.

57. Turner, D. H. (1978) Fluorescence detected circular dichroism. Meth. Enzymol., 49, 199–214.

58. Turner, D. H., Tinoco, I. and Maestre, M. (1974) Fluorescence detected circular dichroism. J. Am. Chem. Soc., 96, 4340–4342.

59. Turner, D. H., Tinoco, I. and Maestre, M. F. (1975) Fluorescence detected circular dichroism study of the anticodon loop of yeast tRNA[Phe]. Biochemistry, 14, 3794–3799.

60. Bayley, P. M. and Anson, M. (1974) Stopped-flow circular dichroism: A new fast-kinetic system. Biopolymers, 13, 401–405.

61. Ferrone, F. A., Hopfield, J. J. and Schnatterly, S. E. (1974) The measurement of transient circular dichroism: A new kinetic technique. Rev. Sci. Instrum., 45, 1392–1396.

62. Tsuda, M. (1975) Apparatus for rapid measurement of optical rotation changes. Rev. Sci. Instrum., 46, 1419–1420.

63. Goodall, D. M. and Cross, M. T. (1975) Polarimetric stopped-flow apparatus. Rev. Sci. Instrum., 46, 391–397.

64. Anson, M., Martin, S. R. and Bayley, P. M. (1977) Transient CD measurements at submillisecond time resolution -- application to studies of temperature-jump relaxation of equilibria of chiral biomolecules. Rev. Sci. Instrum., 48, 953–962.

65. Gruenewald, B. and Knoche, W. (1978) Pressure jump method with detection of optical rotation and circular dichroism. Rev. Sci. Instrum., 49, 797–801.

SPECTROSCOPY AND PHOTOPHYSICS. I. RADIATIVE AND NON-RADIATIVE DECAY PROCESSES[1]

David Phillips

Department of Chemistry, The University

Southampton, SO9 5NH

INTRODUCTION

This lecture is concerned very briefly with the fundamentals of absorption and emission of electromagnetic radiation through electric dipole transitions, and where appropriate factors determining the rate of depopulation of excited electronic states through non-radiative decay paths. Discussion is concentrated upon radiative decay, since the study of luminescence of molecules in biological systems is of both fundamental and applied importance.

ELECTRIC DIPOLE TRANSITIONS

Electric dipole transitions arise through the interaction of the oscillating electric field associated with a light wave and the electrons in the atomic or molecular system. Quantum mechanics leads to the mathematical expression of the above qualitative explanation for the probability of an electronic transition as proportional to R_{if}^2 where R_{if}, termed the transition moment integral between the initial state i and final state f is given by Equation(1)

$$R_{if} = \int_{-\infty}^{+\infty} \psi_{ef} \hat{M} \psi_{ei} d\tau_e \int_{-\infty}^{+\infty} \psi_{nf} \psi_{ni} d\tau_n \quad \cdots (1)$$

where the ψ_e represent electronic wavefunctions, the ψ_n represent vibrational wavefunctions, M is the electronic dipole moment operator and where the Born-Oppenheimer principle of separability of electronic and vibrational wavefunctions has been invoked. The first term in Equation(1)is an integral involving only the electronic wavefunctions of the system, and the second terms, which are simply vibrational overlap integrals, when squared are the familiar Franck-Condon factors.

For an atomic system we are concerned only with electronic wavefunctions, and we should enquire what determines the magnitude of the electronic integral. The exact evaluation of the electronic transition moment integral is extremely difficult, but general statements can be made about the conditions under which the value is zero or finite, which corresponds to a forbidden and an allowed transition respectively; in other words selection rules are generated. Firstly we can say something about transitions between different spin states by separating electronic wavefunctions into space and spin parts. The electronic part of the transition moment integral $(R_{if})_e$ then is given by (2) in which the final term represents the overlap of the spin wavefunction of the initial and final states:

$$(R_{if})_e = \int_{-\infty}^{+\infty} \psi_{esf}\hat{M}\psi_{esi}d\tau_{es} \int_{-\infty}^{+\infty} \psi_{sf}\psi_{si}d\tau_s \qquad (2)$$

where ψ_{es} represents the space part of the electronic wavefunctions and ψ_s the electronic spin wavefunctions.

Pure spin wavefunctions are orthonormal, i.e.

$$\int_{-\infty}^{+\infty} \psi_s(\alpha)\psi_s(\alpha) = \int_{-\infty}^{+\infty} \psi_s(\beta)\psi_s(\beta) = 1 \qquad (3)$$

$$\int_{-\infty}^{+\infty} \psi_s(\alpha)\psi_s(\beta) = 0 \qquad (4)$$

Thus an electric dipole transition is spin-forbidden if the initial and final states have different multiplicities. Absorption corresponds to the radiative transition from a state of lower energy to one of higher energy. Fluorescence is exactly the reverse process between states of the same multiplicity. Emission processes may however be stimulated (induced by a radiation field, as is absorption) or spontaneous. Figure 1 shows the rates of the absorption, of stimulated emission of spontaneous emission.

In a two-level system as shown given that $B_{if} = B_n$, the net rate of change in the system undergoing both stimulated and spontaneous emission process is given by the left-hand side of Equation (5) and is:

$$n_i B_{if}\rho - n_f B_{if}\rho - n_f A_{if} = 0 \qquad (5)$$

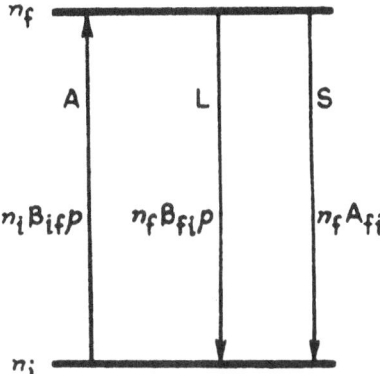

Figure 1. Rates of absorption, stimulated emission, and spontaneous
           emission processes between levels i and f of molecular
           systems of populations n_i and n_f respectively. A is the
           absorption process, L is stimulated emission, and S is
           spontaneous emission. A_{fi} is the Einstein A coefficient,
           B is the Einstein B coefficient, and ρ is the radiation
           density.

set equal to zero in an equilibrium situation. Initially the population in the upper state is given by the Boltzmann distribution: $g=1$

$$\frac{n_f}{n_i} = \frac{g_f}{g_i} e^{-(\varepsilon_f - \varepsilon_i)/kT} = e^{-h\nu/kT} \qquad (6)$$

for singlet states. For electronic transitions, this means that the
initial population in the upper state n_f is negligible, and there
is in the presence of a radiation field a net stimulated absorption
of radiation followed by spontaneous emission. The relationship
between the Einstein A and B coefficient is:

$$A_{if} = \frac{8\pi h\nu^3}{C^3} \cdot B_{if} \qquad (7)$$

from which it can be seen that the probability of spontaneous
emission depends upon ν^3. This spontaneous emission is likely for
electronically excited states, but for vibrational transitions, or
rotational transitions, where ν^3 becomes very small, spontaneous
emission processes have very low probability. Except in the presence of high radiation fields, spontaneous emission processes are
dominant.

Equation (7) is of importance since it relates the Einstein A
coefficient, analogous to a rate constant for spontaneous radiative
decay k_R, to the B coefficient which is related to the easily

measured molar decadic extinction coefficient (for molecules) for absorption, ε (7) is only quantitatively correct for atoms, but Strickler and Berg[2] have given an expression (8) which is often used to relate these quantities in molecules, where n^2 is

$$k_R = 2.88 \ 10^{-9} \ n^2 \left\langle \bar{\nu}^{-3} \right\rangle^{-1}_{AV} \int \frac{\varepsilon(\bar{\nu})}{\bar{\nu}} \ d\bar{\nu} \qquad (8)$$

he refractive index of the medium in which the measurements are made, and $\left\langle \bar{\nu}^{-3} \right\rangle_{AV}$ is an elaborate way of fixing the centre of gravity of fluorescence (see below). There exists a more approximate relationship (9) less rigorous than (8) but easily remembered.

$$10^4 \ \varepsilon \ (1.mol^{-1} \cdot cm^{-1}) \simeq k_R \ (s^{-1}) \qquad (9)$$

Thus for a fully allowed transition, $\varepsilon \sim 10^5$, the corresponding value of k_R would be $\sim 10^9 s^{-1}$.

SELECTION RULES

Manipulation of equation (1) permits evaluation of selection rules for radiative transition in a variety of cases, as was shown above for the case of the spin selection rule. For light atoms (i.e. first two rows in the periodic table) and molecules made up of three atoms, the following selection rules are generated.

Atoms   State notation $^{2S+1}L_J$ e.g. $^1S_0$, $^3P_1$ etc.

Selection rules $\Delta S=0$, $\Delta l=\pm L$, $\Delta L=0$, ± 1
$\Delta J=0$, 1 plus $J=0 \not{/} J=0$

Diatomic Molecules   State notation $^{2S+1}\Lambda_\Omega$; g,u; +, −.

e.g. $^1\Sigma_g^+$, $^3\Sigma_g^-$, $^1\Delta_g$ etc.

Selection rules $\Delta S=0$, $\Delta\Lambda=0,\pm1$, $\Delta\Sigma=0, \Delta\Omega=0,\pm1$
$g \leftrightarrow u$, $g \not\leftrightarrow g$, $u \not\leftrightarrow u$, $+ \leftrightarrow +$, $- \leftrightarrow -$
$+ \not\leftrightarrow -$

In addition to the electronic selection rules given above for diatomic molecules, the observed intensity in radiative transitions is dictated by the Franck–Condon distribution. These are illustrated for a hypothetical diatomic molecule in Figure 2. The difference in energy between most intense absorption and fluorescence bands is a measure of the change in bond length between ground and excited states, and is termed the Stokes loss (see below).

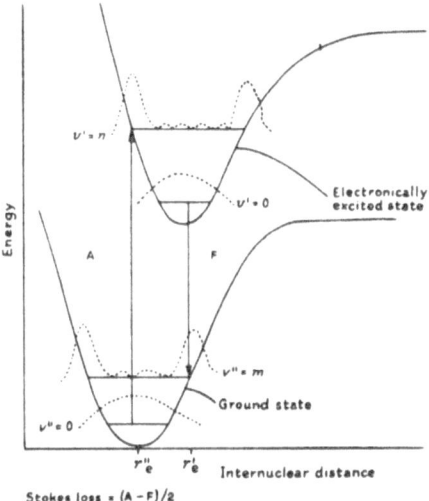

Stokes loss = (A - F)/2

Figure 2.   The significance of the Stokes loss in terms of potential energy curves of upper and lower electronic states
of a hypothetical diatomic molecule.

Complex polyatomic molecules

State notation At its simplest the
orbital whence the electron came and
that to which it is excited are
identified, e.g. n→π* excitation
gives rise to ^{1}nπ* (and ^{3}nπ*) states
π→π* excitation gives singlet and
triplet ππ* states, σ→σ gives ^{1,3}σσ*
states etc.

For molecules possessing reasonably
high degrees of symmetry, group
theoretical notation is used, e.g.
^{1}A$_{1g}$ (ground state of benzene), ^{1}B$_{2u}$
(first excited singlet state of ben-
zene) etc.

Selection rules. Apart from the
ΔS=0 and g→u (for centrosymmetric)
molecules) selection rules, those
for polyatomics are not easily written
in shorthand. The rules can, however,
be summarised in terms of oscillator
strengths using the following factors
with approximate order of magnitudes
shown in brackets; fully allowed
transition, F_A=1; spin-forbidden,
$f_s \sim 10^{-5}$;

$$F = F_A \cdot f_s \cdot f_o \cdot f_p \cdot f_m \qquad (10)$$

overlap forbidden, $f_o \sim 10^{-2}$; parity forbidden, $f_p \sim 10^{-1}$ symmetry
(or momentum) forbidden, $f_m \sim 10^{-1} - 10^{-3}$. These factors apply equally
to radiative rate-constants (k_R) as molar decadic extinction coefficients (ε) (see above).

The distribution of absorption and fluorescence in complex polyatomic molecules often contains no resolvable vibrational features
due to the high density of vibrational levels accessible in the
radiative transition. In symmetry forbidden transitions, however,
the stringent symmetry requirements may dictate that few of the
dense manifold of vibrational levels carry any oscillator strength,
leading to very sharp bands and easily identifiable progressions.
This is nicely illustrated in the case of benzene in Figure 3. For
many polyatomic molecules an approximate mirror symmetry relationship between vibrational distribution in absorption and emission
is obeyed (cf. Figure 3).

ENVIRONMENTAL EFFECTS

We consider here the general effect of solvent relaxation which
is universally observed. Figure 4a shows that the vapour phase
0-0 bands in absorption and fluorescence of a molecule are identical, whereas in solution with solvent of static dielectric constant
ε, refractive index n, the bands are no longer coincident. The
differences can be rationalized as follows. From Onsager theory,
a solute molecule of dipole moment μ in a spherical cavity of
radius a polarizes the dielectric of the solvent, producing a
reaction field. This is given for the ground-state of the solvent
molecule (of dipole moment μ_o), by (11). Upon excitation, and invok-
ing the Franck-Condon principle, the electronic-excitation is much

$$R_o = \frac{2\mu_o}{a^3} \frac{(\varepsilon-1)}{(2\varepsilon+1)} \qquad (11)$$

more rapid than the dielectric relaxation time of the solvent, so
the reaction field of the Franck-Condon excited state S_1' is given
by (12), in which the solvent still experiences the ground-state

$$R_1' = \frac{2\mu_o}{a^3} \frac{(n^2-1)}{(2n^2+1)} \qquad (12)$$

solvent dipole moment, and the high frequency dielectric constant
$(=n^2)$ is used in place of ε. If the decay time of the excitation
decay time of the excited state of the solvent exceeds the dielectric

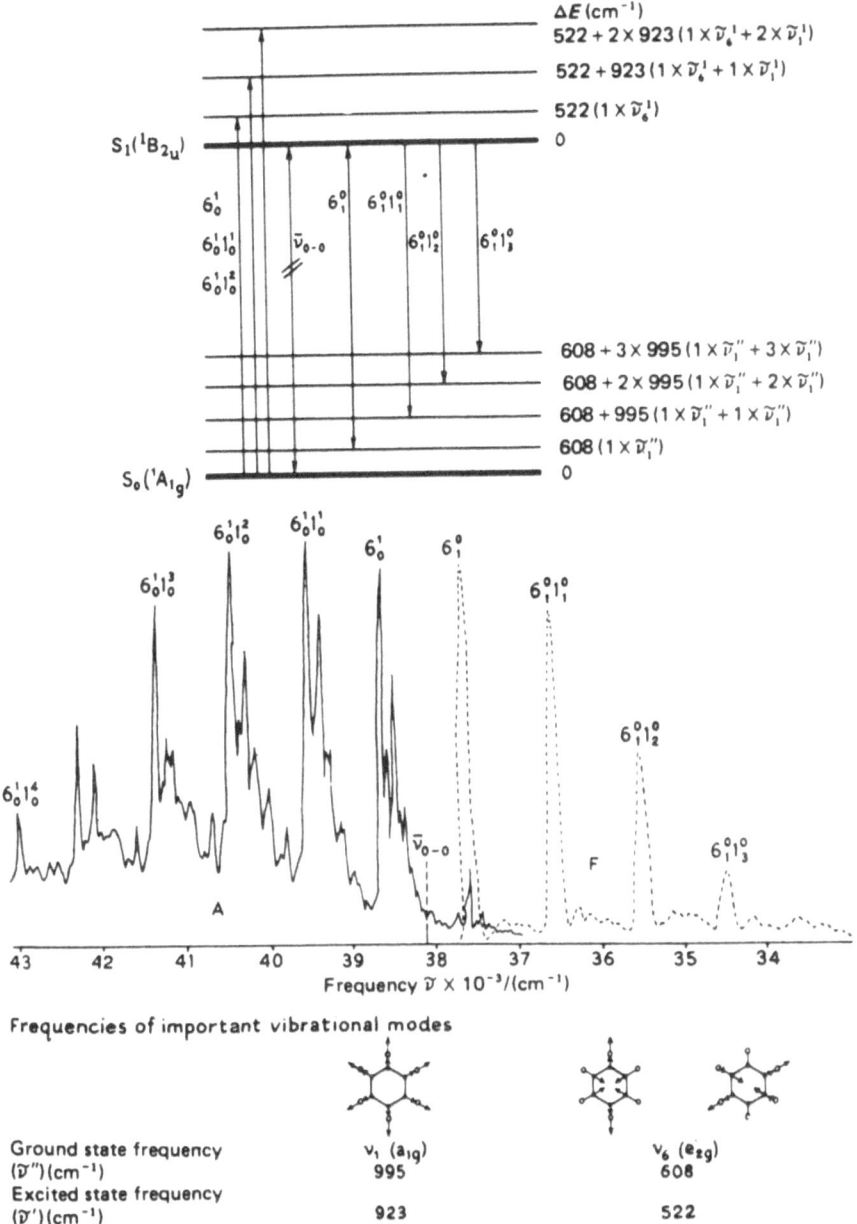

Frequencies of important vibrational modes

	Ground state frequency $(\tilde{\nu}'')(cm^{-1})$	Excited state frequency $(\tilde{\nu}')(cm^{-1})$
ν_1 (a$_{1g}$)	995	923
ν_6 (e$_{2g}$)	608	522

Figure 3.   Absorption and fluorescence of benzene vapour. The
zero-zero transition is absent because of symmetry restrictions. The spectral features are built upon the
ν_1 (a$_{1g}$) totally symmetric ring breathing vibration and
the doubly degenerate ν_6 (e$_{2g}$) in-plane distortion mode.
(based upon Fig. 1 in J. Chem. Phys., 52, 5366 (1970),
and Proc. Roy. Soc. A., 259, 499(1966).

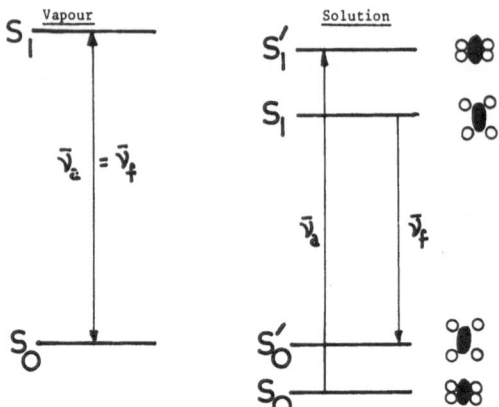

<u>Figure 4</u>   Schematic diagrams showing solvent relaxation.

relaxation time, then the equilibrated excited state system of
dipole moment μ, has a reaction field given by (13).

$$R_1 = \frac{2\mu 1}{a^3} \frac{(\varepsilon -1)}{(2\varepsilon +1)} \qquad (13)$$

Following fluorescence, the non-equilibrium ground-state   has
a reaction field R_0' Manipulation of these equations shows that the

$$R_0' = \frac{2\mu 1}{a^3} \frac{(n^2 -1)}{(2n^2 +1)} \qquad (14)$$

difference in energy of the 0-0 transitions in absorption and
emission is given by (15). For non-polar solvents $\varepsilon \approx n^2$, and

$$\Delta \bar{\nu} = \frac{2(\mu 1 - \mu_0)^2}{hca^3} \left[\frac{(\varepsilon -1)}{(2\varepsilon +1)} - \frac{(n^2 -1)}{(2n^2 +1)} \right] \qquad (15)$$

no shifts are observed. For polar solvents shifts can be large.
Thus fluorescence spectral position can be very sensitive to the
environment of the fluorescent molecule, and such observations are

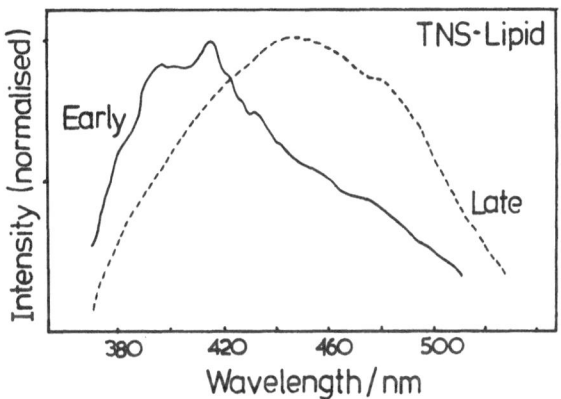

Figure 5. Time-resolved emission spectra for TNS bound to egg phos-
phatidyl choline. Early spectrum Δt=0, gate width δt=
2.5 ns. Late spectrum Δt=31ns, δt=2.5 ns.

widely used in biological systems to identify different sites in
heterogeneous systems such as cell membranes.

Where solvent relaxation is slow compared with the
decay time of the fluorescent molecule, time-dependent fluorescence
spectra will be observed. This is illustrated in Figure 5 for the
case of the aromatic probe molecule 3,6-toluidinyl naphthalene
sulphonate (TNS) bound to a lipid (egg phosphatidylcholine) observed
by ns time-dependent fluorescence spectroscopy.[3] When photons
detected shortly after pulsed excitation only are recorded, the
spectrum is blue-shifted, typically of emission from the Franck-
Condon (unrelaxed) excited state. At later times a red-shifted
spectrum is observed which is typical of that from a state about
which solvent relaxation is complete.

In addition to the general interaction outlined above, specific
interactions may be observed with particular chromophores. For
example, $n \rightarrow \pi^*$ transitions of carbonyl compounds exhibit a blue-shift
in hydrogen bonding, protic, or polar solvents. In addition to such
effects, long-lived species (e.g. triplet-states) may suffer diffusional quenching by small concentrations of dissolved impurities
unless diffusion is suppressed. Such effects are discussed in the
succeeding paper.

NON-RADIATIVE DECAY

Whether or not any excited state of a molecule is emittive
depends upon the magnitude of the radiative rate-constant depopula-
ting that state relative to the sum of those for competing non-
radiative decay processes, including chemical reactions, as can be
seen for the expression for the quantum yield of fluorescence given
by (16). The most critical factors in determining the size of non-
radiative rate constants in complex polyatomic molecules are the

$$\Phi_F = \frac{k_R}{(k_R + \Sigma k_{NR})} \tag{16}$$

proximity of lower-lying electronic states (inverse energy gap law).
and for spin-forbidden processes, the presence of atoms of high
nuclear charge (heavy atom effect). Deuterium substitution reduces
the magnitude of non-radiative rate-constants. The reasons for
these observations cannot be discussed here, but result in the
following generalisations concerning polyatomics. With the notable
exception of azulene, only the lowest excited singlet and triplet
states (S_1 and T_1) are strongly luminescent. Aromatic molecules,
with relatively large $S_1 \rightarrow T_1$ and $T_1 \rightarrow S_0$ energy gaps are in general
strongly luminescent. By contrast carbonyl compounds, for which
the $S_1 \rightarrow T_1$ energy gap is very small are only weakly or non-fluorescent
and also weakly phosphorescent. Simple olefins are non-luminescent.

Thus aromatic molecules are widely used as fluorophores in biological systems. Thus for example naphthalene derivatives can be used as extrinsic probes, tryptophan derivatives as intrinsic probes in fluorescence studies.

POLARIZATION OF EMISSION

Referring back to equation (1) we can recognise that the transitional dipole is a vector quantity, or in other words the change in position of the electron during the transition is in a fixed direction with respect to some system of co-ordinates in which the molecular frame is fixed. Thus the dipole moment operator is resolvable into three directions such that:

$$\hat{M}^2 = \hat{M}^2_x + \hat{M}^2_y + \hat{M}^2_z \tag{17}$$

and in general only one of these components will be non-zero in value. Thus if a perfectly ordered system such as a fixed molecular single crystal is observed in absorption, using plane-polarized light, there will generally be one orientation of the crystal axes with respect to the plane of the polarization of the light which maximizes the absorption probability. Normally fluorescence involves the transition between the same two states as are observed in absorption and thus the fluorescence is usually polarized parallel to the absorption. For system oriented such that the transitional dipole is aligned at an angle other than zero degrees from the electric vector of the exciting light however, two components of fluorescence can be observed, one parallel to the exciting light $I_{\parallel}$, the other perpendicular to it, $I_{\perp}$. In these cases a quantity termed the degree of polarization, P,

$$P = (I_{\parallel} - I_{\perp})/(I_{\parallel} + I_{\perp}) \tag{18}$$

can be defined which has the limits of +1 for a system with transition moment perfectly aligned with electric vector of exciting light, and tends to -1 for the perpendicular orientation. For a perfectly random array of molecules, the degree of polarization takes the value of +0.5 for emission whose transition moment is parallel to that for absorption (as in fluorescence) and -0.33 for a transition polarized perpendicularly to the moment for absorption, as is often observed in phosphorescence.

The above discussion assumes that the molecule under observation remains stationary during the observation of fluorescence. Clearly if rotational motion of the molecule occurs on a time-scale short compared with the decay time, any polarizations in emission will be lost before the experimental measurement is made. Many fluorophores have decay

times in the ns region, and thus observation of loss of polarization
on this time-scale can lead to information about the rotational
motion of molecules in media for which the relaxation time is of
the same time domain. For fluorophores with nanosecond decays
this is confined to macromolecules in fluid media, such as proteins
and other biological systems in non-viscous media,synthetic polymers in
fluid solution or smaller molecules in viscous media. For smaller
fluorescen   molecules in non-viscous media, rotational relaxation
occurs on a ps time-scale, and fluorophores such as dyestuffs with
decay times of this order are required for such studies. Since
rotational relaxation times are related to the shapes and conformations adopted by molecules, the observations of partial depolarization is of great potential use. It should be noted that pulsed
excitation is not a requirement for these measurements except in
the sense that the decay time of the probe fluorophore needs to be
known, otherwise static steady illumination suffices (see below).
However, much more information can be gleaned if time resolved
polarization measurements are made[4], and in this regard high intensity sources of tunable short-duration pulses are desirable. Both
synchrotron storage ring and mode-locked lasers could be used for
these purposes.

EXPERIMENTAL TECHNIQUES

Emission Spectra

In brief fluorescence and phosphorescence spectra under continuous illumination can be obtained on a spectrophosphofluorimeter
shown schematically in Figure 6, and comprising the following.

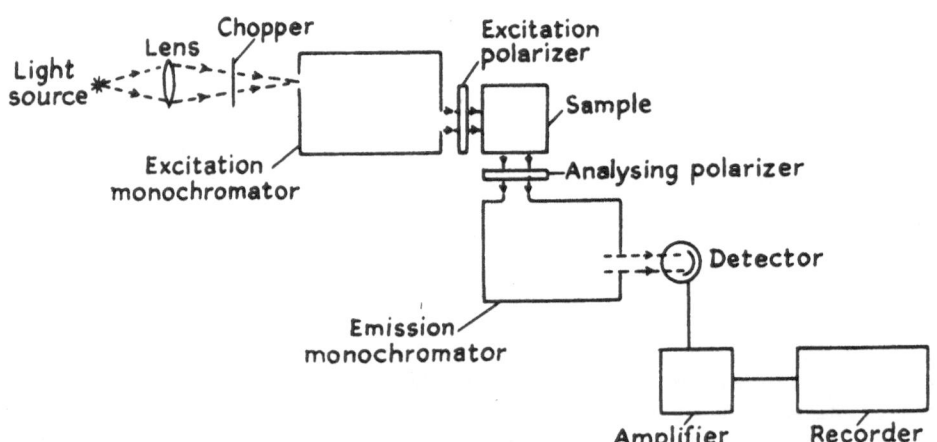

Figure 6.   Block diagram of typical spectrofluorimeter

(a)   Light source to give excitation wavelengths typically between
      200 and 800 nm. A xenon arc lamp of 150-500W power run from
      a d.c. stabilized power supply is a favourite source.

(b)   An excitation monochromator for the selection of excitation
      wavelengths, usually of high resolution grating type.

(c)   An emission monochromator for the spectral analysis of the
      luminescence with similar or better characteristics to those
      of the excitation monochromator.

(d)   A cell-housing in which a variety of cells, often of fused
      silica, may be accommodated, together with temperature control
      systems.

(e)   Detection system, usually in the form of a sensitive photo-
      multiplier, the electrical output from which is amplified
      and displayed on a strip-chart recorder. For greater sensi-
      tivity single photon-counting detection may be employed.

(f)   Correction system. Emission spectral profiles obtained from
      a simple instrument of the above type will be distorted from
      their true shapes owing to the fact that sensitivity of the
      photodetector and the transmission characteristics of the
      optical system are non-linear functions of wavelength of the
      light emitted. The simplest method of correcting spectra is
      to compare the instrumental response for a standard compound
      of known corrected spectral characteristics with that of the
      sample of interest, although spectrofluorimeters are now avail-
      able which compensate electronically for the wavelength response
      of the system.

(g)   Phosphorimeter attachment. This is used for the measurement
      of delayed emission characteristics, and consists of a mech-
      anical or electronic device which interrupts the excitation
      beam periodically. If the detection system is gated so that
      emission time is viewed only after a time has elapsed after
      the cut-off of the excitation, then short-lived emission such
      as prompt fluorescence will have decayed to zero intensity,
      and thus only long-lived emission will be observed. The decay
      time of the luminescence can also be measured using this
      technique, and for a mechanical device such as a rotating
      toothed chopper, the limit of measurable lifetime is of the
      order of 1 ms. Electronic devices can be used to measure
      decay times much shorter than this (see below).

(h)   Polarizers. If polarized spectra are required, a prism polar-
      izer can be inserted at the exit slit of the excitation mono-
      chromator which transmits plane polarized light. An analysing

polarizer is then inserted on the emission side of the optical
system, and usually this can be easily set to measure light
which has the electric vector parallel to that of the excitation radiation, and perpendicular to it (see above).

Emission Decay Times

Phosphorescence. The use of mechanical devices such as that just
described to obtain emission life-times limits lifetime measurements to those emissions having lifetimes longer than a few ms.
This is adequate however for many phosphorescent samples.

Fluorescence. Two methods have been found to be particularly useful: phase modulation of the excitation source coupled with
phase sensitive detection and single photon counting techniques.

(a) Phase modulation. Phase modulation involves the modulation
     of the exciting beam at a high frequency. The phase of the
     luminescence is then compared with that of the exciting light.
     For an emission decaying exponentially with a lifetime τ, the
     phase lag p relative to that of the exciting beam is given by:

$$\omega\tau = \tan p$$

     where ω is the modulation frequency. Thus by feeding the
     excitation and emission into an electronic circuit capable of
     determining the out-of-phase angle of the two signals, τ may
     be obtained. This method has found much favour in laboratories
     studying luminescence of biological systems, and has a claimed
     short lifetime limit of $\sim$ 100 ps. For analysis of multi-
     component decays the method has some difficulties however.

Time-correlated single-photon counting methods. This is now widely
used, and a typical set-up using a conventional excitation source
is shown in Figure 7. Here light from a nitrogen flash-lamp excites
the sample, after passing through a monochromator. This flash is
also used to start a counting sequence in a time-to-amplitude converter (TAC). In simple terms the start pulse triggers a voltage
ramp where the voltage is proportional to time. The voltage ramp
is stopped by a signal from the photomultiplier on observation of
a single emitted photon. The voltage attained by the ramp is
stored as one count in a multichannel analyser and the TAC reset
for the reception of the next start pulse. In this way the multichannel analyser accumulates a histogram of probability of emission
as a function of time after excitation. If the experiment is continued over a long data accumulation time, the histogram resolves
itself into a smooth curve, which for simple single-component decay
will be a single exponential. With suitable computational procedures, the method can be used (with caution) to analyse multi-
component decays. There are considerable advantages to be gained

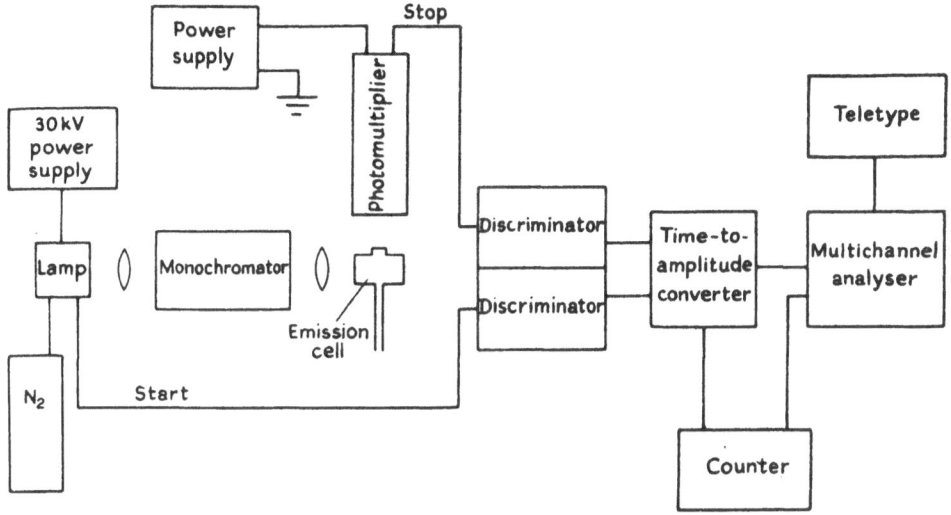

Figure 7. Block diagram of time-correlated single-photon counting apparatus.

by replacing the conventional flash discharge lamp shown in Figure 7 by higher powered faster repetition-rate sources such as the synchrotron storage-ring source and pulsed lasers. These developments are discussed elsewhere in this volume.

REFERENCES

1. These notes are based in part upon an article entitled "Fluorescence and Phosphorescence Spectroscopy" by D. Phillips and K. Salisbury in Spectroscopy Vol 3 pp 161-198, eds. B.P. Straughan and S. Walker, Chapman and Hall, London 1976.

2. S.J. Strickler and R.A. Berg., J. Chem. Phys., 1962, 37, 814.

3. K.P. Ghiggino, D. Phillips and A.G. Lee "Time-resolved fluorescence properties of binding-site probes", Photochem. Photobiol. (submitted).

4. T. Tao, Biopolymers, 1969, 8, 609.

SPECTROSCOPY AND PHOTOPHYSICS. II. BIMOLECULAR PROCESSES

David Phillips

Chemistry Department, The University, Southampton SO9 5NH

INTRODUCTION

The bimolecular processes by which electronically excited states may be deactivated are listed below

(a) Chemical reaction
(b) Enhancement of non-radiative decay
(c) Electronic energy transfer
(d) Complex-formation, including charge-transfer complexes.

These will be discussed briefly, followed by a discussion of other phenomena of interest to those using photoluminescence techniques to study biological systems, namely delayed fluorescence and phosphorescence measurements, and quenching of phosphorescence. There follows first a discussion of static and dynamic quenching processes and kinetic models.

It should be recognised that collisional processes are governed by spin-conservation laws due to Wigner which differ from those for radiative processes. Stated briefly the Wigner conservation law states that for the collisional reaction between A and B to give products C and D, with the molecules having spins S_A, S_B, S_C and S_D respectively, a complex between A and B may have any spin given by the series I. Likewise, the products correlate with spins in series II.

$$A + B \rightarrow C + D$$

Series I $S_A + S_B$, $(S_A + S_B - 1)$, ... $|S_A - S_B|$

Series II $S_C + S_D$, $(S_C + S_D - 1)$, ... $|S_C - S_D|$

The following rules then apply

Spin-allowed process All terms in Series I are found in Series II

Spin-forbidden process No terms in Series I are found in Series II

Spin-restricted process Some terms in Series I are found in
                          Series II

In the case of spin-restricted process, a spin statistical
factor may be easily calculated from the probabilities of forming
a reactant complex in a particular spin state. Thus in the
formation of $O_2(^1\Delta_g)$ sensitised by triplet states of many molecules
(below), the product spin (Series II) must be

$$^3M + O_2(^3\Sigma_g^-) \longrightarrow M + O_2(^1\Delta_g)$$

singlet, whereas the reactant complex spins (Series I) may be
quintet, triplet or singlet. Only the singlet reactant complex
may proceed to products, and thus the reaction can have a maximum
rate given only by the collision frequency of ^3M and $O_2(^3\Sigma_g^-)$
modified by the probability of producing the reactant complex in a
singlet state, in this case $^1/_9$.

KINETIC MODELS FOR BIMOLECULAR INTERACTIONS

Diffusional Model

To illustrate this model, a generalised kinetic scheme showing
the fate of an excited singlet chromophore and the involvement of
a quencher has been given below in which the quenching process is
not specified, and could be any of the four listed above.

		Rate Constant
$D + h\nu \longrightarrow {}^1D^*$	absorption	
${}^1D^* \longrightarrow D + h\nu$ fl.	fluorescence	k_F
${}^1D^* \longrightarrow {}^3D^*$	intersystem crossing	k_{ISC}
${}^1D^* \longrightarrow D + heat$	internal conversion	k_{IC}
${}^1D^* + Q \longrightarrow$	quenching	k_Q

The quantum yield of fluorescence of excited species (D^*) in the absence of quencher (Q) is defined as

$$(\phi_F)_o = \left(\frac{k_F}{k_F + k_{ISC} + k_{IC}}\right) \tag{1}$$

whilst that in the presence of quencher is

$$(\phi_F) = \frac{k_F}{\left(k_F + k_{ISC} + k_{IC} + k_Q [Q]\right)} \tag{2}$$

Manipulation of equations (1) and (2) yield the familiar Stern-Volmer relationship (3)

$$\frac{(\phi_F)_o}{(\phi_F)} = 1 + k_Q \tau_o [Q] \tag{3}$$

where τ_o is the lifetime of the excited molecule in the absence of quencher. For pulsed excitation, the measured decay time obeys the relationship (4), from which it can be seen that static and

$$\tau_F^{-1} = k_F + k_{ISC} + k_{IC} + k_Q [Q] \tag{4}$$

dynamic measurements produce the same measured value of k_Q. Many rate constants for bimolecular interaction in solution approach the diffusional rate, given by (5)

$$k_{diff} = \frac{8\ RT}{3000\ \eta} \tag{5}$$

where η is the viscosity of the solvent.

Static interactions

In this extreme case, confined practically to rigid media, total absence of diffusion is supposed. Two cases may be considered, those of close-range interactions and of long-range interaction, and are illustrated with respect to <u>electronic energy transfer</u> phenomena.

Short-range. The simplest model devised for relating the observed donor luminescence as a function of acceptor concentration for those exchange interactions in rigid media where diffusion is limited, is that of Perrin[1], which expressed mathematically is:

$$\phi_o/\phi \;=\; e^{VN'[A]} \tag{6}$$

where $N' = 6.02 \times 10^{20}$, ϕ_o is the luminescence quantum efficiency in the absence of acceptor, ϕ that in the presence of a concentration of acceptor $[A]$ and V the volume of a quenching sphere (cm^3). This model implies that an excited state is instantaneously deactivated without emitting light if a quenching molecule is within a sphere of quenching action of volume V whereas a molecule of the acceptor located outside this volume has no influence on the donor.

The theory is limited, however, in that the implication of a

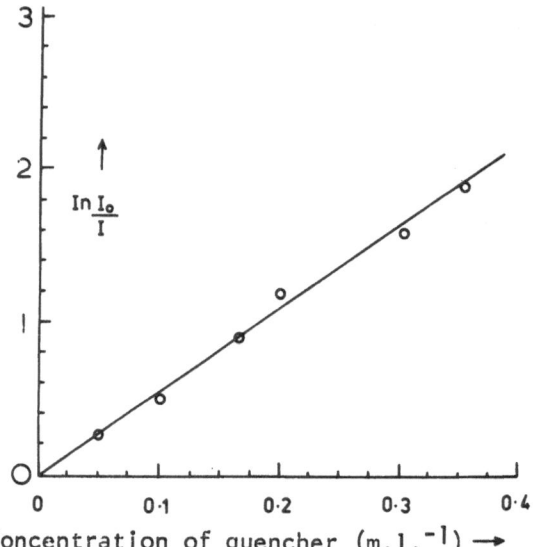

Figure 1.   Perrin plot of the phosphorescence emission quenching at 77°K of thermally oxidated PBD in 1:1 THF:DEE by COD.

sharp spherical boundary does not in reality describe the observed more gradual fall off in quenching action as a function of donor-acceptor separation. A further obvious discrepancy is that donor lifetime is in fact a function of acceptor concentration, whereas the Perrin model predicts the complete independence of donor life-time upon presence of acceptor.

Nevertheless, the treatment is attractive because of its simplicity and has been applied to a variety of systems, exemplified by Figure 1, in which the quenching of triplet excited $\alpha\beta$-unsaturated carbonyls by cyclo-octadiene at low temperatures in a synthetic polymer poly(butadiene) is depicted in Perrin form[2].

A more elaborate method of characterising the efficiency of triplet energy transfer by an exchange mechanism has been developed by Hirayama and Inokuti[3], which recognises that the donor lifetime will be influenced by the acceptor. The efficiency of non-radiative transfer f_{NR} is given by: (6)

$$f_{NR} = 1 - \left[\tau_o^{-1} \int_o^\infty \phi(t) dt \right] \tag{6}$$

$$\text{where } \phi_t = \exp\left[-t/\tau_o \ ^{-\gamma-3}[A]/[A_o] \ g(e^{\gamma t}/\tau_o) \right] \tag{7}$$

and is a decay function of the excited state, τ_o is the lifetime of the same state in the absence of quenching, $[A_o]$ is the critical concentration, γ is a parameter related to R_o, τ_o and the spectral overlap whilst

$$[A_o] = 3/(4 \pi R_o^3) \tag{8}$$

The method of evaluation involves comparing experimental transfer efficiencies as a function of concentration with the theoretical plot of f_{NR} versus $[A]/[A_o]$ for a certain value of γ. The value of γ is obtained from an experimental determination of the decrease of donor lifetime and luminescence intensity due to increasing acceptor concentration. The best fit will provide a value of $[A_o]$ from which R_o may be calculated. The variation of τ/τ_o with ϕ/ϕ as a function of γ is illustrated in Figure 2 for various theoretical values of γ, and the variations expected on the basis of diffusional (Stern-Volmer) and Perrin models is also illustrated.

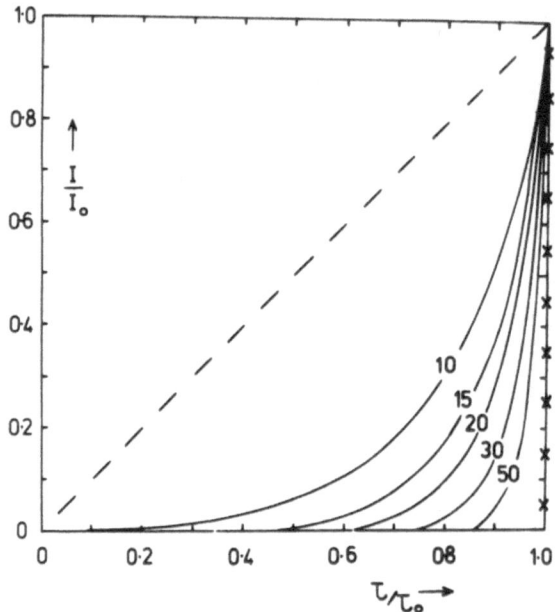

Figure 2. Theoretical relative yield vs. decay time plots for steady-state excitation: — HI (numeral on the curve denotes value of γ), —×— Perrin and ---- Stern-Volmer models.

The application of the Hirayama-Inokuti treatment to the same experimental quenching data as was used to illustrate the Perrin model shown in Figure 1 is shown in Figure 3[2]. R_0 values obtained from the treatments are similar, being $R^{HI} = 12.7$ A, $R_0^P = 12.8$ A.

Long-range. The coulombic interaction between chromophores can in some circumstances be dominant, and through induced dipole interactions (neglecting other multipolar terms), can lead to efficient energy transfer over large distances. The expression for induced dipole, commonly referred to as Forster[4] or long-range

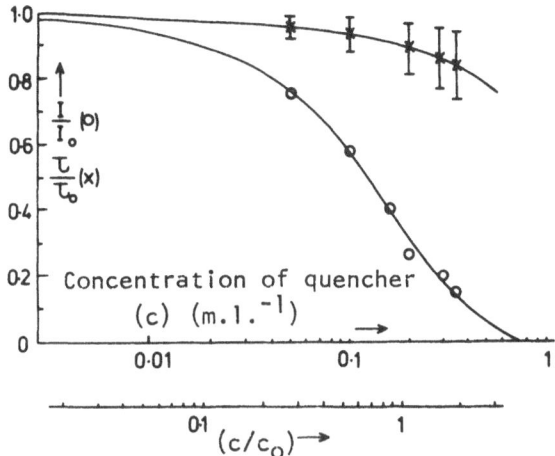

Figure 3. Hirayama-Inokuti plot of the phosphorescence emission
and lifetime quenching at 77°K of thermally oxidized PBD in 1:1
THF:DEE by COD (_____ represents the theoretical curves for the
H-I value of γ = 20).

energy transfer can be written as

$$k_{ET} = \frac{8.8 \times 10^{-25} K^2 \phi_D}{n^4 \tau_D R^6} \int_0^{\infty} F_D(\nu) \epsilon_A(\nu) \nu^{-4} d\nu \qquad (9)$$

where $F_D(\nu)$ is the normalised donor fluorescence spectrum, $\epsilon_A(\nu)$ the
acceptor absorption spectrum, ϕ_D the donor fluorescence quantum
yield, τ_D the donor decay time, n the medium refractive matrix,
R the interchromophoric distance, and K^2 an important term which
contains the orientational dependence of the chromophores. The

last parameter arises because of the requirement of correct
orientation of the transitional dipoles in donor and acceptor
molecules, and may take values anywhere from zero to 4. For a
completely random distribution, χ^2 takes the value of $2/3$, and this
is widely used in quantitative studies on energy transfer. It is
evident from equation (9) that provided all other parameters are
known, including χ^2, observation of energy transfer efficiencies
by this mechanism can yield values of R, the interchromophoric
distances, and this is of some importance as a structural probe in
biological systems. However, since in many biological systems,
e.g. in proteins random orientation of chromophores cannot be
assumed, but equally cannot be defined, interchromophoric distances
calculated by these means are subject to the inaccuracies imposed
by arbitrary choice of χ^2 as equal to $2/3$[5]. It is of interest
to note that energy transfer by the electron exchange (close-
range) mechanism as obserbed in so-called triplet-triplet energy
transfer does not appear to have a strong orientational dependence[6].

Intermediate kinetics

The two cases of static and diffusional kinetics discussed
above can be defined in terms of a molecular energy transfer
distance r where

$$r = (2(D + \Lambda)\tau_0)^{\frac{1}{2}} \tag{10}$$

in which D is the relative molecular diffusion coefficient of the
donor and acceptor, Λ is the donor excitation energy migration
coefficient, and τ_0 is the donor luminescence decay time. If
$r \ll R_0$, the molecules remain stationary during transfer, and static
kinetics will be appropriate. When $r \gg R_0$, Stern-Volmer kinetics
apply. For intermediate cases which may well occur in biological
systems, two theories have been developed.

Voltz applied the diffusion theory to diffusion controlled
singlet energy migration and developed a relationship between the
rate constant for energy transfer (k_t) and R_0 such that

$$k_t = 2 \pi N_0 (D + \Lambda) R_0 \, 10^{-3} \; L\,mol.^{-1}sec.^{-1} \tag{11}$$

Here N_0 is Avogadro's number, D and Λ are as defined above, R_0 is
either the Forster or Perrin critical transfer distance dependent
on whether or not the migration is singlet-singlet or triplet-
triplet, whilst k_t can be obtained from a modified Stern-Volmer
plot.

Birks[8] compared the experimental data for the naphthalene/
9,10-diphenyl anthracene system in various solvents with the Voltz
theory and found a considerable discrepancy between actual and
expected behaviour. This is perhaps due to the incorrect boundary
assumptions employed by Voltz, i.e.

$$p = 0.5 \text{ for } r < R_o$$

$$p = 0 \quad \text{ for } r > R_o$$

where p is the transfer probability and r is the energy migration
distance, whereas in fact $p \propto \dfrac{R_o}{r}$.[6] For this reason the treatment
described below is often
preferred.

Yokota and Tanimoto[9] developed a statistical representation of
Forster's theory and gave an expression for dipole-dipole transfer
in a fluid medium in which the distribution of donor and acceptor
is determined both by diffusion and by the decay transfer of the
donor excited state, such that

$$I_x(t) = I_x(o) \exp\{-t/\tau_o + {}^{-2B\gamma} (t/\tau_o)^{\frac{1}{2}}\} \tag{12}$$

where $I_x(t)$ is the donor fluorescence response function, τ_o is the
donor fluorescence lifetime in the absence of acceptor,

$$\gamma = \frac{[A]}{[A_o]} \quad B = (\frac{1 + 10.87x + 15.5x^2}{1 + 8.743x})^{\frac{3}{4}}$$

$$x = D \alpha^{-\frac{1}{3}} t^{\frac{2}{3}} \text{ and } \alpha = (\frac{R^o}{\tau_o})^6$$

It is then possible to relate the above expression to non-
radiative transfer efficiencies. Comparison of the two
intermediate theories, along with that of Forster, for a synthetic
polymer system of poly(N-vinylcarbazole) in poly(methylmethacrylate)
incorporating anthracene as the quencher. The study of the singlet
energy transfer showed similar conclusions as those reached above
i.e. that the Yokota-Tanimoto model gives a much better fit to
experiment than that of Voltz.

Of the quenching mechanisms outlined in the introduction to
this chapter, (a), (b) and (d) are essentially short-range

interactions, whereas as has been shown above, electronic energy
transfer can be long-range. Further brief comments will be made on
bimolecular quenching processes.

(a) Chemical Reaction

Photochemical reactions may of course proceed in polymeric
systems as in smaller analogues. However, the structural
constraints present in a solid bio-polymer may render such
interactions inefficient.

(b) Enhancement of non-radiative decay

The proximity of a molecule containing an atom of high
nuclear charge or unpaired spin may cause a dramatic enhancement
of spin-forbidden intra-molecular non-radiative decay rates. The
presence of molecular oxygen (discussed elsewhere in this volume)
and metal ions may perturb the luminescence of biological systems
by such processes.

(c) Electronic energy transfer

In addition to the induced-dipole and exchange mechanisms
discussed fully above, radiative transfer (emission of the donor
followed by reabsorption by the acceptor) is possible. In highly
ordered systems excitation energy may diffuse rapidly, leading
to the concept of 'excitons'.

(d) Complex formation

This is discussed fully elsewhere in this volume.

BIPHOTONIC PROCESSES

The absorption of a second-photon by an excited state
may be used to monitor the excited state concentration, to
identify it by its absorption spectrum, and to follow the kinetics
of decay by monitoring concentration as a function of time. The
nature of the experiment is dictated by the decay time of the
species in question. For the time-domain longer than μs,
conventional flash photolysis suffices, whereas in the sub-
microsecond region extending nowadays down to the ps region solid
state laser flash photolysis methods are required.

TRIPLET STATE PROCESSES

Equation (3) derived above for the diffusional quenching of
fluorescence has its analogue for the quenching of the triplet
state emission, i.e. phosphorescence given by (13)

$$(\phi_p)_0/_{\phi_p} \ = \ 1 \ + \ k'_Q \ \tau'_p \ [Q] \tag{13}$$

where τ'_p is now the phosphorescence decay time in the absence of quencher Q, given usually by $(k'_p + k'_{ISC})^{-1}$ from the simple scheme below, and k'_Q is the second order rate constant for quenching. Comparison of equations (3) and (13) show that the respective

$$^3A \ \longrightarrow \ A \ + \ h\nu p \qquad k'_p$$

$$^3A \ \longrightarrow \ A \qquad k'_{ISC}$$

$$^3A \ + \ Q \ \longrightarrow \ quenching \qquad k'_Q$$

decay times dictate the fate of singlet and triplet decay times. Since τ_F is usually many orders of magnitude shorter than τ'_p, modest concentrations of adventitions quencher such as oxygen do not usually drastically reduce the fluorescence intensity, whereas they can cause total quenching of phosphorescence emission. In many biological systems it is of considerable use to study phosphorescence, but from the above argument it can be seen that in order to carry out such studies diffusion must be prevented to avoid total quenching by impurities such as oxygen. This necessitates the use of rigid matrices, and a variety of solvents which give clear glasses at 77K are frequently employed[11].

DELAYED EMISSION

Delayed emission is that which persists long after the excitation process, and clearly includes phosphorescence. The techniques for recording delayed emission spectra are the same as for fluorescence, except that the excitation radiation is no longer continuous, but is 'chopped' by a mechanical or electronic device which interrupts the excitation beam periodically.

If the detection system is gated so that emission is viewed only after a time has elapsed after the cut off of the excitation, then short-lived emission such as prompt fluorescence will have decayed to zero intensity, and thus only long-lived emission will be observed. The decay time of the luminescence can also be measured using this technique, and for a mechanical device such as a rotating toothed chopper, the limit of measurable lifetime is of the order of 1 ms. Electronic devices can be used to measure decay times much shorter than this.

Using the above methods it is possible to observed delayed emission which by definition is long-lived, but which has the same spectral distribution as normal (very short-lived) fluorescence.

Such emission is termed delayed fluorescence, and is of two main types, E-type, first seen in eosin, and P-type, first seen in pyrene.

E-type delayed fluorescence

E-type delayed fluorescence arises from thermal reactivation of a triplet state molecule to the singlet state from which fluorescence occurs. Thus the spectral distribution is that of normal fluorescence, but the lifetime is associated with the triplet state. Since the reactivation process is thermal this phenomenon is confined to species in which the singlet-triplet energy separation is small. The mechanism is indicated below.

$$
\begin{array}{lll}
 & & \text{Rate constant} \\[4pt]
M + h\nu \longrightarrow {}^1M & & I_a \\[4pt]
{}^1M \longrightarrow M + H\nu_F & & k_R \\[4pt]
{}^1M \longrightarrow {}^3M & & k_{ISC} \\[4pt]
{}^1M \longrightarrow M & & k_{IC} \\[4pt]
{}^3M \longrightarrow M + h\nu_p & & k'_p \\[4pt]
{}^3M \longrightarrow M & & k'_{ISC} \\[4pt]
{}^3M \longrightarrow {}^1M & & k'_T
\end{array}
$$

Now $\Phi_F = k_R/(k_R + k_{ISC} + k_{IC})$, as usual,

$$\Phi_{DF} = \Phi_F \cdot k'_T/(k'_{ISC} + k'_p + k'_T) \tag{14}$$

$$\Phi_p = k'_p/k'_{ISC} + k'_p + k'_T) \tag{15}$$

Thus

$$\Phi_{DF}/\Phi_p = \Phi_F k'_T/k'_p$$

$$= \Phi_F \tau'_R k'_T \tag{16}$$

Where τ'_R is the radiative lifetime of the triplet state, Φ_{DF} the quantum yield of delayed fluorescence. Note that k'_T is a thermal rate constant, and can be expressed in Arrhenius form,

Thus
$$\Phi_{DF}/\Phi_p = \Phi_F \tau'_R Ae^{-\Delta E/RT} \tag{17}$$

plots of log ϕ_{DF}/ϕ_P against $1/T$ should thus be straight lines with slope related to ΔE, which is just the energy separation of the S_1 and T_1 states. This has been validated for many systems. Note that the intensity of delayed fluorescence in this case is directly proportional to I_a.

<u>p-type delayed fluorescence</u>

p-type delayed fluorescence arises from the mutual annihilation of <u>two</u> triplet states to produce one excited singlet and one ground-state molecule. This energy pooling process results in the initial formation of an excited dimer (excimer) which if emittive, will give rise to excimer emission as well as normal monomer emission which have the same spectral characteristics as the excimer and monomer fluorescence produced by continuous illumination, but again, which have lifetimes associated with the triplet state.

$$^3P + {}^3P \longrightarrow {}^1(D)^* \longrightarrow {}^1P^* + P$$

$$\downarrow \qquad\qquad \downarrow$$

$$P+P+h\nu \qquad P+h\nu_M$$

In the prompt spectrum (continuous illumination), the monomer is the precursor of the dimer. In the delayed spectrum, the dimer is the precursor of the monomer. Thus we expect

$$(I_m/I_d)_{\text{prompt}} > (I_m/I_d)_{\text{delayed}}$$

This behaviour is observed.

Note in this case, the intensity of delayed emission is proportional to the square of the exciting intensity. The process is often referred to as <u>triplet-triplet annihilation</u>.

The mechanism of P-type delayed fluorescence can be expressed by the additional steps shown below[11]

Rate

$$^3M + {}^3M \longrightarrow {}^1D \qquad k_A'[{}^3M]^2$$

$$^1D \longrightarrow {}^1M + M \qquad k_S[{}^1D]$$

$$^1D \longrightarrow M + M \qquad k_D[{}^1D]$$

Now if I_a is very small, the $k_A' [^3M]^2$ term is very small, and steady-state analysis gives

$$I_a \Phi_{ISC} = (k_{ISC}' + k_p') [^3M] \tag{18}$$

$$= [^3M]/\tau_p$$

$$\Phi_{DF} = 1/2\ k_A' [^3M]^2 \ \frac{\Phi_F}{I_a} \ \ k_S/(k_s + k_D) \tag{19}$$

Substituting we get

$$\Phi_{DF}/\Phi_F = 1/2 k_A' I_a (\Phi_{ISC} \tau_p)^2 k_S/(k_D + k_S) \tag{20}$$

Thus the delayed fluorescence efficiency depends upon I_a, unlike the case for E-type delayed fluorescence. Φ_{DF} also depends linearly upon the rate constant for triplet-triplet annihilation, k_A'. This is often diffusion controlled (see above) and thus depends upon the viscosity of the solvent. One might thus expect that use of a higher viscosity solvent should cause a reduction in Φ_{DF}, but more often an increase in the delayed fluorescence efficiency is observed. This is because τ_p^2 also appears in equation (20), and triplet lifetimes are often controlled by the difusion rate of impurities, particularly molecular oxygen to the triplet state (see above). Use of a higher viscosity solvent, while it will certainly reduce the magnitude of k_A', will increase the value of τ_p in such cases, and since the latter term appears to the power 2 in equation (20), a net increase in Φ_{DF} results.

Sensitized delayed fluorescence

With a donor-acceptor pair, such as phenanthrene, P, and anthracene, A, we could get

$$P + h\nu \longrightarrow {}^1P \longrightarrow {}^3P$$

$$^3P + A \longrightarrow {}^3A + P$$

Thus we could have $^3P + {}^3P \longrightarrow {}^1P$

$$^3P + {}^3A \longrightarrow {}^1A + P$$

$$^3A + {}^3A \longrightarrow {}^1A + A$$

Energy pooling of this sort occurs in biological systems, and can lead to an 'upgrading' of electronic energy content, such as is observed in sensitized anti-Stokes delayed fluorescence (Figure 4) Here the photon emitted is more energetic than the photon absorbed, although of course two of the latter are required to produce one of the former.

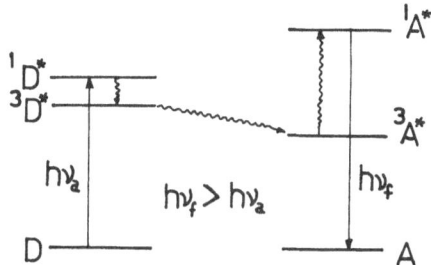

Figure 4   Schematic of sensitized anti-Stokes delayed fluorescence.

Delayed fluorescence measurements are widely used as an investigative tool in biological systems. One further type is that arising from photo-ionization (usually a two-photon process) followed by recombination of ion and electron to produce an excited state. This has been observed in indole-type compounds in biological systems.

References

1. F. Perrin, Comptes Rendues Ser. C., 1924, 178, 1978.

2. J.S. Hargreaves and D. Phillips (unpublished results).

3. F. Hirayama and M. Inokuti, J. Chem. Phys., 1965, 43, 1978.

4. T. Forster, Disc. Faraday Soc., 1939, 27, 1; Naturwissenschaft 1946, 33, 166.

5. J. Eisinger and R.E. Dale in 'Excited States of Biological Molecules' ed. J.B. Birks, Wiley 1976, London, p579.

6. A. Adamczyk and D. Phillips, J.C.S. (Faraday II), 1974, 70, 537.

7. R. Voltz, G. Laustriat and A. Cocher, J. Chimie Physique,
    1963, 63, 1255.

8. J.B. Birks and M.S.C.P. Leile, J. Phys. B, 1970, 3, 513.

9. M. Yokota and O. Tanimoto, J. Phys. Soc. Japan, 1967, 22, 779.

10. A.M. North and M.F. Treadaway, Eur. Polym. J., 1973, 9, 609.

11. C.A. Parker "Photoluminescence in Solution", Elsevier, London,
    1969.

PHYSICAL-ORGANIC PHOTOCHEMISTRY. I. INTRODUCTION AND σ → σ*, π → π*

EXCITATIONS

David Phillips

University of Southampton

Department of Chemistry, The University, Southampton

INTRODUCTION

In the limited time and space available for these topics a co-herent resume of organic photochemistry in terms of detailed molecular orbital descriptions is inappropriate. Instead what is offered here are selected topics in physical-organic photochemistry which may be of some interest. As a brief introduction, Figure 1 (over) shows some of the important <u>unimolecular</u> physical and chemical decay paths in a complex polyatomic molecule following $S_0 \rightarrow S_1$ excitation. The quantum yield of any process is defined as the rate of that process relative to the rate of light absorption. For steady-state illumination it is easily shown that, for example, the quantum yield of fluorescence for a molecule undergoing the processes shown in Figure 1 is where rate-constant subscripts refer to the processes shown in Figure 1.

$$\Phi_F = \frac{k_{NF}}{(k_{NF} + k_{RE} + k_{ISC} + k_{IC})} \tag{1}$$

Whether or not a molecule is fluorescent thus depends upon the relative magnitude of the individual rate-constants in equation (1). For spin-allowed transitions k_{NF} may have values between 10^5 and 10^9 s^{-1}. Where k_{RE} is small (as for example in aromatic systems) the fate of the excited singlet state molecule is often dictated by the magnitude of k_{ISC} relative to k_{NF}. In many systems k_{RE} is very large, and thus photochemistry dominates (e.g. in σ → σ* excitations, see below).

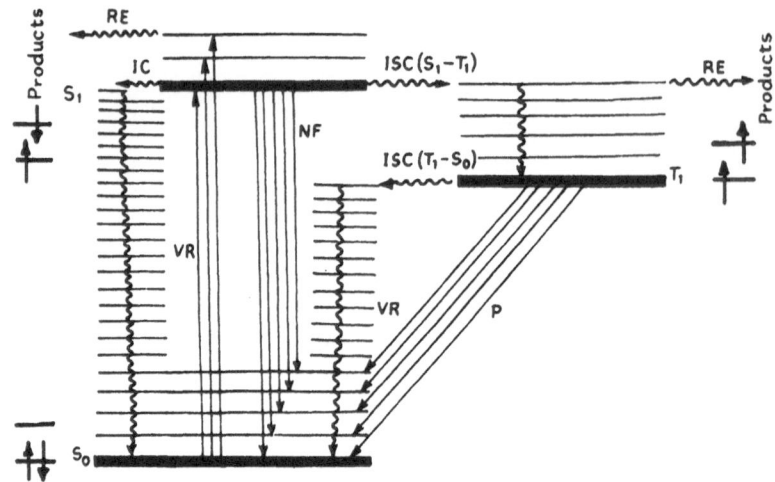

Figure 1. A state diagram showing the important electronic and vibrational relaxation processes in competition with emission.

Key to processes

A = absorption	IC = internal conversion
NF = normal fluorescence	ISC = intersystem crossing
VR = vibrational relaxation	RE = photochemical reaction
P = phosphorescence	

In high pressure gases, and condensed phases, bimolecular fates of excited states tend to, be dominant, and quenching of excited states can result. The magnitude of a quenching interaction by an additive Q in fluid media is often expressed in terms of the Stern-Volmer relationships (equation 2).

$$\frac{(\phi)_o}{\phi} = 1 + k_Q \, \tau \, [Q] \qquad \dots\dots\dots\dots\dots\dots\dots \ (2)$$

where $(\phi)_o$ is the quantum yield of any process when $[Q] = 0$, ϕ that at any concentration $[Q]$, k_Q is the second-order rate-constant for the bimolecular interaction, and τ is the decay time of the excited state when $[Q] = 0$. Since the decay times of triplet states usually are very long compared with those of corresponding excited singlet states, bimolecular interactions are particularly significant in triplet state chemistry. Equation (2) is invalid in possible bimolecular interactions which are listed below.

(a) Complex formation, including charge-transfer ('exciplex') interactions.

(b) Electronic energy transfer

(c)   Enhancement of non-radiative decay, usually promoted by atoms of high nuclear charge ('heavy   atom' effect) or by paramagnetic species, notably molecular oxygen.

(d)   Chemical reaction, which may be preceeded by any of the first three.

PHOTOCHEMICAL FATES OF ORGANIC MOLECULES

σ-σ^* Excitations

These are high energy excitations occurring in the vacuum ultraviolet region of the spectrum (where synchrotron radiation is a good light source). Since however all organic molecules will exhibit such electronic transitions, <u>selective</u> excitation is not possible. Moreover, since no 'inert' solvent is available in this wavelength region, any detailed studies on the photochemistry consequent upon such excitations have of necessity been carried out in the gas phase.

<u>Alkanes</u>. Figure 2 shows some uv absoption profiles of simple saturated hydrocarbons. Whereas in methane only $\sigma \rightarrow \sigma^*$ excitations associated with the C-H bond are possible, the   lowest energy $\sigma \rightarrow \sigma^*$ transition is clearly that localised in the C-C bond in ethane, since the respective bond dissociation energies are 83 k cal mole^{-1} (τ-C) and 102 k cal mole^{-1} (C-H). Since the photon energy greatly exceeds the bond-dissociation energy, it is not surprising that fragmentation is the major fate of photoexcited alkanes. Typical reaction pathways are given below for simple alkanes.

<u>Methane</u> $CH_4 + h\nu_{123.6} \rightarrow CH_2 + H_2 \quad \phi = 0.8$ (3)

$CH_4 + h\nu_{123.6} \rightarrow CH_3\cdot + H\cdot \quad \phi = 0.2$ (4)

$CH_4 + :CH_2 \rightarrow C_2H_6^+$ (5)

<u>Ethane</u> $C_2H_6 + h\nu \rightarrow CH_3CH: + H_2$ (6)

$C_2H_2 + H_2 \leftarrow CH_2{=}CH_2^+ \rightarrow CH_2{=}CH_2$

$C_2H_6 + h\nu \rightarrow CH_2{=}CH_2 + 2H$ (7)

$C_2H_6 + h\nu \rightarrow CH_4 + :CH_2$ (8)

$\rightarrow 2CH_3\cdot$ (9)

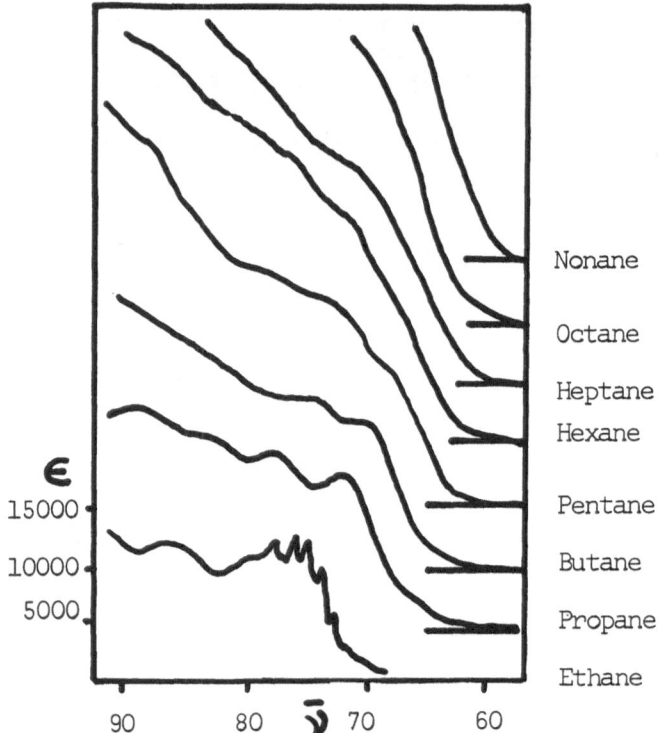

Figure 2   Absorption spectra of some linear alkanes

(7) is of increasing importance as photon energy increases. Reactions (8) and (9), C-C bond cleavages, are of minor importance only. Recently, it has been found that upon C-C $\sigma \rightarrow \sigma^*$ excitation, relatively strong fluorescence is observed in many alkanes ($\phi_F \sim 10^{-2}$). In all cases the fluorescence is very red-shifted ($\lambda_{max}^F \sim$ 206-250 nm) with respect to the absorption, indicating a considerable change in geometry between ground and emitting states.

Rydberg transitions. A Rydberg orbital is a linear combination of atomic orbitals which spatially resembles a pure atomic-type orbital. In the vacuum ultra-violet region of the spectrum one thus expects to observe transitions involving electrons initially in σ (bonding) orbitals (and π-(bonding) and n-(non bonding) where appropriate) promoted to Rydberg orbitals also.

Saturated amines. Since most systems of biological interest contain nitrogen, it is appropriate to consider the photochemical fates of the very simplest aliphatic nitrogen-containing compounds as models. In ammonia the longest wavelength transition has been ascribed to a Rydberg type transition (n → 3s) although it might more correctly be described as a combination of this Rydberg transition and an intravalence n → σ* excitation. Whereas the ground-state of ammonia is pyramidal, the excited state is planar, and thus optical excitation is accompanied by a severe geometry change. In ammonia, primary and secondary aliphatic amines, the consequence of photoexcitation is almost exclusively dissociation. In <u>tertiary</u> amines however, this decay channel is absent, at least

$$RR'NH + h\nu \ \to \ RR'N. + H. \qquad\qquad \dots\dots\dots\dots\dots\dots \quad (10)$$

at low photon energies, and surprisingly strong fluorescence is observed (ϕ_F = 1.0 in the vapour phase). The dramatic difference in behaviour can be understood of quantum-mechanical tunnelling from the excited state surface to the dissociative surface is required in these compounds. For compounds containing a hydrogen atom tunnelling is facile, but bulky alkyl groups cannot achieve this, thus tertiary amines are non-photodissociative.

Since simple amines have very low ionization potentials, they are frequently encountered as electron donors in complex formation.

$\pi \to \pi^*$ Excitations

These can occur in the near ultra-violet region of the spectrum, being of lower energy than σ → σ* excitations.

<u>Aromatic molecules</u>. The physical fates of these molecules has been considered in detail elsewhere, and tends to dominate. Thus unimolecular photochemical reactions in for example benzene to give valence isomers (11) have a maximum quantum yield of only a few percent. Bimolecular reactions with unsaturated compounds to give cycloaddition compounds are often observed.

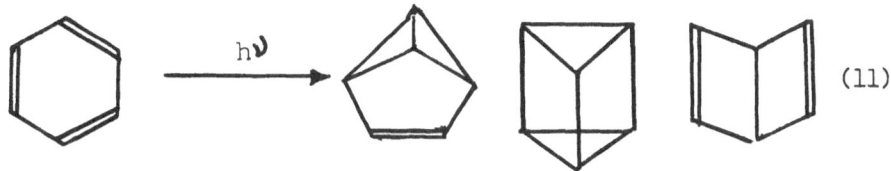

(11)

<u>Olefins</u>. In simple olefins, e.g. ethylene, excitation in the π → π* singlet-singlet transition (overlapped by the Rydberg π → 3s transition) results in fragmentation.

$$CH_2=CH_2 + h\nu \rightarrow CH_2=C: + H_2 \tag{12}$$

$$CH\equiv CH + H_2 \tag{13}$$

$$CH\equiv CH + 2H \tag{14}$$

$$CH_2=C: + 2H \tag{15}$$

By contrast, sensitization of the triplet state results in rapid cis-trans isomerization, commonly met in all olefins, due to the intersection of ground and excited potential surfaces, Figure 3.

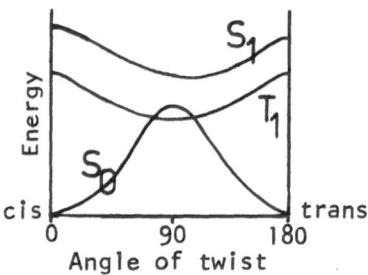

Figure 3   Potential energy surfaces for typical olefin as function of rotation about double bond

In polyenes, direct excitation results in electrocyclic reactions exemplified below which are governed by orbital symmetry rules.

Suggested Further Reading

'Excited States in Organic Chemistry', J.A. Barltrop and J.D. Coyle, Wiley, (Interscience) London, 1975.

'Fluorescent Properties of Saturated Systems Hydrocarbons and Amines', A.M. Halpern, Mol. Photochem., 1973, 5, 517.

PHYSICAL-ORGANIC PHOTOCHEMISTRY. II. n → π* EXCITATIONS

David Phillips

University of Southampton

Department of Chemistry, The University, Southampton

Transitions involving non-bonding electrons promoted to π* antibonding orbitals are most frequently encountered in carbonyl compounds and nitrogen containing systems.

ALIPHATIC CARBONYLS

Transitions possible in a simple aliphatic ketone, acetone, are shown in Figure 1, from which it can be seen that the lowest energy transition is of n → π* character, and occurs in simple carbonyls at around λ_{max} 280 nm. It should be noted that by analagy with formaldehyde the transition is a weak one, with ε_{max} ∿ 10 1 mol^{-1} cm^{-1}, being symmetry and overlap forbidden and is promoted by the out-of-plane bending vibration taking the methyl substituents from the planar configuration to a pyramidal geometry with respect to the carbonyl group. Since the S_1-T_1 (nπ*) energy gap is small in simple ketones, intersystem crossing to the ^{3}nπ* state is facile. Such compounds are weakly fluorescent and phosphorescent, and undergo photochemical reactions with high quantum yield. The most important of these are:

Norrish Type 1 Reaction

This is a homolytic split to give free radicals and occurs very widely in aliphatic ketones.

$$CH_3COCH_3^* \rightarrow CH_3CO. + .CH_3 \tag{1}$$

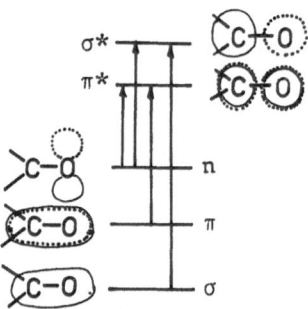

Figure 1   Electronic transitions and orbitals in acetone
Norris Type II Reaction

For aliphatic ketones with a hydrogen atom in the γ position,
a molecular split can occur giving an olefin and further carbonyl,
e.g. in pentan-2-one. This reaction

$$CH_2 \cdots CH_2 - CH_2 \quad O = C - CH_3 \quad + \ h\nu \longrightarrow$$

$$CH_2 \cdots CH_2 - CH_2 \quad O = C - CH_3$$

$$CH_2 = CH_2 \ + \ CH_2 = \overset{OH}{\underset{}{C}} - CH_3$$

$$CH_3COCH_3$$

is of particular importance in higher acylcic ketones.

Reaction with Olefins

Of the many bimolecular reactions undergone by simple aliphatic ketones, that with simple olefins (3) to form oxetans is of interest, since it probably procedes through exciplex formation, and produces a species of some current interest in that the thermal reverse dissociation of the oxetan can give rise to chemiluminescence. Dissociation of dioxetan is an efficient clean source of triplet acetone with which to carry out photochemistry without light.

$$\begin{array}{c} CH_3 \\ \diagdown \\ \diagup \\ CH_3 \end{array} C{=}O* \ + \ CH_2{=}CH_2 \ \rightarrow \ \begin{array}{c} CH_3 \\ \diagdown \\ CH_3 \diagup \end{array} \begin{array}{c} C - O \\ | \quad | \\ CH_2 - CH_2 \end{array} \tag{3}$$

AROMATIC CARBONYLS

There are complications here due to the presence of $^1\pi\pi*$, $^3\pi\pi*$ states as well as the $^1n\pi*$, $^3n\pi*$ states. The ordering of the states is important, since intersystem crossing between states of different configuration is faster than that between states of the same configuration (El-Sayed's rule). Figure 2 shows a typical ordering of states for an aromatic ketone (e.g. benzophenone). Very fast intersystem crossing from $^1n\pi*$ to $^3\pi\pi*$ occurs, followed by rapid internal conversion to $^3n\pi*$. In alkyl phenyl ketones the $^3\pi\pi*$ state lies higher in energy than $^1n\pi*$ and thus the intersystem crossing is slower. The energy levels of the $n\pi*$ states are highly solvent dependent, and thus the fates of these carbonyl compounds are also very solvent dependent. In all cases however, triplet state chemistry dominates.

An important reaction of aromatic ketones is a hydrogen abstraction with hydrocarbon substrates, e.g. with benzophenone

$$\begin{array}{ccc} O & & OH \\ {}^3(\phi{-}C{-}\phi) + RH & \rightarrow & \phi{-}\underset{\bullet}{C}{-}\phi \ + R. \end{array} \tag{4}$$

UNSATURATED KETONES

These are of biological importance since steroids contain these chromophoric groups. Woodward's rules allow a simple identification of structure from uv-absorption data. Taking the basic αβ unsaturated ketone structure below, the uv absorption maximum

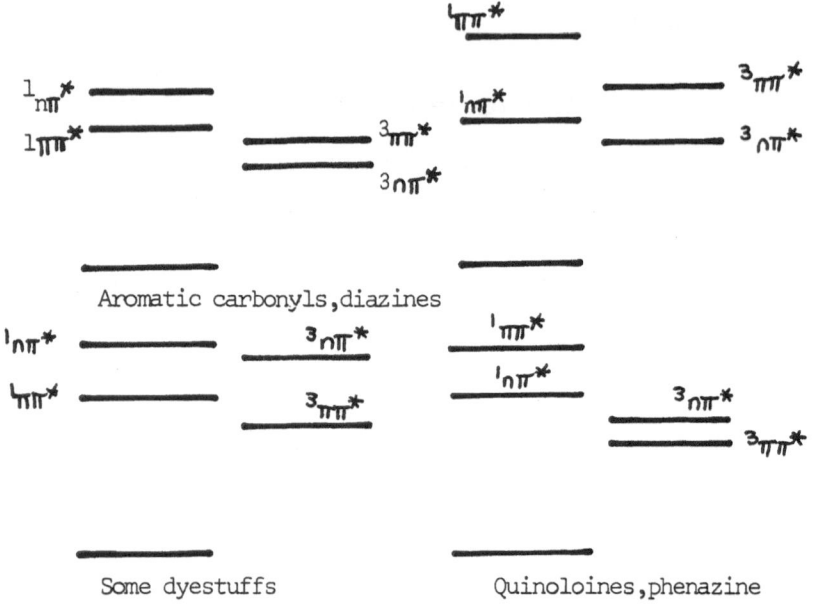

Figure 2   Energy levels of some organic molecules

can be calculated according to the simple rules shown.

$$C=C-C=O$$

Parent αβ unsaturated λ_{max}	215 nm
For each extra double bond in conjugation	Add 30nm
For each α alkyl substituent	Add 10nm
For each β alkyl substituent	Add 12nm
For each α hydroxyl substituent	Add 35nm
For each α chloro-substituent	Add 15nm
For exo location of double bond	Add 5nm
If a homo-diene	Add 39nm

Some examples of the application of these rules to keto-steroid structures are shown in Figure 3.

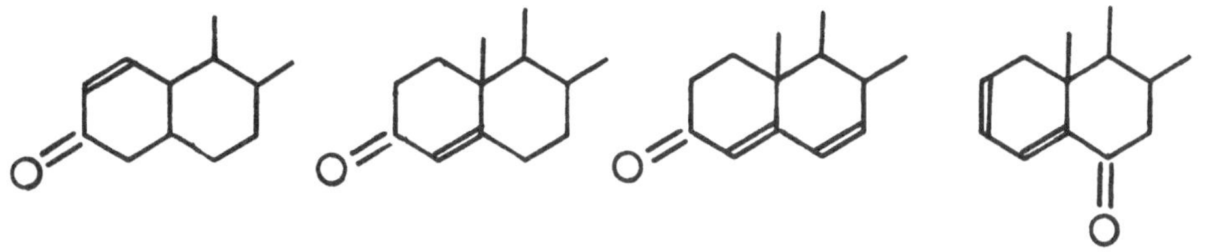

$\lambda_{max} = 230$ nm $\lambda_{max} = 241$ nm $\lambda_{max} = 284$ nm $\lambda_{max} = 315$ nm

$\lambda_{cal} = 227$ nm $\lambda_{cal} = 244$ nm $\lambda_{cal} = 280$ nm $\lambda_{cal} = 317$ nm

Figure 3   Experimental and calculated absorptions

Simple αβ unsaturated ketones are non-luminescent, but bi-cyclic enones exhibit phosphorescence, the nature of the emitting state depending upon positions of substituents.

AZA-AROMATICS

Here again n→π* excitations are possible which in this case often overlap the stronger π→π* transitions. Thus in pyridine a weak n→π* absorption is overlapped by the stronger π→π* aromatic transition. Different excitations may be characterized by their polarization with respect to absorption of plane-polarized light. Thus the π→π* transition in simple benzenoid hydrocarbons is polarized in the plane of the aromatic ring system, whereas the n→π* transition is polarized perpendicularly. Solvent shifts can also be used to distinguish n→π* from π→π* excitations.

Suggested Further Reading

"Excited States in Organic Chemistry", J.A. Barltrop and J.D. Coyle, Wiley(Interscience), London, 1975.

Chemical Society Specialist Periodical Reports "Photochemistry", Senior Reporter, D. Bryce-Smith, Volumes 1-9, 1969-1978.

PHYSICAL-ORGANIC PHOTO-CHEMISTRY. III. PHOTOOXIDATIONS

David Phillips

Southampton University

Department of Chemistry, The University, Southampton

ELECTRONICALLY EXCITED OXYGEN

We are concerned here with the reactions of electronically excited states with oxygen, and those of electronically excited states of oxygen with substrates. Figure 1 shows the electronic energy levels of the three lowest states of molecular oxygen along with those of other molecules of interest, and Figure 2 the distribution of electrons in the $\pi_g^*(2_p)$ antibonding orbitals in the three states. The $^3\Sigma_g^-$ ground state is radical in character, whereas the $^1\Delta_g$ state, some 22.5 k cal mol^{-1} above the ground state, can be considered to be a dienophile because of the vacant π_g orbital. $O_2(^1\Delta_g)$ is reactive, and has a typical solution phase lifetime of 10^{-4}-10^{-6} s. The second excited singlet state ($^1\Sigma_g^+$), 37.5 k cal mol^{-1} above $^3\Sigma_g^-$ does not have the vacant orbital as does the $^1\Delta_g$ state, and thus cannot participate in two-electron processes. In any case, the solution phase lifetime of O_2 ($^1\Sigma_g^+$) is $\sim 10^{-8}$ s, the species usually decaying to O_2 ($^1\Delta_g$), and thus it is the chemistry of this species which tends to dominate in photooxidations.

Population of $O_2(^1\Delta_g)$

Although direct excitation and microwave discharges are possible, the most widely available source of $O_2(^1\Delta_g)$ in condensed media is through sensitization. In principle, for aromatic sensitizers such as naphthalene (Figure 1), both singlet (Equation 1) and triplet state (Equation 2) sensitizations are possible, since

$$^1N* + O_2(^3\Sigma_g^-) \rightarrow {}^3N* + O_2(^1\Delta_g) \tag{1}$$

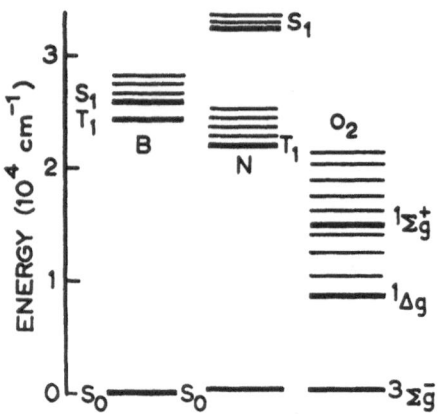

Figure 1   Energy levels of oxygen,benzophenone,naphthalene

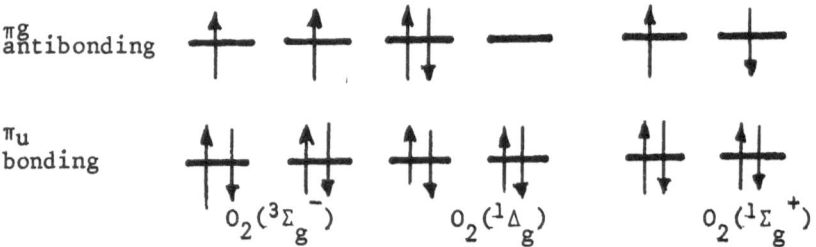

Figure 2 π-electron distribution in states of oxygen

$$^{3}N* + O_{2}(^{3}\Sigma_{g}^{-}) \rightarrow N + O_{2}(^{1}\Delta_{g}) \tag{2}$$

(1) is energetically feasible, and does not violate the Wigner spin
conservation rules. However, there is no evidence that (1) occurs
to any extent, and in the case of sensitizers such as benzophenone,
(Figure 1), is not possible energetically. Triplet sensitization
through reactions analagous to (2) is thus the main source of $O_{2}(^{1}\Delta_{g})$
and the reaction occurs at 1/9 collision efficiency for many
sensitizers. There are many naturally occurring sensitizers
(chlorophyll, porphyrins, etc) and subsequent damaging reactions of
$O_{2}(^{1}\Delta_{g})$ with living organisms is called 'photodynamic action'. A
further solution phase source of $O_{2}(^{1}\Delta_{g})$ is reaction (3).

$$H_{2}O_{2} + OCl^{-} \rightarrow O_{2}(^{1}\Delta_{g}) + Cl^{-} + H_{2}O \tag{3}$$

Reactions of $O_2(^1\Delta_g)$

These are summarized very briefly below.

With olefins

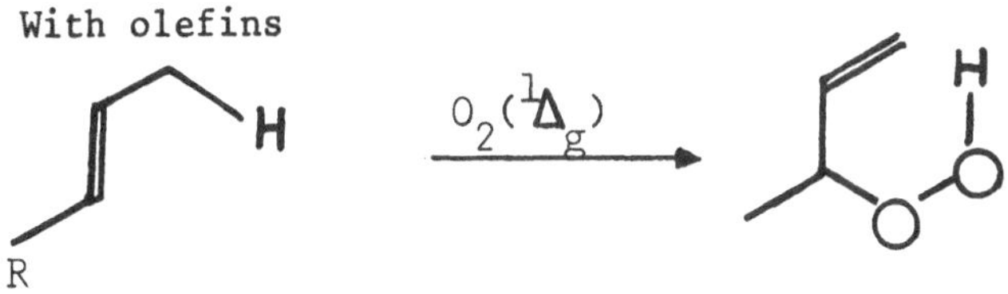

$$\xrightarrow{O_2(^1\Delta_g)}$$

(4)

('ene' reaction)

Reactions are stereospecific.

With dienes

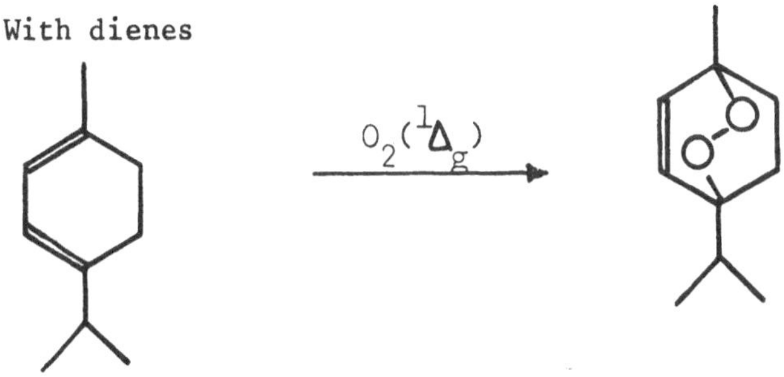

$$\xrightarrow{O_2(^1\Delta_g)}$$

(5)

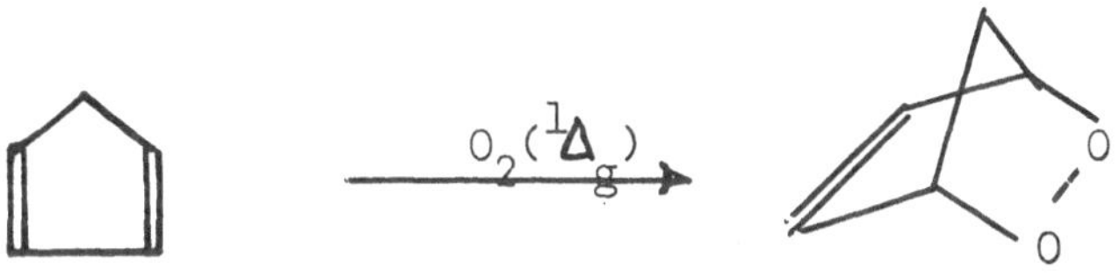

$$\xrightarrow{O_2(^1\Delta_g)}$$

(6)

peroxide formation

With aromatics

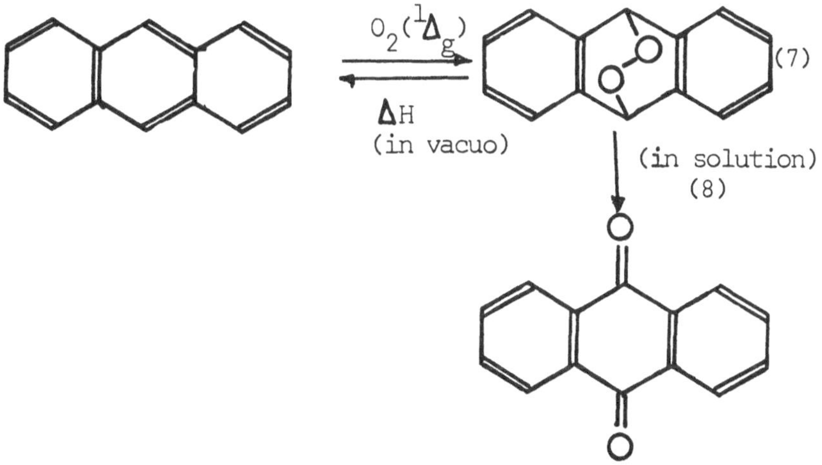

$$\xrightleftharpoons[\substack{\Delta H \\ (in\ vacuo)}]{O_2(^1\Delta_g)}$$

(7)

(in solution)
(8)

With sulphides

$$2R_2S + O_2(^1\Delta_g) \longrightarrow 2R_2SO \tag{9}$$

suggested mechanism

$$\text{example} \quad MeSCH_2CH_2\underset{\underset{NH_2}{|}}{C}HCO_2H \xrightarrow{O_2(^1\Delta_g)} Me\overset{O}{\overset{||}{S}}CH_2CH_2\underset{\underset{NH_2}{|}}{C}HCO_2H \tag{11}$$

With amines

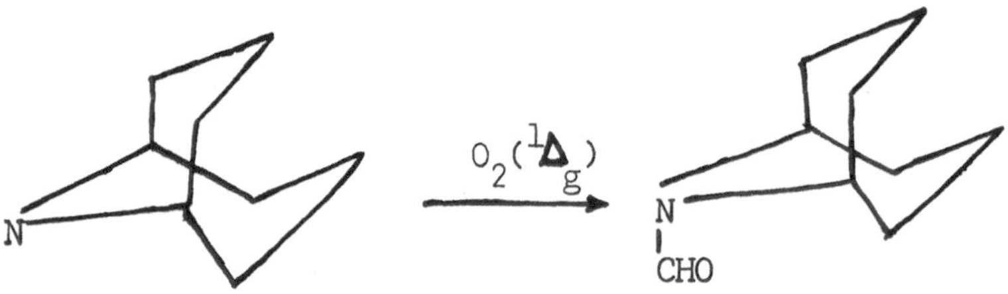

$$\tag{12}$$

CHAIN OXIDATION PROCESSES

Where radicals are produced in a system in the presence of ground-state molecular oxygen, a radical chain oxidation process may occur. This is illustrated in Figure 3 with respect to the oxidation of a hydrocarbon RH through steps 5,6. Radicals can be produced photochemically through steps 2, or 3 and 4 via some excited state X* of an absorbing chromophore X. X* may also dissociate to give molecular fragments (step 9) for example through the Norrish Type II reaction when X is a carbonyl compound. As we have seen above, X* may also interact with substrates to produce $O_2(^1\Delta_g)$ (step 10) which react further to give peroxides (step 11) which may then initiate the chain oxidation process by thermal or photochemical dissociation to produce alkoxy and hydroxyl radicals (7), followed by (8), (5), (6). The reaction scheme shown in Figure 3 is very familiar to those concerned with the photooxidation of synthetic polymers, but is appropriate for consideration here, since it also explains the oxidation of foodstuffs and many biological systems in secondary stages, when secondary breakdown of hydroperoxides formed through $O_2(^1\Delta_g)$ reactions initiates chain oxidation. Chain oxidation is frequently accompanied by low-level visible chemiluminescence,, generally arising through termination reactions of peroxy and hydroperoxy radicals.

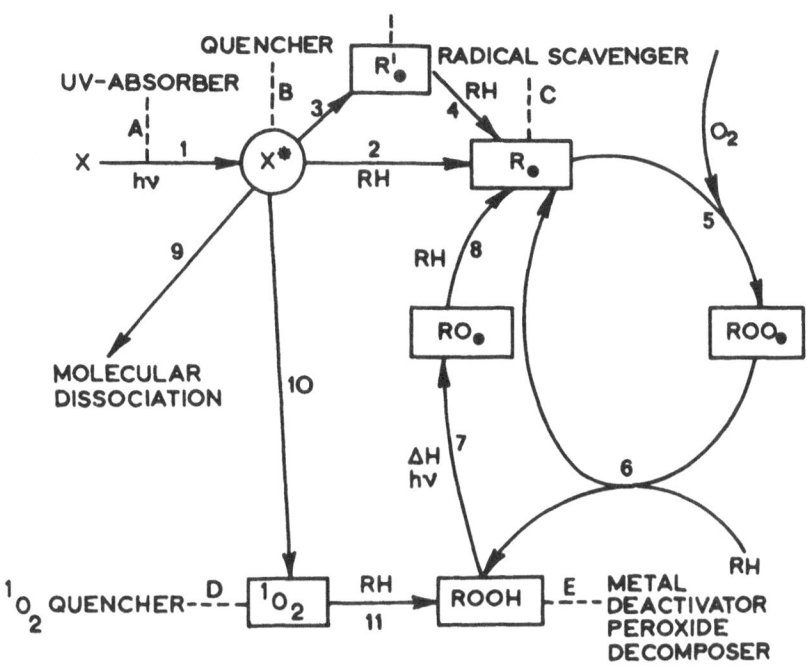

Figure 3   Possible mechanisms for oxidation of hydrocarbon RH

Prevention of oxidation may be achieved by any of the following steps individually or in combination. (A) Use of uv-absorbers or screening agents to prevent absorption. (B) Quenching of electron-ically excited states of X*. (C) Provision of radical scavengers to interrupt the chain oxidation process. (D) Provision of a sink for $O_2(^1\Delta_g)$. (E) Removal of hydroperoxides and peroxides. In nature, D is frequently met in the form of β-carotene.

Oxidation in biological systems may be brought about by other oxygenated species such as the superoxide radical $O_2^{\cdot-}$ produced enzymatically, hydrogen peroxide, and OH. radicals produced for example through reaction (13).

$$O_2^{\cdot-} + H_2O_2 \rightarrow .OH + OH^- + O_2 \tag{13}$$

Suggested Further Reading

'Proceedings of International Conference on Singlet Oxygen and Related Species in Chemistry and Biology', Pinawa, Canada, 1977.

PHYSICAL-ORGANIC PHOTOCHEMISTRY. IV. COMPLEX FORMATION

David Phillips

Southampton University

Department of Chemistry, The University, Southampton

CHARGE-TRANSFER INTERACTIONS

Charge-transfer interactions between electron donors (D) and acceptors (A) in the ground-states can result in the formation of complexes (EDA complexes) for which absorption characteristics are different from those of the uncomplexe D and A molecules. If the equilibrium constant for complex formation K_c^{AD} is given by (1), where the subscript zero refers to initial concentrations, recognizing that the absorbance of the complex A for unit path length is given by $\varepsilon_\lambda^{AD} \, [AD]$, then the equation can be rearranged to give the

$$K_c^{AD} = \frac{[AD]}{[D][A]} = \frac{[AD]}{([D]_o - [AD])([A]_o - [AD])} \qquad \cdots \cdots \cdots \cdots (1)$$

Benesi-Hildebrand equation, (2) provided $[D]_o \gg [A]_o$

$$\frac{[A]_o}{A} = \frac{1}{K_c^{AD} \varepsilon_\lambda^{AD}} \cdot \frac{1}{[D]_o} + \frac{1}{\varepsilon_\lambda^{AD}} \qquad \cdots \cdots \cdots \cdots (2)$$

Thus plots of $[A]_o/A$ against $[D]_o$ yield straight lines, from which ε_λ^{AD} and K_c^{AD} can be evaluated. The frequency of charge-transfer absorption depends upon the stabilization energy of the complex, which depends upon the ionization potential of donor, I_D, electron affinity of acceptor E_A. Such charge-transfer interactions are frequently observed in ground-state molecules.

Electronic excitation of an EDA complex produces an excited state complex which may emit fluorescence. There is, however, a related phenomenon in which interaction of an excited state

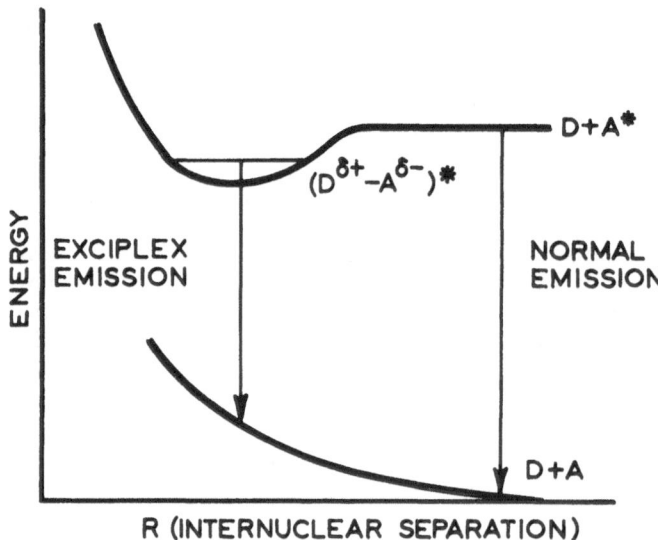

Figure 1   Potential energy surfaces   for Exciplex formation and
emission

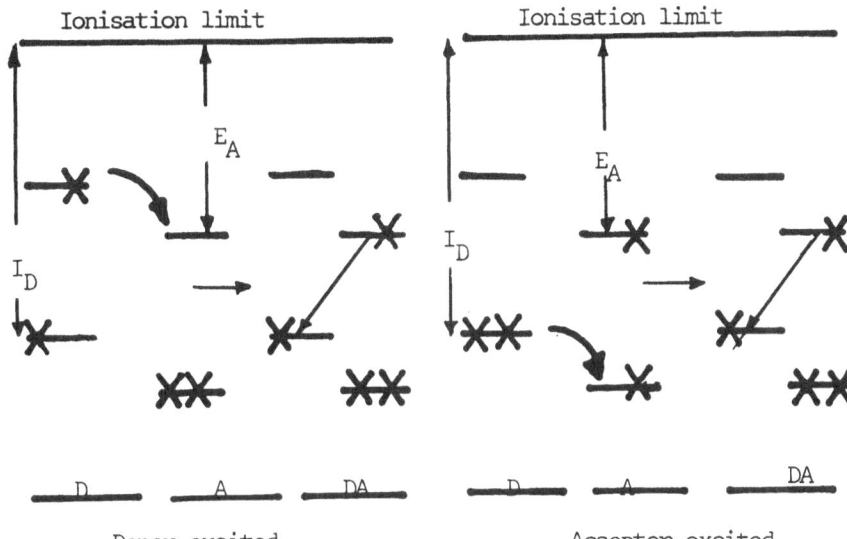

Donor excited   Acceptor excited

Figure 2   Orbital arrangement in exciplex formation

partner with a ground state molecule produces an excited state
complex ('exciplex') which again may emit fluorescence, but for
which there exists no ground-state interaction, i.e. the potential
energy surface between D and A is purely repulsive (See figure 1).
Emission from such systems is thus redshifted with respect to
monomer fluorescence, and is structureless.

A simple molecular orbital picture of what occurs in exciplex
formation and emission is given in Figure 2, from which it can be
seen that for either donor excitation or acceptor excitation, the
free energy change associated with exciplex formation is given by
equation (3) where $h\nu_{oo}$ is the (singlet) excitation energy, C is the

$$\Delta G = I_o - E_A - h\nu_{oo} - C - D \qquad\qquad (3)$$

coulombic energy term ($-e^2/r$, where r is the equilibrium separation
of A^- and D^+) and D a polarization term. Assuming C and D are constant, it can be seen that the frequency of exciplex emission will
depend linearly upon the magnitude of I_D-E_A, neglecting solvent
interactions. This has been verified for a large number of pairs
of molecules.

There are other fates of exciplexes than emission, some of which
are indicated in the scheme below (which assumes donor excitation).
There are many bimolecular photochemical reactions which are now
believed to be preceded by exciplex formation. Whether or not
exciplex emission is observeable depends upon the magnitude of the
exciplex radiative rate constant k_E with respect to those for the

Scheme 1

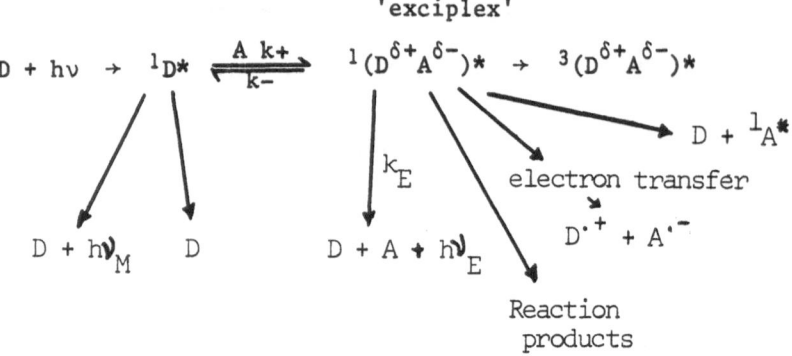

competing decay paths. Evidence for participation of non-fluorescent exciplexes in any photochemical system can be adduced in two ways.

Indirect Evidence

From equation (3) it can be seen that for a single donor and range of acceptors (varying E_A), or single acceptor and range of donors (varying I_D), the free energy of complex formation will vary linearly with the parameter E_A or I_D being changed. If the overall quenching from excited monomer species (say D*) by various additives can be measured, a range of k_Q values can be obtained. These are related to the rate constant values shown in Scheme 1 by where k_q is the sum of all rate-constants depleting the exciplex

$$k_Q = \frac{k + k_{q-}}{(k + k_q)} \tag{4}$$

concentration. Using the thermodynamic relationship between equilibrium constant and free energy, $-\Delta G = RT\ln K_p$, it can be shown that plots of $\ln k_Q$ against ΔG measured from (3) should be straight lines. The success of such plots is usually taken as an indication of charge-transfer interactions, and one such correlation is shown in Figure 3.

Direct Evidence

In favourable cases, even though the exciplex is non-fluorescent careful monitoring of the decay characteristics of the excited monomeric species (D* or A*) using pulsed excitation techniques (such as synchrotron radiation sources) can yield information about exciplex formation. It can be seen from Scheme 1 that the molecule D* is populated by two processes, namely direct excitation by the light pulse and by reverse dissociation of the exciplex. The decay of the monomer fluorescence with time $I_{D^*}(t)$ can thus be fitted to dual exponential decay function of the form

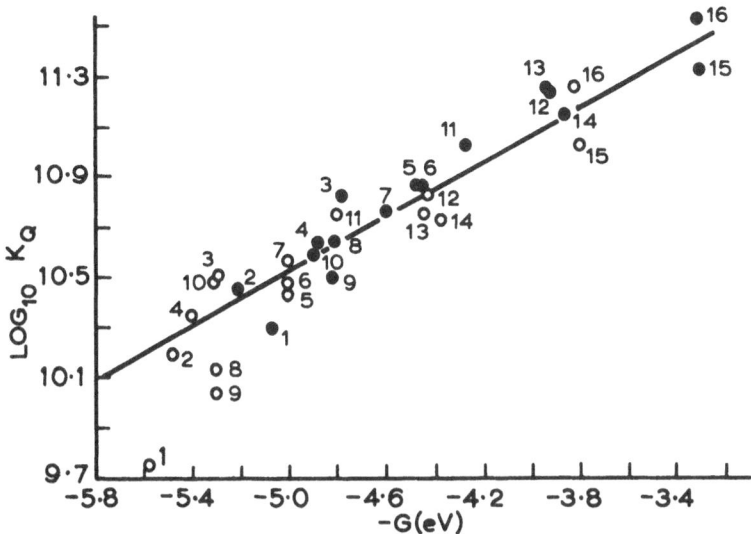

Figure 3   Plots of ln k_Q against $-\Delta G$ for a series of electronically excited benzenoid hydrocarbons quenched by conjugated diolefins in the vapour phase (J. Amer. Chem. Soc., For key to compounds see original article).

$$I_D^*(t) = A_1\, e^{-\lambda_1 t} + A_2\, e^{-\lambda_2 t} \qquad\qquad (5)$$

where λ_1 and λ_2 are related to the rate constants for monomer and exciplex decays.   This has been achieved in a number of important cases.

Exciplex formation is of such general occurrence in photo-chemical reaction pathways that an exhaustive coverage is impossible here.   Typical model systems are listed below with brief examples.

(a) Aromatic hydrocarbons - olefins.   The formation of cyclo-addition compounds between benzene and simple mono and diolefins is considered to occur via exciplex formation (cf. Figure 3).

(b)   Ketones - olefins. e.g. oxetane formation occurs via exciplex formation.

(6)

(c) Aromatic ketones - amines. (Used as photocuring initiator industrially)

$$Ar_2CO^* + RCH_2CH_2NR_2 \rightarrow \left[exciplex \right]$$

$$\downarrow \qquad\qquad (7)$$

$$ArCOH + RCH_2CHNR_2 \rightarrow (Ar_2C=O)^{\bar{.}} (RCH_2CH_2NR_2)^{\dot{+}}$$

Further examples will be discussed.

Excimers

Excimers of course are related to exciplexes except that complexes are formed between excited electronic state and ground state partners of the same species. Excimer emission resembles exciplex emission spectrally, and fluorescence of dinucleotides, polynucleotides and DNA is of this type. Excimer stability is given by both exciton resonance and charge-transfer terms, whereas resonance terms are absent in the case of exciplexes. Excimers may also be implicated in many photochemical reactions e.g. (reaction 8), although instances are fewer than is the case for exciplexes.

(8)

Suggested Further Reading

'Excited States in Organic Chemistry' J.A. Barltrop and J.D. Coyle, (Wiley Interscence), London 1975.

'The Exciplex' Ed. M. Gordon and W.R. Ware, Academic Press, N.Y. 1975

For discussion of analysis of dual exponental decay curves see C. Lewis and W.R. Ware, Mol. Photochem., 1973, 5, 261.

'Photoassociation in Aromatic Systems', B. Stevens, Advances in Photochemistry, 1971, 8, 161.

TIME-RESOLVED FLUORESCENCE STUDIES ON DRUG BINDING SITES

David Phillips , Kenneth P. Ghiggino and Anthony G. Lee

Chemistry Department, The University, Southampton SO9 5NH

INTRODUCTION

Fluorescence spectroscopic techniques have been widely used in the past to obtain information about simple lipid bilayers and purified lipid-protein complexes. The same techniques have also been used to study intact biological membranes but the results of such studies have been much more difficult to interpret, simply because the membrane is a highly heterogeneous structure. In general a single composite spectrum is obtained either from all components of the membrane or from probe molecules dissolved in a variety of sites in the membrane when excited with continuous wave radiation.

The fluorescence properties of anilino-naphthalene sulphonate derivatives make it a useful probe molecule used to investigate, for example, the nature of the binding sites in various macro-molecules and cell constituents[1]. Changes in fluorescence yields, absorption profiles and fluorescence spectra accompanying the binding of these derivatives to biomolecules from aqueous solution are used in many studies as a sensitive probe of the polarity of the binding site[2], although a number of recent investigations have suggested that solvent cohesion and structure[3], the nature of the electronic states involved[4] and environment rigidity[5] may also play an important role in determining the photophysical properties of these molecules. A detailed knowledge of the influence of these is vital in interpreting the usefulness of fluorescent probes and dyes for biochemical studies. Any such studies should include temporal resolution of fluorescence, since this adds another dimension to the investigative armoury of the research worker.

The technique of time-correlated single-photon counting decay
fluorimetry with pulsed discharge lamp excitation has certainly
permitted such studies to be carried out, but the low repetition
rate and intensity of such light sources, together with their
variable pulse duration, intensity and repetition rate make them
less than ideal, particularly in the short wavelength uv region.
Improved light sources for such studies would require high
repetition rate and intensity, tunability and stable operation.
Clearly synchrotron radiation fulfils these requirements, and will
thus be particularly useful in those wavelength regions where
alternative sources are not available, e.g. below 260 nm. We
describe here, however, an alternative system, namely a mode-locked
gas laser which also has many of the desirable features of the ideal
pulsed excitation source, (including location in the experimenters
own laboratory), and which is particularly useful in the following
wavelength regions 257.3 nm, 280-320 nm, 488 nm, 514.5 nm,
550-650 nm.

INSTRUMENTATION

        The time-resolved fluorescence apparatus employed[6], shown
schematically in Figure 1 uses either the second harmonic output
of a mode-locked, cavity dumped argon-ion laser (λexc 257.25 nm) or
the second harmonic of tunable radiation from a dye laser using
Rhodamine dyes in the 550-650nm region (λexc 280-320nm) as the
excitation source. Spectral dispersion of fluorescence is obtained
with a Rank Precision Monospek 1000 grating monochromator and
detection is by the time-correlated single photon counting technique.
Features of this system are the very high laser repetition rate
(5 MHz) enabling the rapid recording of both fluorescence lifetimes
and time-resolved emission spectra and the exact reproducibility
of pulse shape and intensity allowing accurate computer analysis
of the fluorescence decay lifetimes.

        The argon-ion laser can be operated in the mode-locked, cavity
dumped, or cavity dumped mode-locked configuration to produce a
train of narrow, highly stable, intense light pulses with repetition
rates selectable from single shot to 100 MHz. The dye laser can
also be synchronously pumped and cavity dumped to provide a source
of tunable picosecond light pulses. Second harmonic generation of
UV pulses is achieved with temperature and angle tuned ADA and
ADP crystals. The cavity-dumping method of pulse selection has
certain advantages over other methods in that either narrow
sub-nanosecond pulses can be obtained when operated in the mode-
locked and dumped mode, or wider but highly reproducible and
intense ns pulses may be obtained using the cavity-dumped
output alone. The exact reproducibility of pulse shape throughout

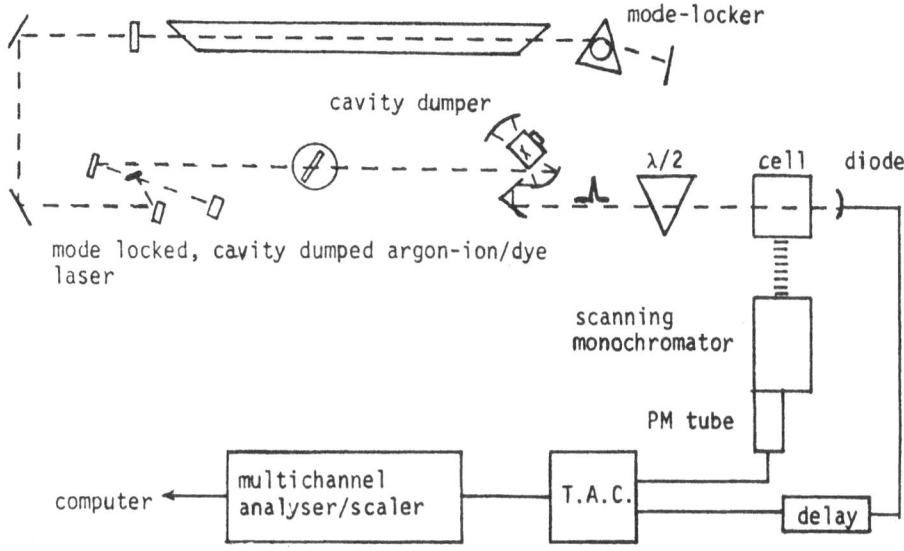

Figure 1   Schematic of laser excited time-resolved emission
apparatus

the experiment allows accurate computer analysis of the
fluorescence decay kinetics. Examples of lifetimes of a number of
compounds recorded on the apparatus are given in Table 1.   In all
cases the lifetimes obtained by a non-linear least squares
convolution fitting of the data to a single exponential decay were
identical within the quoted errors using both mode locked cavity
dumped pulses (FWHM ∿ 300 psec) and cavity dumped pulses (FWHM
∿ 7 nsec) for excitation. Accuracy of fitting is tested using the
method of inspection of weighted residuals. An example of these
is shown in Figure 2 for fitting (a) a single exponential, (b) the
weighted sum of two exponentials to an experimental decay curve
obtained in a heterogeneous system (see below). Best fit is
indicated when residuals are random as in (b).

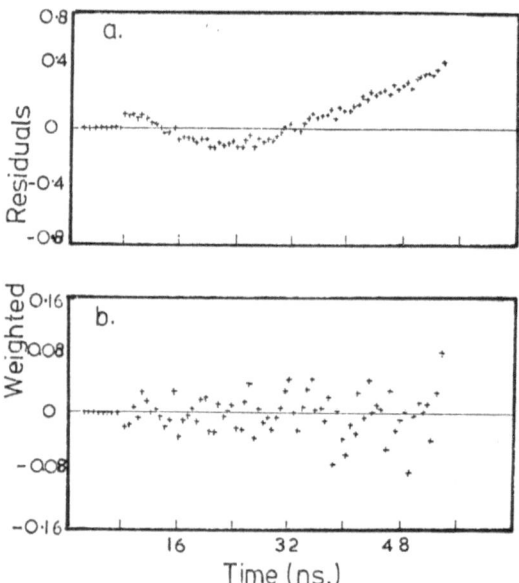

<u>Figure 2</u> Weighted residuals for fluorescence decay of probe
molecule ANS (see below) bound to microsomal
membrane at 480 nm. (a) Single exponential fit.
(b) Double exponential fit. The random residuals in
(b) clearly indicate a superior fit to the data.

Table 1

Lifetimes of compounds in degassed room temperature solutions

Compound/Solvent	λ_{F1}^{a} (nm)	τ^{b} (ns)	τ^{c} (ns)
Acridine/Ethanol	440	-	0.33±0.03
Rose Bengal/Ethanol	600	0.71±0.02[d]	-
Anthracene/Cyclohexane	440	4.76±0.05	4.79±0.05
Poly(styrene)/Dichloromethane	280[e]	0.74±0.05	0.76±0.05
	340[f]	13.4±0.3	13.9±0.2
9-Cyanoanthracene/Cyclohexane	440	12.7±0.1	12.8±0.1
N-acetyltryptophan methyl amide/H$_2$O	330	2.95±0.03	3.00±0.03

a Emission wavelength monitored
b Using cavity-dumped pulses from second harmonic (257.25nm) of the ion laser only, FWHM ∿ 7 ns
c Using mode-locked cavity-dumped pulses from ion laser, second harmonic (257.25nm) FWHM ∿ 300 ps
d Using cavity-dumped dye laser pulses (580nm) FWHM ∿ 7ns
e Corresponding to monomer emission only
f Corresponding to excimer emission only

A further indication of the system stability and the accuracy of the deconvolution technique is obtained from an analysis of the fluorescence decay from a dilute mixture of anisole (τ = 5.42 nsec) and aniline (τ = 2.49 nsec) in undegassed ethanol. The application of a dual exponential function to the decay curves recorded at wavelengths where both species fluoresce yields a random residual distribution and lifetime values of 5.57±0.7 nsec and 2.26±0.4 nsec with a pre-exponential weighting ratio of 0.35 thus temporally isolating the mixture components, It should be recognised that these measurements were made with excitation pulses (cavity-dumped second harmonic ion laser, FWHM ∿ 7ns) which are longer than both lifetimes measured. The success of the experiments is due to the inherent stability of pulse shape, intensity, and interval between pulses. Similar stability should be realisable with synchrotron radiation sources. Both sources (laser and synchrotron) lend themselves to the technical trick of inverting the normal mode of operation of the single-photon counting detection equipment such that the emitted photon acts as the start signal, the next

excitation pulse as the stop signal. This permits the full
repetition rate of laser or storage ring to be utilised without
difficulty, drastically reducing data acquisition times. Thus in
the present experiments, in all cases the fluorescence decay data
was accumulated in less than thirty seconds.

A further application of the apparatus is the recording of time
(optimal 200 psec) and wavelength (optimal 1 Å) resolved emission
spectra. By using appropriate gating electronics it is possible to
directly record the spectral distribution of photons emitted at
different times after the excitation pulse. This will be illustrated
below for biological systems.

APPLICATION TO BIOLOGICAL SYSTEMS

Modern concepts concerning the biological membrane are
summarized in the fluid-mosaic model (figure 3) of S. J. Singer
and G. L. Nicolson[7]. The structure is assumed to consist of a
bilayer of lipid molecules containing a variety of protein molecules,
some of which penetrate right through the lipid bilayer, whilst
others do not. The lipid bilayer provides the basic permeability
barrier, while the proteins are concerned both with the transport of
molecules from one side of the membrane to the other, and with the
various enzymatic activities of the membrane.

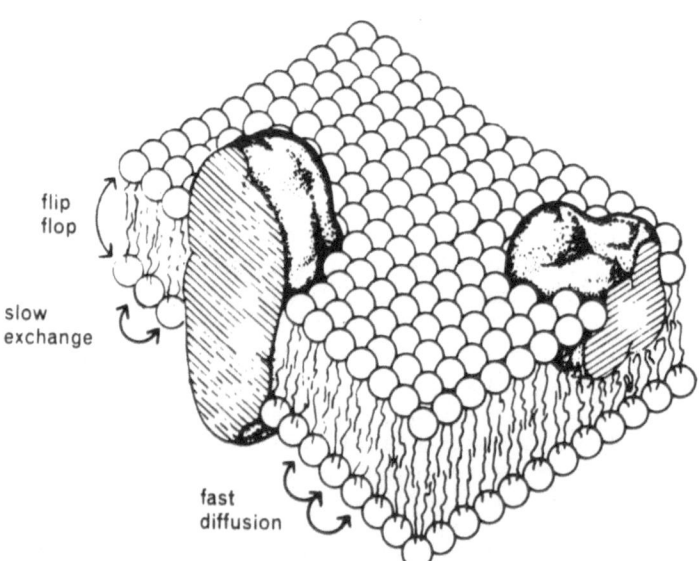

Figure 3   The Singer-Nicolson model of a membrane.   From A.G. Lee
           Endeavour, 1975, 34, 67.

8-Anilino-1-naphthalene sulphonate (ANS) has been much used as a fluorescence probe in such systems since it exhibits a large increase in fluorescence intensity and spectral shifts when bound to certain proteins, lipids or cellular membranes. However, it is uncertain to which components of cellular membranes the ANS binds. Although ANS can bind to protein solubilized from some membranes[8], it is generally thought that the major binding sites are lipid in nature. Thus although proteolysis of intact myelin or erythrocyte membranes has little effect on ANS binding, phospholispase treatment has a large effect[1]. On the other hand, Yguerabide[9] have shown that the fluorescence decay for ANS bound to red blood cells can be fitted to a double exponential, one component of which is characteristic for the fluorescence decay of ANS bound to lipid and the other for ANS bound to a protein such as apomyoglobin.

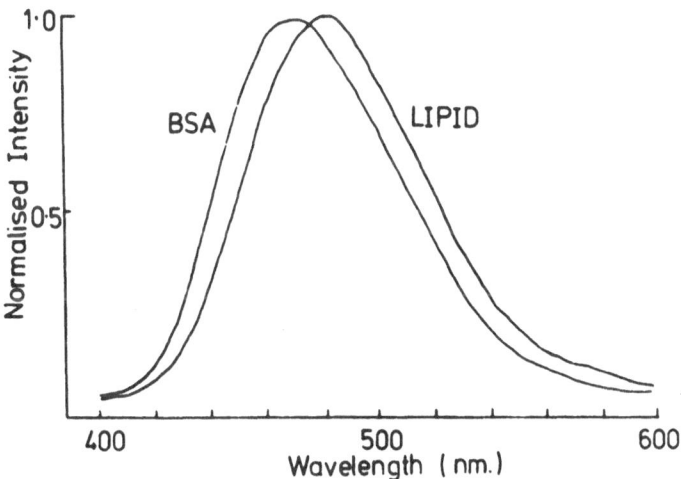

Figure 4   Total fluorescence emission spectra for ANS bound to liposomes of egg yolk phosphatidylcholine and bovine serum albumin.

In the present experiments as shown in Figure 4, there is a detectable difference in emission maxima for ANS bound to a model lipid system, liposomes of egg phosphatidylcholine, and for ANS bound to a model protein, bovine serum albumin, BSA, under continuous illimination. However, in a microsomal membrane (from Tetrahymena pyroformis) under continuous illumination, fluorescence from ANS shows only a single peak at 470nm, and thus different binding sites, if present, are unresolvable (Figure 5). Note that because of excitation conditions a significant background emission due to protein excitation is also observed.

Time-resolved measurements can provide much more information. Thus fluorescence decays for ANS bound to liposomes of egg phosphatidylcholine or bovine serum albumin can be well fitted to

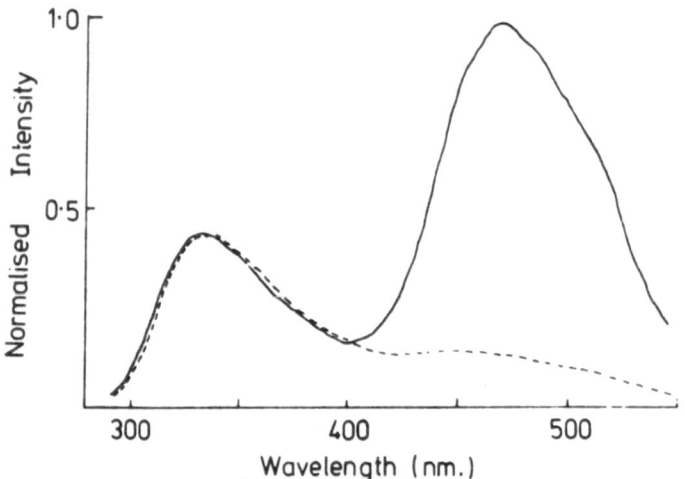

Figure 5   Total fluorescence emission spectra for ANS bound to micro-
            somal membranes, (----) Fluorescence from blank
            microsome sample; (———) Fluorescence from microsomes
            plus ANS (1.7 x 10^{-5}M).

a single exponential, with fluorescent lifetimes listed in Table 2.
Fluorescence decays for ANS bound to microsomal membranes can no
longer be fitted to a single exponential but can with confidence
be fitted to a double exponential, (see Figure 2(b)) with decay
components shown in Table 2.

Fluorescence lifetimes decrease slightly with increasing ANS
concentration, probably as a result of fluorescence quenching at
high concentrations of bound ANS, as also observed in measurements
of fluorescence intensity. The close correspondence of these two
decay components to values obtained with the model lipid and protein
bound ANS systems lead to the speculation that the time resolution
permits identification of lipid and protein binding sites in the
microsomal membrane. Further support for this view comes from
time-resolved spectroscopy. Thus, the significant differences in
fluorescence lifetimes and emission maxima of the probe in the two
environments allows one to isolate the emission from ANS bound
to lipid and protein components by time resolved emission
spectroscopy. The spectral distribution of photons emitted
coincident with the excitation pulse ($\Delta t = 0$, $\delta t = 2.5$ns) will be
weighted towards the component of fluorescence of shorter lifetime

TABLE 2

Fluorescence decay characteristics for ANS

Systems	Decay time (nsec) Laser excitation
Egg-phosphatidylcholine	4.22±.1
Bovine serum albumin	15.6±.1
Microsomal membranes (1mg protein/ml)	
(ANS 6.3 x 10^{-6}M)[a]	4.51±0.4 ($A^{[b]}$=.056)
	16.4±0.8 (A=.017)
(ANS 5.2 x 10^{-5}M)[a]	3.72±0.4 ($A^{[b]}$=.065)
	14.3±0.8 (A=.019)

[a] data fitted to a double exponential,
with a X^2 statistic of 1.3

[b] Pre-exponential weighting factors

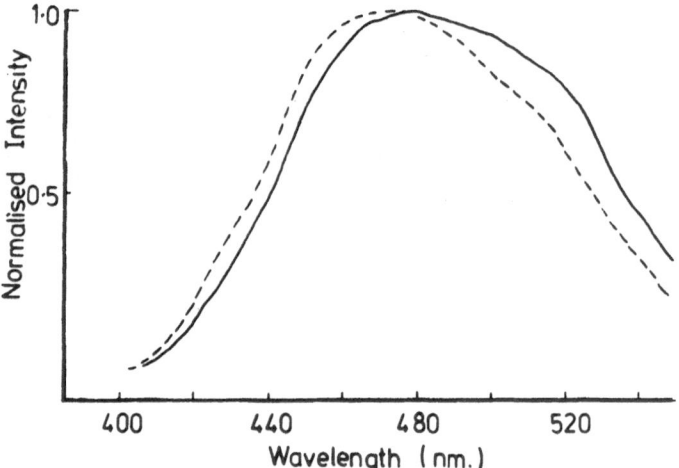

<u>Figure 6</u>   Time resolved fluorescence emission spectra for ANS bound to microsomal membranes (————) Early gated spectrum Δt = 0 δt = 2.5ns (————) Late gated spectrum Δt = 49ns, δt = 3.2ns.

while with the time gate placed at a considerable delay after
excitation (Δt = 49 ns, δt = 3.2 ns) the emission spectrum will
arise solely from the long-lived component of fluorescence. The
time-resolved spectra recorded are shown in Figure 6.

The early-gated spectrum has a fluorescence emission maximum
the same as that for lipid bound ANS whereas the component of long
fluorescence lifetime has a spectrum comparable to that of ANS bound
to BSA (compare Figs 6 and 4). It has therefore been possible to
resolve the spectra of ANS into two broad categories of binding
site in the membrane, although the conventional steady-state
fluorescence spectrum gives no indication of two components (Fig. 5).
It is, of course, not possible to conclude from these results that
only two distinct types of binding site exist, but the time-resolved
spectra together with the fluorescence lifetime data do suggest that
the majority of binding sites under these conditions do fall into
the broad categories of lipid - like or protein (BSA) - like.

Having confirmed the assignment of the two components of the
fluorescent decay curves for ANS binding to microsomes, it is now
possible to use the fluorescent decay curves to study environmental
effects of ANS binding. Figure 7 shows the effect of addition of
increasing amounts of barbiturate. Clearly, barbiturate causes
an increase in fluorescence decay time for ANS bound to the
membrane, attributable to an increased weighting for the longer-
lived protein-bound component in the composite decay (Table 1).
Since steady state measurements show that addition of barbiturate
leads to a decrease in fluorescence intensity, this means that
barbiturate must preferentially displace ANS from its lipid binding
site.

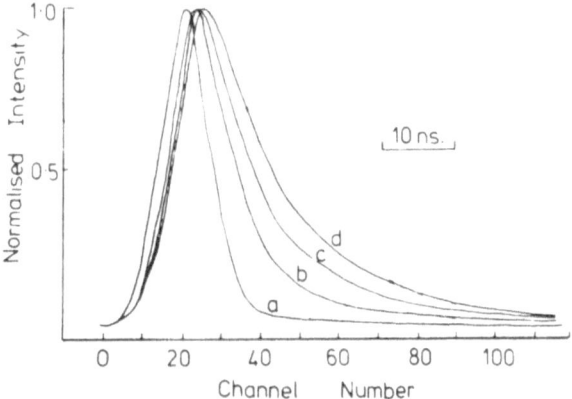

Figure 7 Fluorescence decay curves for ANS (1.7×10^{-5}M) bound to
microsomal membranes (1mg/ml) recorded at 480nm, in the
presence of pentobarbitone. (a) excitation pulse (b)
without pentobarbitone, (c) with 10^{-6}M pentobarbitone,
and (d) with 5×10^{-6}M pentobarbitone.

In conclusion, the technique of nanosecond time-resolved fluorescence spectroscopy is able to resolve spectra for fluorescent molecules bound in lipid and protein environments in a membrane, whenever the fluorescence emission maxima and fluorescence lifetimes differ significantly in the two enviroments. Time-resolved fluorescence spectroscopy also, of course, has potentially important applications in other areas of biochemistry. Thus, for example, in proteins containing two Trp. residues in different environments, fluorescence spectra for these two residues will be resolvable.

Although the experiments described here were carried out with pulsed laser excitation, synchrotron radiation should provide an alternative and in some respects superior excitation source for such studies, although there are disadvantages also to be considered.

REFERENCES

1. L. Stryer, Science, 1968, 162, 526; J. Vanderkooi and A. Mantonosi, Arch. Biochem. Biophys., 1971, 144, 37; M.B. Feinstein and H. Felsenfeld, Biochemistry, 1975, 14, 3041.

2. W.O. McClure and G.M. Edelman, Biochemistry, 1966, 5, 1908; C.J. Seliskar and L. Brand, J. Amer. Chem. Soc., 1971, 93, 5405.

3. R.L. Reeves, M.S. Maggio and L.F. Costa, J. Amer. Chem. Soc., 1974, 96, 5917.

4. E.M. Kosower and H. Doduik, J. Amer. Chem. Soc., 1974, 96, 6195.

5. S.K. Chakrabarti and W.R. Ware, J. Chem. Phys., 1971, 55, 5494; R.P. DeToma, J.H. Easter and L. Brand, J. Amer. Chem. Soc., 1976, 98, 5001.

6. K.P. Ghiggino, R.D. Wright and D. Phillips, J. Polym. Sci. Polymer Physics, 1978, 16, 1499; K.P. Ghiggino, D. Phillips, K. Salisbury and M.D. Swords, J. Photochem., 1977, 7, 141.

7. S.J. Singer and G.L. Nicolson, Science, 1972, 175, 720.

8. R.B. Freedman and G.K. Radda, Febs Lett., 1969, 3, 150.

9. J. Yguerabide in "Fluorescence techniques" Ed. A.A. Thuer and M. Sernetz, Springer-Verlag, NY, 1973.

THE USE OF SYNCHROTRON RADIATION IN FLUORESCENCE STUDIES ON

BIOCHEMICAL SYSTEMS

David M. Jameson and Bernard Alpert

Lure, Universite de Paris-Sud
Centre D'Orsay, Batiment 200
91405 Orsay, France

Our lectures will concern the application of fluorescence spectroscopy, utilizing synchrotron radiation as an excitation source, to the study of biochemical systems. At present, in conjunction with the LURE (Laboratoire d'Utilisation du Rayonnement Synchrotron) group at Orsay, France, we are establishing a dedicated light port with complete spectrofluorometric instrumentation on the ACO (Anneau de Collision d'Orsay) electron storage ring. When completed our facility, which will be available to a number of researchers from various disciplines, will have the capability for measuring excitation and emission spectra, polarizations, lifetimes and time-resolved spectra. In the course of our lectures we will describe various aspects of our facility in more detail. Perhaps some of the participants of this meeting will propose worthwhile and interesting experiments which our ACO light port can accomodate.

Before one can appreciate the specific advantages (and limitations) of synchrotron radiation in the wavelength region we plan to utilize one must necessarily possess a knowledge of the information available from fluorescence spectroscopy. The participants of this meeting have quite divergent backgrounds and areas of specilization, hence it is difficult to present material challenging yet comprehensible to all. Since certain of the other lecturers plan to discuss some of the fundamentals of absorption and fluorescence processes we shall concentrate on the philosophy of specific applications of fluorescence techniques to biochemical problems describing what appear to us to be the particularly interesting areas of development. Our central questions will be :

1) What information of interest to biochemists can we expect

to obtain from fluorescence spectroscopy?

2) What aspects of synchrotron radiation lend themselves to fluorescence techniques?

We shall try to illustrate our salient points with references to the recent literature.

Fluorescent probes are generally divided into the categories intrinsic and extrinsic. By intrinsic we mean those natural components of the system which demonstrate sufficient fluorescence yield to be of practical utility. These include such molecules as NADH, FMN and the aromatic amino acids, specifically tryptophan and tyrosine. Extrinsic probes encompass all those fluorophors such as the aminonaphthalene sulfonates, the fluoresceins, etc., which are normally foreign to the system and are introduced by the investigator. The discovery and implimentation of a host of fluorescence Probes in the last two or three decades have done much to advance our knowledge of the fluorescence phenomenon and biochemistry. It is beyond the scope of our lectures to present a comprehensive review on fluorescent probes; interested parties are referred to an excellent article by Brand and Gohlke (1972).

It is interesting to note that the discovery of new probes is due sometimes not to the astuteness and skill of the organic chemist or biochemist but to improvements in electronics and instrumentation. For example, as late as the early 1950's it was not known with certainty if the aromatic amino acids fluoresced. Improvements in instumentation, however, eventually permitted investigators to detect and quantitate the fluorescence from tryptophan, tyrosine and phenylalanine (Shore and Pardee (1956), Teale and Weber (1957) and Weber (1959)) which led to the wide application of these fluorophors in the study of proteins. More recent instrumentation has permitted the study of the intrinsic fluorescence at room temperatures of aqueous solutions of nucleotides and nucleosides, molecules with notoriously low ($\sim 10^{-4}$) quantum efficiencies (Daniels and Hauswirth (1971) and Vigny and Duquesne (1974)). Experiments conducted at Urbana, Illinois and Orsay, which we shall discuss in more detail later, have convinced us that given present day instrumentation the intrinsic, very weak, tryptophan fluorescence from hemeproteins may prove to be a useful tool in the study of protein dynamics and energy transfer processes.

One of the major advantages of fluorescence over absorption spectroscopy is its sensitivity. In conventional absorption spectroscopy one measures the difference in light intensity between two relatively large signals; conventional fluorescence spectroscopy, on the other hand, monitors a light level above a dark background (ideally). Thus we find that while absorption spectroscopy is useful in the range from 10^{-3} to 10^{-7} molar for most biologically

interesting molecules, fluorescence techniques permit the extension
to the range of 10^{-8} to 10^{-12} molar fluorophor.

An example of the sensitivity of fluorescence spectroscopy is
given in figure 1. These spectra were acquired on a computer inter-
faced, photon-counting spectrofluorometer (Jameson et al (1976))
very similar to the instrumentation being established at ACO. Figure
1A shows the solvent, 0.2N H_2SO_4, excited at 340 nm and scanned for
approximately three hours from 370 to 560 nm (excitation and emission bandwidths of 4 nm). The principal feature in the spectrum is
the raman peak at 384 nm due to the O-H stretch of water. Solvent
raman peaks can serve as excellent internal standards for sensitivity
and wavelength calibration yet they may cause some confusion for the
inexperienced fluorescence practitioner. One should remember that
raman signals are quite weak and are only evident if the fluorescence
signal under scrutiny is poor due to low quantum efficiency or low
fluorophor concentration. Raman peaks can be distinguised from
fluorescence peaks by changing the excitation wavelength which shifts
the raman peak, since it remains at a fixed energy relative to the
excitation wavelength, while the fluorescence peak maximum remains
constant (with a few exceptions). Figure 1B shows the results from
a similar after the addition of quinine sulfate to a concentration
of 10^{-11} molar. The subtraction of spectrum 1A from spectrum 1B is
shown, with a change in the vertical scale, in figure 1C. To demonstrate that this spectrum does indeed correspond to quinine sulfate
this spectrum and one obtained at 10^{-8} molar concentration are shown
overlaid in figure 1E.

An important consequence of this sensitivity is that fluorescence techniques permit us to study systems, even protein ensembles,
under conditions of virtually ideal thermodynamic behavior. One
must realize that the high dilution attainable in fluorescence
studies may affect the system by the law of mass action, eg., multi-
subunit proteins are subject to dissociation at low concentrations.

We also note that various emission parameters such as intensity
changes and spectral shifts are often much more sensitive to the
environment than the corresponding absorption parameters. The
classic example of this effect would be the large increase in quantum efficiency and dramatic blue shift in the emission of 1- anilino-
naphthalene-8- sulfonate (ANS) upon binding to bovine serum albumin
demonstrated by Weber and Laurence (1954). A more recent example is
the work of Li et al (1976) on the reversible denaturation of proteins under high pressure. These investigators demonstated that the
intrinsic protein fluorescence or the fluorescence of a specifically
bound probe, eg., FMN, could be monitored to follow the reversible
pressure denaturation of a protein yielding large changes (in the
FMN case, a 30 fold change) in the signal as the pressure was increased. Comparable high pressure absorption studies on proteins
exhibited signal changes of only a few percent.

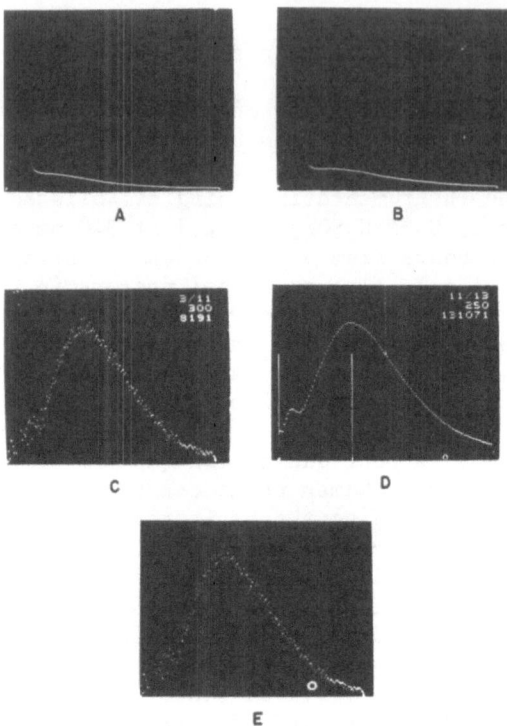

Figure 1.   Emission Spectrum of 10^{-11} M Quinine Sulfate
  A.   Background emission of 0.2 N H_2SO_4 scanned from 371 to 560
       nm.   Excitation at 340 nm.   Raman peak appears at 385 nm.
  B.   Background plus 10^{-11} M quinine sulfate.
  C.   Spectrum B minus Spectrum A : 10^{-11} M quinine sulfate.
       Vertical scale expanded.
  D. 10^{-8} M quinine sulfate.
  E.   Overlay of Spectra C and D.

        Finally, absorption measurements are generally limited to deter-
minations of peak maximums and extinction coefficients (excluding
techniques such as circular dichroism) whereas fluorescence measurements include peak positions, intensities, polarizations, lifetimes
and time-resolved spectra.   Thus, although interpretation of this
wealth of data is not always straightforward, the potential information content of fluorescence experiments is impressive.

        We may briefly summarize the most important principles of
fluorescence as follows :

1) The absorption or emission of a photon occurs instantaneously relative to the motions of the nuclear framework - this is the
Franck-Condon principle.

2) After excitation, a rapid loss of energy through vibrational
coupling modes brings the system to its lowest excited level in a
few picoseconds (we refer to condensed phase systems). Direct measurements on this vibrational relaxation process have in fact been
made by Lauberau et al (1972). A consequence of this rapid loss of
excess energy is Vavilov's law which states that the quantum efficiency of a fluorophor is independent of the excitation wavelength;
a number of exceptions to this rule are now known, though - see
Birks (1976).

3) A further reduction in the energy of the system can occur
during the lifetime of the excited state by such processes as solvent relaxation, proton dissociation or various quenching interactions.

This last point expresses implicitly the utility of fluorescence spectroscopy in biochemistry for we can only observe the
effects of those processes which occur during the lifetime of the
excited state. It seems remarkably fortuitous, in fact, that a
technique exists which is able to probe exactly the time scale of
most critical interest for the motions and interactions of biological molecules. Since the typical lifetimes of fluorescence probes
extend from 10^{-7} to 10^{-10} second, the processes which can be monitored include rotations of small molecules in media of low viscosi-
ties as well as the complex mobilities of macromolecules such as
proteins, even in relatively viscous media such as biological membranes. By implication then, fluorescence is a technique primarily
suited to study dynamics rather than structures. Indeed, fluorescence techniques have been important in establishing the dynamical
picture of proteins that has emerged in recent years (an excellent
review by Lumry and Hershberger (1978) covers the literature on
protein dynamics thoroughly; also see Weber (1976) for a marvelous
exposition on the nature of the information obtainable from fluorescence techniques).

To emphasize the importance of dynamics in fluorescence measurements we may consider the case of solvent relaxation around an
excited dipole. If the fluorophor exhibits a different dipole moment in its excited state than in its ground state then we may expect the solvent molecules to reorient during the lifetime of the
excited state. The reoriented or relaxed solvent molecules confer
an energetic stability to the system which is reflected in a red
shift of the emission spectrum relative to an unrelaxed state. The
classical theory developed by Lippert (1957) relates this spectral
shift to the dielectric constant and refractive index of the solvent
as well as the difference between the excited and ground state
dipole moments.

In principle, by determining the spectral shifts for a fluorophor in a series of solvents of known dielectric constants one can calculate the difference between the excited and ground state dipole moments and determine the dielectric constant - refractive index term for an unknown environment. Such a measurement holds an obvious interest for the researcher who wishes to probe the surroundings of a fluorophor bound to a protein or embedded in a membrane since the nature of the media will impose its effect on the biological function of the system. We now understand, however, that it is the total dynamics of the fluorophor - solvent system which determine the final emission properties (Bakhshiev (1964)). Evidently if the radiative rate (reciprocal of the lifetime) is rapid compared to the solvent relaxation rate there may be negliable spectral shift regardless of the nature of the solvent. Clearly the important parameter to consider is the ratio of the solvent relaxation rate to the radiative rate. Modification of either term in this ratio may be expected to result in altered emission properties. The solvent relaxation rate can be altered quite simply by changing the temperature or viscosity. For example, the emission maximum of the fluorophor 2-diethylaminonaphthalene sulfonate in propylene glycol demonstrates a strong dependence upon the temperature - viscosity term over the range -60° C to +50°C (Weber (1976)). The elegant time-resolved spectroscopy experiments of Ware et al (1971) and Brand and Gohlke (1971) on various systems have demonstrated directly the dynamical aspects of solvent relaxation. The technique of time-resolved spectroscopy, which we will discuss in more detail later, has permitted these investigators to obtain directly emission spectra characteristic of a fluorophor at specified time intervals after the excitation process. The observations of Brand and Gohlke (1971) on the TNS - bovine serum albumin system demonstrated clearly a time dependence in the spectral shifts which may be correlated to the relaxation of the protein residues around the bound fluorophor. Thus we must apparantly proceed cautiously before attempting to characterize the environment of a fluorophor bound to a protein by consideration of the spectral shift alone.

An alternative approach to this problem was adopted by Weber and Lakowicz (1973) who chose to shorten the lifetime of the excited state by addition of oxygen, a highly effective dynamic quencher. Since the ratio of the solvent relaxation rate to the radiative rate is the important parameter we may expect that shortening the lifetime, i.e., increasing the radiative rate, will have the equivalent effect as increasing the solvent viscosity. Weber and Lakowicz observed blue shifts in the emission of tryptophan in water and indole and ANS in ethanol upon the addition of oxygen. By these means they were able to estimate the relaxation rate of ethanol around excited state indole and ANS to be 13 and 54 picoseconds respectively at 25°C and 1500 psi of oxygen.

It is important to realize that these oxygen quenched spectra, obtained under steady state conditions, contain essentially the same information as the standard time-resolved spectra but without the complications of the instrumental convolution functions which at present restrict time-resolved spectra to durations greater than one nanosecond after excitation.

Direct measurements of emissive lifetimes, in addition to the insight they offer to the molecular spectroscopist, are of prime importance for the correct interpretation of a variety of fluorescence data. Polarization and quenching experiments are but two examples wherein lifetime determinations may be essential and we should consider these cases in more detail.

If a fluorophor in solution is excited with linearly polarized light then the emission may also be polarized. Within a coordinate system assigning the vertical laboratory axis to the parallel polarized component one can derive various relations demonstrating the dependence of the observed polarization upon several parameters. (For a more complete discussion of the polarization phenomenon the reader is refered to Weber (1966) and Weber (1952)).

The observed polarization, P , is given by:

$$P = \frac{I_{\parallel} - I_{\perp}}{I_{\parallel} + I_{\perp}}$$

where $I_{\parallel}$ and $I_{\perp}$ are the intensities of the parallel and perpendicular components of the emitted light observed at 90° to the propagation direction of the exciting light. Perrin (1929) derived the expression:

$$\frac{1}{P} - \frac{1}{3} = \left\{ \frac{1}{P_0} - \frac{1}{3} \right\} \left\{ 1 + \frac{3\tau}{\rho_h} \right\}$$

P_0 corresponds to the intrinsic or limiting polarization, i.e., the polarization observed in the absence of rotation or energy transfer. In solution P_0 can assume a value between + 1/2 and - 1/3 depending upon the angle between the absorption and emission dipoles. This angle, λ, may vary with the excitation wavelength and is given by:

$$\frac{1}{P_0} - \frac{1}{3} = \frac{5}{3} \left\{ \frac{2}{3\cos^2\lambda - 1} \right\}$$

τ is the radiative lifetime of the fluorophor and ρ_h is the harmonic mean of the rotational relaxation times about the principal axes of rotation. For a sphere we find:

$$\rho_h = \rho_o = \frac{3\eta V}{RT}$$

where η corresponds to the solvent viscosity (we shall not discuss
here the distinction between macro and microviscosity but the reader
is advised that interpretation of η is not always unequivocal),
is the molar volume of the kinetic unit being monitored by the probe,
R is the universal gas constant and T is the temperature.

The polarizations observed for fluorophors embedded in membranes
or bound to macromolecules such as proteins may thus provide information on the hydrodynamics of the rotating unit or the viscosity
of the medium. Complete interpretation of the data, however, requires a knowledge of τ since, as the Perrin equation demonstrates,
the extent of depolarization will depend upon the lifetime of the
excited state.

Discussions of quenching experiments usually begin with the
Stern-Volmer equation (Stern and Volmer (1919)).

$$\frac{I_o}{I} = 1 + k_+ \tau_o Q$$

where I_o and I are the fluorescent intensities in the absence and
presence of quencher, k_+ is the bimolecular quenching constant,
the lifetime in the absence of quencher and Q the quencher concentration. Hence, by measuring the fluorescent intensity as a
function of quencher concentration and knowing τ_o , one can determine the accessibility of the quenching molecule to the fluorophor.
In such experiments, however, one must be careful to distinguish
between the dynamic and static quenching modes. In the case of
static quenching a nonfluorescent complex, predating excitation,
may form between quencher and fluorophor. One finds then that, although the fluorescent intensity decreases as quencher is added,
the lifetime is unchanged. Direct measurement of τ at each quencher concentration permits one to determine the existence of a static
component. In this way Lakowicz and Weber (1973) were able to demonstrate, for a number of proteins, that the quenching molecule
oxygen is able to diffuse through a protein matrix and dynamically
quench excited tryptophan residues, even those buried in a protein's
interior.

One of the major emphasis in fluorescence spectroscopy in the
last ten years has been the development of increasingly sophisticated instrumentation and theory for lifetime determinations. The bulk
of this effort has been directed towards the development of the
pulse technique in which the sample is illuminated by a light pulse
of short duration as opposed to the phase/modulation technique in
which the illumination is continuous but modulated. At the present

neither technique appears to hold a decisive advantage over the
other especially since the commercial availability of multifrequen-
cy phase/modulation fluorometers (SLM Instruments Inc., Urbana,
Illinois, U.S.A.).

To date, synchrotron radiation has been coupled only with the
single photon counting technique for lifetime measurements (although
the application of phase/modulation techniques for lifetime determinations utilizing synchrotron radiation is perhaps not far off -
R. Lopez-Delgado and E. Gratton; to be published). We shall briefly
review the methodology of the single photon counting technique with
reference to the experimental arrangement utilized at ACO. Our
review cannot be exhaustive and those with a sustaining interest
are refered to the exceptionally good reviews by Birks and Munro
(1967), Knight and Selinger (1973) and Ware (1971).

An excellent technical description of ACO itself has been given
by Guyon et al (1976). One will find in this paper a detailed discussion of the ring operating conditions and various technical considerations on the beam lines, transfer optics and vacuum equipment.
Each time the electron bunch in the storage ring passes the light
port (figure 2) a burst of radiation passes down the light transport
system. Clearly, the nature of the window material (saphire in our
case) and other elements of the light transport system will delineate
the range of frequencies available (see Sutherland, J. this volume).
The synchrotron radiation then excites the sample; an emitted photon
reaches the photomultiplier giving rise to an anode pulse which,
properly amplified and discriminated, starts a voltage ramp in the
time-to-amplitude converter (TAC). The voltage ramp is stopped upon
receipt of a pulse from an electrode embedded in the ring which de-
tects passage of the electron bunch providing a constant temporal
reference for the sample pulse. We note that this arrangement is
reversed from the conventional setup where the reference pulse, triggered by flashlamp ignition or an excitation photon, starts a TAC
and the sample signal stops the resultant ramp. The very high re-
pitition rate of the ACO storage ring (13.62 MHz with one bunch in
the ring) compared to the fastest available flashlamps (around 50
KHz) precludes the conventional arrangement as the TAC could burn
out due to the extremely high duty cycle. It is more sensible to
start the TAC only when an emission event is detected (see figure 3).
The resulting decay curve merely appears to be established in the
negative time direction. The voltage ramp from the TAC then addres-
ses the multichannel analyzer (MCA) indicating the channel into which
the count is to be stored. This process, repeated many times to the
level of statistical significance required, develops the direct decay curve in the MCA for display on a CRT and for further data reduction.

A considerable effort has been expended in recent years on the

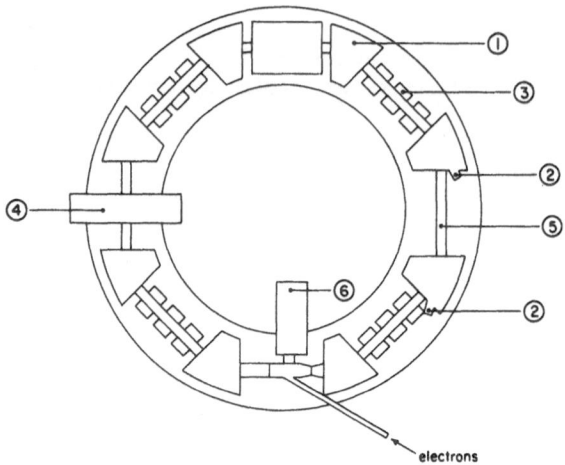

electrons

Figure 2. Simplified Schematic of the ACO Storage Ring (adapted
from Guyon et al (1976)).
   1. Magnets
   2. Light ports
   3. Magnetic quadrupole lenses
   4. RF cavity
   5. High energy physics experimental chamber
   6. Pumps

mathematical analysis of these decay curves. For simple monoexponential decays of a duration long compared to the excitation pulse
width the analysis is straightforward; the slope of the decay curve
plotted on a logarithmic scale yields the lifetime directly. Complications arise, however, when the emission decay curve must be
deconvoluted from the excitation pulse and when the decay is multi-
exponential. A number of methods for data analysis have appeared
in recent years; these include the Laplace transform procedure
(Gafni et al (1975)), the method of nonlinear least squares (Grinvald and Steinberg (1974)), the method of moments (Isenberg and
Dyson (1969)) and the method of modulating functions (Valeur and
Moinez (1973)) to name a few. We cannot discuss here the details
and relative merits of these various methods and the interested
party is refered directly to the sources listed and references
therein. Suffice to say that the accurate analysis of a heterogeneous emitting population (as certainly almost all biological systems
are) is probably the most difficult task facing the fluorescence
practitioner studying a biochemical or biological problem. Assigning weights to the relative contributions from the different emitting
species, for example the tryptophan residues in a protein, presup-

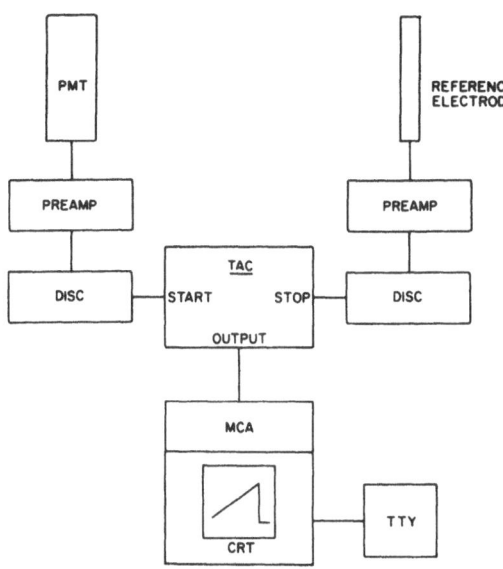

Figure 3. Lifetime Apparatus at ACO
          PMT - photomultiplier observing fluorescence
          Preamp - preamplifiers
          Disc - discriminators
          TAC - time-to-amplitude converter
          MCA - multichannel analyzer

poses a simple model which may not capture the essential reality.
The situation is rendered more complicated in fluorescence spectroscopy by the fact that the various emission components may be weighted differently by the instrument response functions, perhaps in
subtle fashions. These comments are not meant to be discouraging
of the technique but rather to elicite caution in the analysis of
any biological system.

The utilization of synchrotron radiation for fluorescence lifetime determinations offer a number of advantages over conventional
pulse fluorometry. Lopez-Delgado, Tramer and Munro (1974) have
written a classic paper describing in detail one of the first applications of synchrotron radiation to single photon counting lifetime
determinations and time resolved spectroscopy. For a wealth of
descriptive data on ACO and typical results interested readers are
refered to this work (discussions of other early work utilizing
synchrotron radiation for lifetime determinations may be found in
the papers by Lindqvist et al (1973) and Lopez-Delgado (1976)).

The spectral range accesible with flashlamps depends upon the
type of gas in the bulb. Hydrogen filled lamps yield a continuum
extending into the ultraviolet but the intensity, particularly at

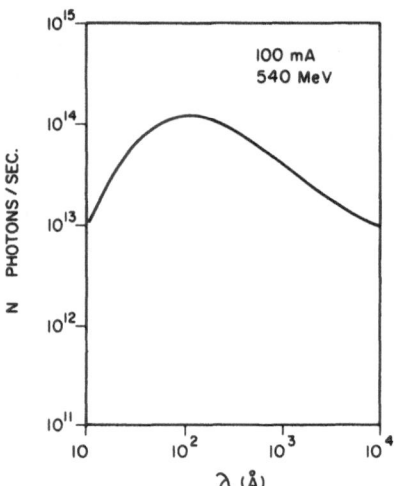

Figure 4. Calculated Photon Flux at ACO (data of Guyon et al (1976))
Constant resolution of $\frac{\Delta\lambda}{\lambda}$ = 0.01

shorter wavelengths is weak. In contrast, synchrotron radiation
provides a tunable source which actually increases in intensity at
shorter wavelengths (figure 4); in our case the window material and
other elements of the light transport system limit the wavelength
region accesible.

One must remember that the single photon technique requires
that only one photon reaches the detector per exciting pulse, ie,
that pulse pileup which would bias the data in favor of the short
lifetime photons must be minimized. This constraint dictates that
the count rate be small (2-4%) compared to the pulse repetition rate
(Ingle and Crouch (1972) and Cova et al (1973)). Thus with a flashlamp operating at 50KHz the permissable count rate is perhaps 2KHz
wheras the ACO pulse rate of 13.62MHz with one electron bunch in
the ring in principle permits a data acquisition rate of several
hundred thousand hertz (the duty cycle of present day MCA's actually
limits us to around one hundred kilohertz). Hence, the acquisition
time for a statistically significant number of counts may actually
be reduced from a matter of hours with the flashlamp to several minutes with the synchrotron radiation. This time factor may be an
important consideration in the case of labile biochemical materials.

The pulse widths of (at half height) of present day flashlamps
are on the order of two nanoseconds or more depending upon the flash
rate and the type of gas utilized. Moreover, since the gas ionization process is a relaxation phenomenon the intensity decays expon-
entially. Hence in the case of fluorescent lifetimes approaching
two nanoseconds one must deconvolute the exponential fluorescence

decay from the exponential flashlamp decay, a procedure which taxes the accuracy of the technique at the shorter lifetime values. Depending upon the ring current, however, the synchrotron radiation pulse at ACO can be less than one nanosecond full width at half height; moreover it is gaussian. The narrow pulse width and gaussian shape mean that one can analyze shorter lifetimes than with the flashlamp method before recourse to deconvolution techniques become necessary.

A persistent problem with flashlamps is the distortion and time jitter of the pulse due to the aging of the lamp. The synchrotron pulse is not subject to significant time distortions and is relatively invariant in intensity and width except for a slow decrease in ring current which can be monitored. A great advantage of the synchrotron radiation is that the pulse shape is identical for all wavelengths and each wavelength is perfectly synchronized with all others. In flashlamps the illumination function may demonstrate a complex wavelength dependency.

A further advantage of the electron storage ring as a light source is that there exists only optical coupling between the source and detector. In the flashlamp technique kilovolt pulses are developed as the lamp discharges; such high voltages invariably result in noise pickup and distortion in the detector response which can complicate analysis. Synchrotron radiation, on the other hand is a quiet source.

We have already refered to the technique of time-resolved spectroscopy which was pionerred by Ware and his collaborators. In this technique a variable time window can be set by interposing a single channel pulse height analyzer (SCPHA) between the TAC and the MCA. In this case the MCA can be synchronized to the monochromator position; the SCPHA upper and lower discriminator settings determine the position and size of the time window. In practice, with flashlamps for excitation, the data acquisition rate is low. The much higher repetition rate of ACO permits us to acquire time-resolved spectra comparatively rapidly (Lopez-Delgado et al (1974)). Another advantage of ACO is that the narrow pulse width and gaussian shape should mitigate the necessity for deconvolution of the instrument response function at the shorter time windows.

The spectrofluorometer at ACO will also permit us to acquire routine emission and excitation spectra and polarizations in addition to lifetimes and time-resolved spectra. Advantages of utilizing ACO for such measurements include (1) the increased light flux over conventional light sources (excluding lasers in the visible region) especially in the ultraviolet region and (2) a precise knowledge of the illumination function at the light port. This last point should facilitate calibration of the instrument's wavelength

dependent response function. Although in principle the synchrotron
radiation is polarized in the plane of the ring, the degree of polarization depends upon the wavelength selected and the vertical
aperture collected (Lopez-Delgado et al (1974)). For our facility
we have elected to utilize a vertical aperature sufficiently large
to yield low polarization exciting light; we will then utilize Glan-
Taylor prism polarizers to select the parallel and perpendicular
components accordingly. The reasons for this approach were that
a large vertical aperature will yield more light and that a strong
polarization in the plane of the ring would require a more complicated instrument geometry for polarization measurements.

One of our principal research interests at Orsay concerns the
tryptophan emission from hemeproteins. The quantum efficiencies
of the tryptophans in hemoglobin and myoglobin are known to be substantially reduced from those of the apoglobins, presumably due to
Forster type energy transfer to the heme moiety (Weber and Teale
(1959)). The low yield has for many years discouraged investigators
from further work on the intrinsic tryptophan emission from these
hemeproteins. Recently, though, the availability of sophisticated
photon-counting instrumentation and access to synchrotron radiation
has brought about a renewed interest in this potentially valuable
area of research.

Alpert and Lopez-Delgado (1976) reported lifetime measurements
utilizing synchrotron radiation on various hemeproteins excited into
the tryptophan absorption band; the proteins investigated included
hemoglobin from ox, horse and adult humans as well as their isolated
α and β subunits. Surprisingly, the results indicated that although
the fluorescence yield was substantially reduced for the hemeproteins relative to their apoglobins the lifetimes were only slightly
different. For example, a lifetime of 3.5 nanoseconds was found
for ferrihemoglobin compared to 4.3 nanoseconds for its apoglobin.
If substantiated these data may offer interesting new information
possibly concerning dynamical aspects of hemeprotein conformation.

A criticism to these initial experiments,however, was that the
spectral distribution of the emission was not obtained, hence the
investigators could not be certain that they were indeed observing
tryptophan fluorescence. Accordingly, the spectra and quantum efficiencies of hemoglobin and its isolated α and β subunits were investigated at Urbana, Illinois in the laboratory of Gregorio Weber,
utilizing highly sensitive photon-counting instrumentation. Pre-
cautions were taken to prepare the best possible samples and to rule
out, as much as possible, the presence of extraeneous proteins or
denatured hemeproteins. The results of these experiments (Alpert,
Jameson and Weber; submitted to Nature) indicate that human adult
hemoglobin A_o and its isolated α and β subunits do exhibit an intrinsic tryptophan emission. Quantum yields for these samples relative to free tryptophan are given in table I.

TABLE I

Quantum Efficiencies Relative to Tryptophan
(280 nm excitation)

	5° C	25° C
α	0.66%, 0.75%*	0.97%
β	0.69%	1.36%
HbO₂	1.19%, 0.90%*	1.34%
APOHb	41.2%, 44.6%*	
APO α	34.7%	
APO β	40.4%	

* Determinations from different preparations

These results are perhaps not surprising since there is no reason to believe, apriori, that the tryptophan emission in hemoglobin should be totally quenched. In a Forster type energy transfer mechanism the quenching efficiency is determined in part by the angle between the emission dipole of the donor and the absorption dipole of the acceptor (Forster (1948)). We know understand that proteins are dynamic structures; detectable fluctuations in the protein matrix occur within the nanosecond timescale (Lakowicz and Weber (1973) and McCammon et al (1977)) and given the appropriate probes, instrumentation, experimental design and interest we may expect to observe fluctuations (appropriately minute) even on the picosecond timescale. Given the fact that hemoglobin contains six tryptophans (α subunit has one, β subunit has two) and that these residues will experience some motion around their mean positions, we should be surprised if transfer to the heme were complete. To date little work has been done on the effects of heterotransfer upon fluorescence lifetimes and the exact relationship between the decrease in intensity and decrease in lifetime remains somewhat obscure (Weber and Teale (1959) and Weber, personal communication (1978)).

The wealth of structural information on hemoglobin affords us some idea of the relative dispositions of the heme and the tryptophans. The tryptophan fluorescence may be an important probe of the dynamics of the protein including subtle changes which may occur upon ligation with oxygen or carbon monoxide. Certainly an important experiment will be to measure the effect of deoxygenation of the hemoglobin upon the fluorescence yield.

One of the major emphases at Orsay will be to measure the yields and lifetimes in the same apparatus and near the same time.

The spectrofluorometer at ACO should be well suited to the task and should permit us to resolve questions concerning the relationship between yields and lifetimes in the hemeproteins.

Besides offering information about protein dynamics, the hemo-globin system may prove to be valuable for the study of energy transfer processes per se. The environment of a fluorophor within a protein matrix is quite different from that available from any solvent. Relative orientations and motions of groups coupled via energy transfer mechanisms are severely constrained compared to the solution case. Hence, instead of utilizing a fluorophor to eluci-date the nature of a protein we may consider the possibility of utilizing a protein to study certain fundamental properties of fluorophors and excited states.

More recently ACO was utilized in a study concerning the effect of variation of excitation wavelength upon the emissive lifetime of tryptophan in aqueous buffer (Alpert, Jameson, Lopez-Delgado and Schooley, submitted to Photochem. Photobiol.). Balcavage and Alvager (1976) have reported that the lifetime of tryptophan changes dram-atically as the exciting wavelength is varied from 290nm to 250nm. A primary reason for our reinvestigation of this problem was the fact that Alpert and Lopez-Delgado, while investigating the anoma-lous lifetime of tryptophan in hemoglobin, had not observed any significant wavelength dependency. Following the publication of Balcavage and Alvager a closer investigation seemed warrented espe-cially since the excellent energy distribution of the synchrotron radiation in the ultraviolet region would expediate the measurement. The phase/modulation fluorometer of Spencer and Weber (1969) was also utilized to determine the lifetimes by a non-pulsed technique.

The ACO and phase data, as a function of excitation wavelength, are compiled in table II along with data from Balcavage and Alvager (1976). Our data indicate that the tryptophan lifetime is constant throughout the spectral region investigated, at neutral pH and 20°C. All decay data, at neutral pH, were also monoexponential. This be-havior persists even in the 220-240nm region where Tatischeff and Klein (1975) and Steen (1974) have reported a decrease of about 50% in the tryptophan quantum efficiency. This work, then, supports the conclusion that L-tryptophan in water solution (at room temp-erature and ambient air) regardless of the excitation wavelength always fluoresces from the same thermally equilibrated excited level. Hence, our data does not support a theory positing the existence, at room temperature, of two electronic excited states directly coup-led to the ground state (and not to each other) through the radiation field. We should note that we have not ruled out the possibility of a different distinct emitting species in the emission of trypto-phan below 320nm, the lower cutoff of our filters, nor have we at-tempted to delineate the effects of very low temperatures upon the singlet manifold distribution.

TABLE II

Tryptophan Lifetimes as a Function of Excitation Wavelength*

Excitation Wavelength (nm)	ACO[1] (nsec)	Phase Data[2] (nsec)	B & A[3] (nsec)
220	3.1± 0.1		
230	3.0± 0.1		
240	3.0± 0.1		
250	3.1± 0.1	3.1± 0.1	14.9
260	3.1± 0.1	3.1± 0.1	
270	3.0± 0.1	3.1± 0.1	14.9
280	3.1± 0.1	3.1± 0.1	
290	3.1± 0.1	3.1± 0.1	2.7/9.0
300	3.0± 0.1	3.1± 0.1	
310	3.1± 0.2		
320	3.1± 0.2		

[1] Single photon data from ACO. Emission observed through Corning 0-54 filter (f 320nm) for excitation from 220nm to 280nm and Corning 0-54 plus Schott WG 345 and WG 360 filters (f 360nm) for excitation from 280nm to 320nm.

[2] Phase-shift data (10 MHz modulation frequency) observed through Corning 0-54 filter

[3] Data of Balcavage and Alvager (1976)

* ACO and Phase data obtained at 20°C and neutral pH.

    Using a conventional flashlamp arrangement, Rayner and Szabo (1978) have studied the emissive lifetime of tryptophan in neutral buffer at 20°C with excitation at 280nm as a function of emission wavelength. They concluded that the lifetime can be described by a two exponential decay function whose components have lifetimes of 3.14 nanoseconds and 0.51 nanosecond which they assign to the solvent equilibrated ^1L_a and ^1L_b states respectively. Furthermore they indicate that the relative proportions of these two components changes dramatically across the emission band, the contribution of the short component decreasing going from the blue to the red edge of the spectrum. A complete comparison between the work at ACO and the work of Rayner and Szabo is not possible since at ACO we observed the emission through filters and Rayner and Szabo utilized a constant excitation wavelength of 280nm. Yet in as much as Rayner and Szabo found the 3.14 nanoseconds component predominate in the longer wavelength region of the emission, the region passed by our filters, we may consider both experiments to be substantially in agreement.

We may say in conclusion that the future of synchrotron radiation in biological research may be bright indeed. However, those utilizing this novel light source are advised to consider carefully their choice of experiments. The constraints on access to storage rings means that the researcher should possess an in depth knowledge of the spectroscopic technique to avoid loss of time due to oversights or artifacts and also that he chooses those experiments which are difficult or impossible to preform with conventional light sources.

REFERENCES

Alpert, B. and Lopez-Delgado, R. (1976) Natrue, 263, 445
Bakhshiev, N. G. (1964) Opt. Spectry. 16, 446
Balcavage, W. X. and Alvager, T. (1976) Mol. Photochem., 7, 309
Birks, J. B. (1976) Jour. Res. Nat. Bur. Stnd. A. 80A, 389
Brand, L. and Gohlke, J. R. (1971) J. Biol. Chem., 246, 2317
Brand, L. and Gohlke, J. R. (1972) Ann. Rev. Biochem., 41, 843
Daniels, M. and Hauswirth, W. (1971) Science, 171, 675
Forster, T. (1948) Ann. Physik., 2, 55
Gafni, A., Modlin, R. L. and Brand, L. (1975) Biophys. J., 15, 263
Grinvald, A. and Steinberg, I. Z. (1974) Anal. Biochem., 59, 583
Guyon, P. M., Depautex, C. and Morel, G. (1976) Rev. Sci. Instru., 47, 1347
Ingle, J. D. and Crouch, S. R. (1972) Anal. Chem., 44, 777
Isenberg, I. and Dyson, R. D. (1969) Biophys. J., 9, 1337
Jameson, D. M., Spencer, R. D. and Weber, G. (1976) Rev. Sci. Instru. 47, 1034
Knight, A. E. W. and Selinger, B. K. (1973) Austr. J. Chem., 26, 1
Lakowicz, J. R. and Weber, G. (1973) Biochemistry, 12, 4161
Lauberau, A., von der Linde, D. E. and Kaiser, W. (1972) Phys. Rev. Lett., 28, 1162
Li, T. M., Hook, J. W., Drickamer, H. G. and Weber, G. (1976) Biochemistry, 15, 3205
Lindqvist, L., Lopez-Delgado, R., Martin, M. M. and Tramer, A. (1974) Opt. Comm., 10, 283
Lippert, E. (1957) Z. Elektrochem., 61, 962
Lopez-Delgado, R., Tramer, A. and Munro, I. H. (1974) Chem. Phys., 5, 72
Lopez-Delgado, R., Miehe, J. A. and Sipp, B. (1976) Opt. Comm., 19, 79
Lumry, R. and Hershberger, M. (1978) Photochem. Photobiol., 27, 819
McCammon, J. A., Gelin, B. R. and Karplus, M. (1977) Nature, 267, 585
Perrin, F. (1929) Ann. Phys. (Paris), 12, 169
Shore, V. G. and Pardee, A. B. (1956) Arch. Biochem. Biophys., 60, 100

Steen, H. B. (1974) J. Chem. Phys., 61, 3997
Stern, O. and Volmer, M. (1919) Z. Phys., 20, 183
Teale, F. W. J. and Weber, G. (1957) Biochem. J., 65, 476
Vigny, P. and Duquesne, M. (1974) Photochem. Photobiol., 20, 15
Ware, W. R. (1971) in "Creation and Detection of the Excited State"
Vol. 1, Ed. A. A. Lamola, p. 239
Ware, W. R., Lee, S. K., Brent, G. J. and Chow, P. P. (1971) J. Chem.
Phys., 54,4729
Weber, G. and Laurence, D. J. R. (1954) Proc. Biochem. J., 56, XXXI
Weber, G. (1960) Biochem. J., 75, 335
Weber, G. (1966) in "Fluorescence and Phosphorescence Analysis" Ed.
D. M. Hercules, Interscience Publishers N. Y., p. 217
Weber, G. and Lakowicz, J. R. (1973) Chem. Phys. Lett., 22, 419
Weber, G. (1976) Horizons Biochem. Biophys., 2, 163

EXCITED STATES OF PROTEINS

Claude HELENE

Centre de Biophysique Moléculaire

45045 Orléans Cedex (France)

INTRODUCTION

Much information can be gained on proteins and on their complexes with ligands by using the excited state properties of their aromatic residues. Not only can one determine the parameters characterizing the environment of these residues and local conformational changes arising from the perturbation of this environment due, e. g., to ligand binding or to denaturation. But one can also obtain information on the processes which take place in the time range of the excited state lifetime (nanosecond). These processes include rotational motion, local fluctuations, relaxation of the fluorophor surroundings, energy transfer processes (1).

ABSORPTION SPECTRA (2)

Among the aromatic amino acids which all have absorption bands below $\sim$ 300 nm, tryptophan has the highest extinction coefficient (figure 1). Peptide bonds give rise to absorption bands below 220 nm. In the wavelength range 240-300 nm, cystein and cystine absorb light with low but non negligible extinction coefficients. This is particularly important in the case of cystin because light absorption by this molecule leads to splitting of the disulfide bridge. Its contribution to the photochemistry of proteins is therefore important.

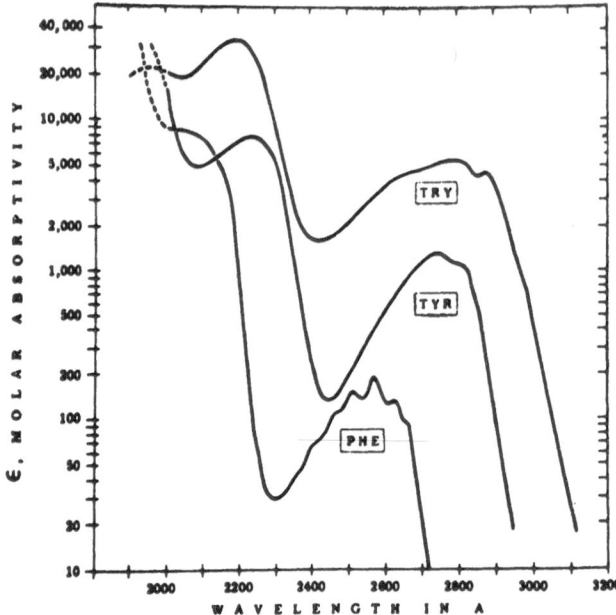

Figure 1 : Ultraviolet absorption spectra of aromatic
amino acids at pH 6 (from reference 2).

Proteins which do not contain cofactors (such as coenzymes)
have an absorption spectrum which is approximately the sum of
the absorption spectra of the constitutive amino acids. However it
must be kept in mind that depending on their respective environments aromatic amino acid residues may have slightly different
spectra.

Also the small changes induced in the absorption spectra of
individual aromatic residues by organic solvents may be used to
obtain information on the accessibility of these residues to solvent
molecules (3). If an aromatic residue is buried inside the protein
structure its absorption spectrum will not be affected if an organic
solvent is added to the protein aqueous solutions. On the contrary
"exposed" aromatic residues will show small changes in absorption which can be recorded by difference absorption spectroscopy
(3).

FLUORESCENCE EMISSION AT ROOM TEMPERATURE

A. Aromatic amino acids

1. Solvent relaxation

Aromatic amino acids (TRP, TYR, PHE) emit fluorescence when they are excited in their near UV absorption bands. The fluorescence of PHE and TYR is not very sensitive to the solvent while that of TRP is strongly dependent on the polarity of the solvent. For the indole ring the wavelength of the fluorescence maximum changes from 300 nm in cyclohexane to 349 nm in water (4, 9). This has been attributed to an interaction of the indole ring with solvent molecules especially in the excited state (exciplex formation) and to a strong reorientation of polar solvent molecules around the excited molecule which has a different charge distribution and dipole moment as compared to the ground state (5). This relaxation process of the excited molecule with its solvent cage is clearly demonstrated by the temperature dependence of the fluorescence spectrum of tryptophan in a mixed solvent water-ethylene glycol (1v/1v) which gives a transparent glass at low temperature (figure 2). This solvent melts over a broad range of temperatures (gradual melting). When the solvent is frozen the fluorescence maximum appears at short wavelengths because solvent relaxation cannot occur. As soon as the solvent begins to melt the fluorescence maximum progressively shifts to longer wavelengths and this is accompanied by a broadening of the fluorescence spectrum. When a completely fluid medium is reached the fluorescence spectrum does not shift any more but the fluorescence quantum yield decreases when temperature increases. This strong temperature dependence requires that any study of tryptophan (or protein) fluorescence must be conducted at a controlled constant temperature.

2. pH dependence

The fluorescence quantum yield of aromatic amino acids depends on the pH (6, 7). Part of this pH dependence is due to a change in the ionization of the amino and carboxyl groups of the free amino acid. Such contribution is not observed in proteins where these groups are involved in the formation of peptide bonds. However some of the transitions in the fluorescence vs. pH curve are due to the aromatic ring itself. For example the indole ring is strongly quenched by H^+ and OH^- ions. In the latter case quenching arises as a result of deprotonation of the NH indole group in the excited state. Strong proton acceptors such as phosphate trianions

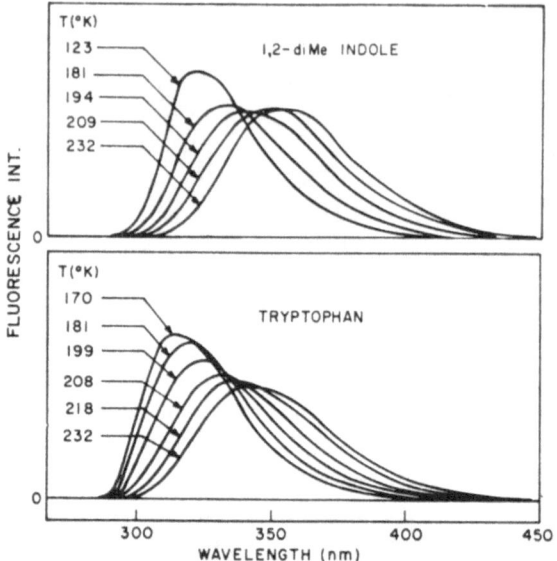

Figure 2 : Fluorescence spectra of 1, 2-dimethylindole
and tryptophan in a mixture of water and ethylene glycol
(1v/1v) at various temperatures. Solvent relaxation occurs
in times comparable to the fluorescence lifetime in the
temperature range covered in these experiments (from
reference 5).

(PO_4^{3-}) also quench the fluorescence of indole by a similar diffusion controlled reaction. The fluorescence of tyrosine is strongly
quenched at high pH as a result of proton dissociation of the phenol
hydroxyl group in the ground state (pK $\sim$ 10). It is known that most
organic molecules have pK values for protonation equilibria which
are different in the excited states as compared to the ground state.
However it is difficult to determine an excited singlet state pK
from the pH dependence of the fluorescence quantum yield because
the protonation equilibrium usually cannot be reached in the excited state due to the too short lifetime of this state (nanosecond
range). The excited state pK can nevertheless be calculated from
the Förster cycle. Such a calculation for the tyrosine phenol ring
leads to a singlet state pK around 5. The fluorescence pH titration
gives a pK of 10 identical to the ground state for the reasons out-

lined above. However a lowering of the pK is observed if the tyrosine OH group interacts with a proton acceptor such as acetate or
phosphate anions which seem to stabilize the excited tyrosinate
anions (an increase in the fluorescence quantum yield of tyrosinate is observed) (8).

3. Excitation spectra and primary photochemical events

The fluorescence quantum yield of tryptophan depends on the
excitation wavelength and changes rather abruptly when excitation
changes from the first to the second absorption band (9). However
the fluorescence lifetime does not change over the same excitation
wavelength range ($\sim$ 3.1 ns at 20°C). This means that a photochemical process takes place in the third excited state (the first
absorption band contains two transitions in indole). This reaction
can be an ejection of electrons which appears to be more and more
efficient when the excitation wavelength decreases. As a matter of
fact it is known that excited indole ejects electrons and gives rise
to a radical cation with a pK of 4.3 (which means that at pH 7 only
the neutral radical is observed) (10). This photoejection process

$$(1)$$

appears to be monophotonic at room temperature (at least when the
incident light intensity is low enough) and occurs from the excited
singlet state. At low temperature this process is clearly biphotonic the indole triplet state being the intermediate species which
absorbs the second photon (11).

Due to the high electron donating ability of the indole ring
tryptophan fluorescence is strongly quenched by electron acceptors (12). In some cases the indole ring can react photochemically
with the electron acceptor. Trichloroethanol is such a compound
which quenches the first excited state of tryptophan and has been
used to photochemically modify in a specific reaction those tryptophan residues in a protein which are fluorescent and accessible
to trichloroethanol (13).

Tyrosine also ejects electron from its first singlet excited state and gives rise to a phenoxyl radical. This process (equation 2) is monophotonic at room temperature and biphotonic at low temperature with the triplet state as the intermediate species which absorbs the second photon.

$$\langle\!\bigcirc\!\rangle\!-OH \xrightarrow{\;h\nu\;} \langle\!\bigcirc\!\rangle\!-O^{\cdot} + H^{\cdot} + e^{-} \tag{2}$$

Tryptophan is a very photoreactive compound. When excited in aqueous solution at room temperature it gives rise to many different photoproducts whose nature and relative proportions depend on the presence of oxygen. One of the most important photoproduct seems to be N-formylkynurenine (NFK in reaction (1)). This compound absorbs light above 300 nm ($\lambda_{max} \sim 330$ nm) and acts as an efficient photosensitizer in many different reactions (14) (including the photooxydation of tryptophan which is thus an autocatalyzed reaction).

In flash photolysis experiments the tryptophan triplet state has been detected. The triplet-triplet absorption is superimposed on the absorption spectra of indole radicals and solvated electrons (15). In the presence of disulfide compounds electrons attach to the disulfide group to give rise to a radical anion $(R-S-S-R)^{\bar{}}$ which has a characteristic absorption spectrum. This reaction is very important in proteins where disulfide bridges can be broken as a result of electron capture.

B. Peptide and Proteins

1. Fluorescence spectra (for a review, see reference 16)

Peptides and proteins which contain aromatic amino acids usually emit fluorescence at room temperature. If all three aromatic amino acids are present in a protein the fluorescence is usually dominated by the emission from tryptophan residues. Some proteins do not contain tryptophan (pancreatic ribonuclease for example) and their fluorescence is characteristic of tyrosine ($\lambda_{max} \sim 305$ nm). In some peptides which contain only phenylalanine the fluorescence spectrum is of course characteristic of the singlet state emission from this residue ($\lambda_{max} \sim 280$ nm). This general behavior of proteins is due to the fact that the extinction coefficients decrease in the order TRP $>$ TYR $>$ PHE and to energy transfer processes which occur in the following direction PHE $\rightarrow$ TYR $\rightarrow$ TRP.

As said above the fluorescence spectrum of tryptophan is very sensitive to the local environment. Consequently depending on the location of tryptophan residues in the protein structure the fluorescence spectrum of a protein may have a maximum wavelength usually varying from 325 nm to 350 nm (17). As expected buried residues fluoresce at short wavelengths whereas exposed residues give rise to the more red-shifted spectra.

A recent study (18) shows however that one has to be very cautious in ascribing the fluorescence of proteins to a particular residue. Two cytotoxins from the venom of Naja naja shows a fluorescence spectrum with a maximum around 345 nm at pH 7. However none of these toxins contains tryptophan. This fluorescence has been shown to arise at pH 7 from a tyrosine residue which is strongly interacting with a carboxylate anion (Glu or Asp) inside the peptide structure. This interaction leads to a proton transfer reaction in the excited state with a concomitant fluorescence emission from tyrosinate anion (which has a much higher quantum yield under these conditions than free tyrosinate in solution).

The circular polarization of protein fluorescence has been recently used to investigate conformational problems (33).

2. Accessibility of tryptophan residues

The accessibility of aromatic residues in proteins, especially tryptophan, can be probed by using external fluorescence quenchers. For a fully accessible molecule in aqueous solution whose fluorescence is usually quenched in a diffusion-controlled reaction the fluorescence intensity F is related to the quencher concentration Q by the Stern-Volmer equation

$$\frac{F_o}{F} = 1 + k_Q \, \tau_o \, Q \tag{3}$$

where F_o is the fluorescence intensity in the absence of quencher, k_Q the bimolecular rate constant for reaction of the excited molecule with the quencher : τ_o is the fluorescence lifetime in the absence of quencher. The rate constant k_Q will depend on the accessibility of the excited molecule. For peptides or proteins containing only one tryptophan residue the determination of k_Q allows one to determine whether this residue is exposed to the solvent or buried inside the peptide structure. For proteins containing more than one tryptophan residue quenching of different residues will be characterized by different rate constants and very often the plot of F_o/F vs Q will not be a straight line. However, in some cases one may distinguish between exposed residues which will be quenched and internal residues which will not be affected by external quencher. In these cases a modified Stern-Volmer equation (4) can be used to determine the fraction of fluorescence which is "accessible" to the quencher (19)

$$\frac{F}{F_o - F} = \frac{1}{fa} + \frac{1}{fa \, k_Q \cdot \tau_o \, Q} \tag{4}$$

where k_Q and τ_o are the average rate constant for quenching and the average lifetime of accessible residues, respectively.

Much information on the dynamic processes which take place in proteins in the nanosecond time range can be obtained from the study of the excited states of their aromatic chromophores and of their accessibility (see references 1a and 1b for reviews).

3. Energy transfer processes

Due to the overlap between the fluorescence spectrum of tyrosine and the absorption spectrum of tryptophan an efficient energy transfer process can take place at the singlet level from tyrosine to tryptophan (20). The Förster critical distance R_o has been calculated to be 12-14 Å (this is the distance at which tyrosine has an equal probability of transferring its singlet energy to tryptophan and of deactivating to the ground state by fluorescence and nonradiative processes).

The rate constant of energy transfer between a donor and an acceptor molecule depends on the 6th power of the distance between the two molecules (equation 5).

$$(5) \quad k_t = \frac{8.8 \times 10^{-25} \, K^2 \, J_{DA}}{n^4 \, \tau_o \, R_{DA}^6} \quad \text{with} \quad J_{DA} = \int_o^\infty \mathcal{E}_A(\nu) F_D(\nu) \frac{d\nu}{\nu^4}$$

$\mathcal{E}_A(\nu)$ is the molar decadic extinction coefficient of the acceptor

$F_D(\nu)$ is the fluorescence intensity of the donor normalized to unity $\left(\int_o^\infty F_D(\nu) \, d\nu = 1\right)$

τ_o is the radiative lifetime of the donor molecule

The Förster critical distance R_o is defined as the distance R_{DA} for which $k_t = 1/\tau_F$ (τ_F = fluorescence lifetime of the donor). Since $\tau_F = \tau_o \times \phi_D$ one obtains

$$R_o^6 = 8.8 \times 10^{-25} \, K^2 \, \phi_D \, n^{-4} \, J_{DA}$$

The transfer rate depends on an orientation factor (K^2) which arises from the fact that singlet energy transfer originates

from a coupling between the transition moments of the two molecules. The value of K^2 varies between 0 and 4. When both donor
and acceptor can adopt all possible relative orientations during
the donor singlet state lifetime an average value of $2/3$ is obtained for K^2. In the case of tryptophan the situation is a little bit
more complicated because the first absorption band of indole is in
fact the superposition of two transitions with nearly perpendicular
transition moments. This in fact increases the probability of energy transfer since the overlap between the fluorescence of tyrosine
and each of the two tryptophan transitions is roughly the same but
the couplings of transition moments (and consequently the values of
K^2) are different for the two transitions. This means that the relative orientations of the tyrosine and tryptophan rings which give
a K^2 value equal to zero are more restricted than in the usual
case where the acceptor has only one transition moment involved
in the coupling process.

Distances between Tyr and Trp residues can be determined
with a relatively good accuracy if the peptide or protein of interest contains one of each type of residues and if the tryptophan
fluorescence polarization can be determined as a function of the
excitation wavelength in a medium of high viscosity to prevent
fluorescence depolarization arising from rotational movements
during the excited state lifetime (21). As a matter of fact energy
transfer from Tyr to Trp should result in a depolarization of Trp
fluorescence when Tyr is excited. This depolarization depends on
the angle between the transition moments of the absorbing donor
and the emitting acceptor and therefore enables one to estimate
the relative orientation of the Tyr and Trp rings.

This energy transfer method has been used to estimate distances between Tyr and Trp chromophores in hormones or toxins
(20, 22). When analyzed in conjunction with other data (circular
dichroism, nuclear magnetic resonance, determination of the accessibility of aromatic residues ...) the results of energy transfer experiments may be very helpful in determining conformational parameters in solution (23).

In alkaline medium ionization of Tyr shifts its absorption
spectrum to longer wavelengths. This leads to an efficient energy transfer from Trp to ionized Tyr (i.e. in the reverse direction
as compared to neutral pH) (16).

4. Phosphorescence spectra of proteins at room temperature

Aromatic amino acids have short-lived triplet states in fluid
solution which are efficiently deactivated by solvent molecules

and oxygen. However if tryptophan residues are sufficiently pro-
tected in the protein structure they may have a longer lived triplet state. This has been recently observed by two different methods : flash spectroscopy and phosphorescence. Not only has it
been possible to measure triplet-triplet absorption by the first
method but it has also been demonstrated that phosphorescence
emission could be observed from the tryptophan triplet state in
proteins (24). The lifetimes may have values of the order of 0.1
to 1 s (i.e. one order of magnitude shorter than at 77 K). This
phosphorescence emission might prove useful in analyzing slow
movements from the depolarization of phosphorescence emission
(24b).

5. Fluorescence characteristics and photochemical modi-
fication of proteins. An example : the lac repressor

There have been many studies devoted to photochemical modifications of proteins. The light used in these experiments is absorbed either by an intrinsic chromophore of the protein (aromatic amino acids, disulfide bridges...) or by an extrinsic photosensitizer (a coenzyme for example or a dye added to the protein
solution). A recent review covers the problems raised by the second type of reactions (photosensitized reactions, photodynamic
action ...) (25).

Excitation of proteins in the near UV results in many different processes : rupture of disulfide bridges, photoejection of
electrons by aromatic residues followed by electron capture by
and sensitized reactions in other amino acid residues and even
peptide bonds, photosensitized processes involving primary photooxidation products such as N-formylkynurenine, etc... It is not
possible in such a review to cover all aspects of protein photochemistry. The behavior of each protein has to be analyzed separately and photochemical modifications have to be related to the biological function(s) of this protein. To give an example of what can
be done using photochemical modifications of proteins the following paragraph will summarize the results obtained in our laboratory with a protein involved in nucleic acid binding.

The repressor of the lactose operon in the bacterium E. coli
is a protein with 4 identical subunits. Each subunit is a polypep-
tide chain of 360 amino acids among which are two tryptophan residues. This protein has a fundamental biological role (26) : it re-
gulates the expression of the three genes involved in the metabo-
lism of lactose by binding to a region of DNA of about 20 base

pairs (called operator). When bound to the lac operator, the lac
repressor prevents the transcription by RNA polymerase of the
three genes of the lac operon. One of the first metabolites of lactose, allolactose, binds to the lac repressor and induces a dissociation of the repressor-operator complex, thereby allowing the
transcription of the lac genes. Thus in the presence of lactose the
bacterium synthesizes the proteins which are required to metabo-
lize lactose. In the absence of lactose these proteins are not needed and their synthesis is blocked at the level of transcription. The
lac repressor molecules has therefore two ligands : the lac operator and the inducer (allolactose). There should exist a link between these two binding sites since inducer binding leads to a dissociation of the repressor from the operator. This is due to a
conformational change induced in the protein by the binding of the
inducer. Each subunit of the lac repressor tetramer has a binding
site for the inducer. The number of subunits involved in operator
binding is not known yet.

The fluorescence quantum yield of the lac repressor decreases when this protein is irradiated with UV light (27). However it
was shown that complete quenching of the fluorescence was observed when only one of the two tryptophan residues per subunit was
photochemically modified. In order to determine whether these two
residues emitted fluorescence the fluorescence decay under pulsed
excitation was measured at different wavelengths by the single photon counting technique (28). This study demonstrated that the two
tryptophans emitted fluorescence with different spectra and diffe-
rent lifetimes (λ^1_{max} = 322 nm, τ_1 = 3.8 ns ; λ^2_{max} = 344 nm,
τ_2 = 9.8 ns). Inducer binding led to a change in the emission parameters of both residues ($\lambda^{1'}_{max}$ = 327 nm, τ'_1 = 3.0 ns ;
$\lambda^{2'}_{max}$ = 333 nm, τ'_2 = 7.6 ns) indicating that the environment of
both tryptophans had changed.

UV irradiation of the lac repressor modifies only one of the
two tryptophans. This tryptophan is transformed into a photoproduct (most probably N-formylkynurenine) which acts as an energy
trap for the excitation energy of the second tryptophan. This means
that as soon as one Trp is modified the other cannot be photochemically affected.

Phototransformation of one Trp residue in lac repressor
inhibits the binding of the inducer and also decreases the binding
constant for non-operator and operator DNA. If these results can
be extrapolated to an in vivo situation this means that it is possible
to induce the translation of the lac operon genes by photochemical

modification of the tryptophan residues of the lac repressor. Therefore attributing all biological effects of UV radiations to DNA
damage might be in error especially at long wavelengths (>300
nm) where proteins (but not DNA) still absorb an appreciable
amount of light.

LUMINESCENCE AT LOW TEMPERATURE

1. Luminescence spectra

Aromatic amino acids and proteins usually emit both fluorescence and phosphorescence at low temperature (77 K) in rigid
glasses (for example a mixture of water and propylene glycol
1v/1v) (16). For free amino acids phosphorescence lifetimes are
1.4 s (tyrosinate), 2.9 s (tyrosine), 6.65 s (tryptophan), 7.7 s
(phenylalanine). In proteins these lifetimes are of the same order
of magnitude. Phosphorescence emitted by tyrosine starts at shorter wavelengths ($\sim$350 nm) than that emitted by tryptophan ($\sim$400
nm). In most cases excitation at 275 nm generates a phosphorescence spectrum in which the contributions of both residues can be
separated. On the contrary excitation at 295 nm leads to a phosphorescence spectrum which is characteristic of tryptophan only
(tyrosine residues are not excited at this wavelength).

As already mentioned above the fluorescence spectrum of
tryptophan at low temperature is shifted to shorter wavelengths as
compared to that at room temperature. This is due to the inhibition of solvent relaxation around the excited residue.

Energy transfer from Tyr to Trp can be demonstrated at
low temperature by measuring the change in fluorescence quantum
yield of Trp with the excitation wavelength and by observing the
depolarization of Trp fluorescence when Tyr and Trp are both
excited. At alkaline pH, Trp is able to transfer its singlet energy
to ionized Tyr. In some cases it has been demonstrated that a reverse transfer occurs at the triplet level from ionized Tyr to Trp.
For example in luliberin, a decapeptide hormone (5- oxo -Pro-His-
Trp-Ser-Tyr-Gly-Leu-Arg-Pro-Gly NH_2) which contains only one
Tyr and one Trp residue the fluorescence and phosphorescence
are emitted by Trp at neutral pH (due to Tyr$\rightarrow$Trp energy transfer
at the singlet level)whereas at alkaline pH the fluorescence is emitted
by ionized Tyr and the phosphorescence by Trp. This last result is
due to Trp$\rightarrow$Tyr$^-$ energy transfer at the singlet level followed (after intersystem crossing in Tyr$^-$) by Tyr$^-\rightarrow$ Trp energy transfer
at the triplet level (29).

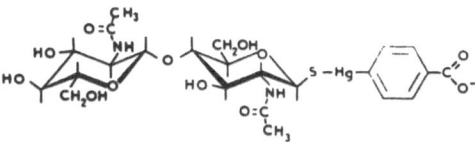

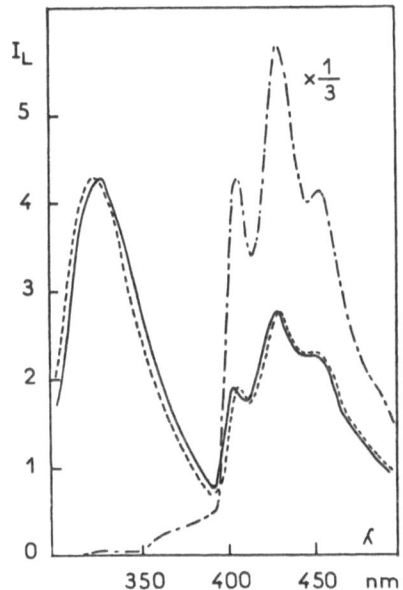

Figure 3 : Luminescence spectra of Wheat Germ Agglutinin at 77 K in the absence (——) and in the presence of
saturating concentrations of the normal ligand di-N-
acetyl-β-chitobioside (- - -) and of the thiomercuribenzoate of di-N-acetyl-β-chitobioside (-•-•-). The formula of the last compound is shown above the figure. Fluorescence is observed at short wavelengths and phosphorescence at longer wavelengths (from reference 31).

2. Heavy atom effects in the analysis of ligand-protein interactions

Heavy atoms are known to enhance spin-orbit coupling in organic molecules (30). This leads to an increase in the intersystem crossing (ISC) probability and in the radiative deactivation from the triplet to the ground state. Three spectroscopic effects result from this property of heavy atoms : fluorescence is quenched (as a result of rapid ISC), phosphorescence is usually enhanced and the phosphorescence lifetime is considerably shortened. The heavy atom may be covalently linked to the organic molecule (internal effect). In the latter case the effect will decrease very rapidly with the distance of the heavy atom from the excited molecule.

We have recently used this property to probe the presence of a tryptophan residue in the ligand binding site of a protein, wheat germ agglutinin (WGA) (31). This protein (lectin) is a dimer of molecular weight of 36, 000 and binds selectively oligomers of N-acetylglucosamine (it agglutinates various types of animal cells, in particular malignant cells). The dimer has four binding sites (2 per subunit). Each subunit contains 3 Trp residues only 2 of which emit fluorescence (one is completely quenched inside the protein structure). Ligand binding induces a change in the fluorescence spectrum and lifetime of the Trp residues. In order to provide evidence for the presence of a Trp residue in each binding site several ligands containing a heavy atom (Hg) were synthesized. Using the thiomercuribenzoate of di-N-Acetylchitobioside (see formula on figure 3) as a ligand it was shown that saturation of WGA was accompanied by the following spectroscopic effects :

   i)   complete quenching of WGA fluorescence
   ii)   enhancement of phosphorescence intensity (x 6)
   iii) shortening of the phosphorescence lifetime from 5. 9 to
         0. 4 s.

These three effects are those expected if the Hg atom of the ligand is in Van der Waals contact with Trp residues of the protein thereby proving that those residues are located in (or near) the binding sites. This example shows how usefull spectroscopic properties can be in analyzing interacting systems.

REFERENCES

(1 a) G. Weber (1976) in Excited States of Biological Molecules"
        J. B. Birks Ed., Wiley, pp. 363-374
    b) R. Lumry and M. Herschberger (1978) Photochem. Photo-
        biol. 27, 819-840
(2 a) G. H. Beaven and E. R. Holiday (1952) Advances in Protein
        Chemistry, 7, 319-386
    b) B. Wetlaufer (1962) Advances in Protein Chemistry, 17,
        303-390
(3) T. T. Herskovits and S. M. Sorensen (1968) Biochemistry, 7,
        2533-2542
(4) M. S. Walker, T. W. Bednar, R. Lumry and F. Humphries
        (1971) Photochem. Photobiol., 14, 147-161
(5) J. Eisinger and G. Navon (1969) J. Chem. Phys., 50, 2069-
        2077
(6) J. Feitelson (1970) Israel J. Chem., 8, 241-252
(7) I. Weinryb and R. F. Steiner (1971) in "Excited States of Pro-
        teins and Nucleic Acids", Mac Millan, pp. 277-318
(8) D. M. Rayner, D. T. Krajcarski and A. G. Szabo (1978) Can.
        J. Biochem., 56, 1238-1245
(9) I. Tatischeff and R. Klein (1975) Photochem. Photobiol., 22,
        221-229
(10 a) L. I. Grossweiner, A. G. Kaluskar and J. F. Baugher (1976)
        Int. J. Radiot. Biol., 29, 1-16
    b) J. F. Baugher and L. I. Grossweiner (1977) J. Phys. Chem.,
        81, 1349-1354
    c) R. F. Evans, C. A. Ghiron, W. A. Volkert and R. R. Kuntz,
        (1976) Chem. Phys. Letters, 42, 43-45 ; Photochem. Photo-
        biol., 24, 3-7
(11) R. Santus, C. Hélène and M. Ptak (1968) Photochem. Photo-
        biol., 7, 341-360
(12) R. F. Steiner and E. Kirby (1969) J. Phys. Chem., 73, 4130-
        4135
(13) J. P. Privat and M. Charlier (1978) Eur. J. Biochem., 84,
        79-85
(14) P. Walrant and R. Santus (1974) Photochem. Photobiol., 19,
        411-417 and 20, 455-460
(15) R. Santus and L. I. Grossweiner (1972) Photochem. Photobiol.
        15, 101
(16) J. W. Longworth (1971) in ref. 7, pp. 319-484
(17) E. A. Burstein, N. S. Vedenkina and M. N. Ivkova (1973)
        Photochem. Photobiol., 18, 263-279
(18) A. Szabo, K. Lynn, D. Krajcarski and D. M. Rayner (1978)
        Intern. Conf. On Luminescence (Paris) pp. 263-264

(19) S.S. Lehrer (1971) Biochemistry, 10, 3254-3263

(20) J. Eisinger (1969) Biochemistry, 8, 3902-3907

(21) R.E. Dale and J. Eisinger (1974) Biopolymers, 13, 1573-1605

(22) P. Marche, T. Montenay-Garestier, C. Hélène and P. Fromageot (1976) Biochemistry, 15, 5730-5737

(23) R.A. Badley and F.W.J. Teale (1969) J. Mol. Biol. , 44, 71-88

(24) a) M.L. Saviotti and W.C. Galley (1974) Proc. Nat. Acad. Sci. USA, 71, 4154-4158
     b) G.B. Strambini and W.C. Galley (1976) Nature, 260, 554-556

(25) J. Spikes and G. Jori (1978) Photochem. Photobiol. Rev., K. Smith Ed. , Plenum Press, vol. 3, ch. 6 (in press)

(26) B. Müller-Hill (1975) Prog. Biophys. Mol. Biol. , 30, 227-252

(27) M. Charlier, F. Culard, J.C. Maurizot and C. Hélène (1977) Biochem. Biophys. Res. Comm. , 74, 690-698

(28) J.C. Brochon, Ph. Wahl, M. Charlier, J.C. Maurizot and C. Hélène (1977) Biochem. Biophys. Res. Comm. , 79, 1261-1271

(29) P. Marche, T. Montenay-Garestier, P. Fromageot and C. Hélène (1976) Biochemistry, 15, 5738-5743

(30) S.P. Mc Glynn, T. Azumi and M. Kinoshita (1969) in "Molecular Spectroscopy of the triplet state", Prentice Hall

(31) M. Monsigny, F. Delmotte and C. Hélène (1978) Proc. Nat. Acad. Sci. USA, 75, 1324-1328

(32) B. Valeur and G. Weber (1977) Photochem. Photobiol. , 25, 441-444

(33) I.Z. Steinberg (1978) Ann. Rev. Biophys. Bioeng. , 7, 113-137.

EXCITED STATES AND PHOTOCHEMICAL REACTIONS IN NUCLEIC ACIDS

Claude HELENE

Centre de Biophysique Moléculaire

45045 Orléans Cedex (France)

INTRODUCTION

Nucleic acids play a central role in the behavior of all living organisms. DNA (and in a few cases RNA) contains all the information which is needed for all cellular reactions and their regulation. Messenger, transfer and ribosomal RNAs are important intermediates in the expression of this information.

Ultraviolet irradiation of microorganisms leads to mutation and death. The wavelength dependence of these biological effects has been shown to be similar to the absorption spectrum of DNA (1). More recently it has been concluded that UV damages to DNA might be responsible for the induction of tumors in animals and in human beings (2,3). It is therefore of great importance to understand what happens to nucleic acids when they are submitted to the action of UV light. This explains the large amount of work which has been devoted to the study of the excited states of nucleic acids and of the photochemical processes which follow the absorption of light.

However nucleic acids never appear as free molecules inside living systems. They are interacting with a large number of molecules. Proteins represent the major constituents of these liganded molecules. These proteins can play different roles i) a

structural role if they maintain the nucleic acid in a given conformation or if they participate in the structural organization of,
e. g. , chromatin or chromosomes ; ii) a regulatory role when they
act as regulators of genetic expression ; iii) an enzymatic role if
they are involved in replication, repair or transcription of DNA,
in aminoacylation of transfer RNAs, in translation of the genetic
message at the level of ribosomes.

Excited state studies have proved to be useful in analyzing
the conformational properties and the interactions of biological
molecules. They are also a prerequesite to the understanding of
the photochemical processes which can take place in the excited
systems. The investigation of excited state properties of nucleic
acid-protein complexes requires a knowledge of the excited state
parameters of the two separated molecules. This series of lectures will be divided into three parts dealing with excited state properties of i) nucleic acids, ii) proteins which interact with nucleic
acids and iii) their associations. Also photochemical reactions in
both systems will be briefly mentionned with special reference to
these reactions which are specific to the associated molecules
(cross-linking, electron and energy transfer).

ABSORPTION SPECTRA OF NUCLEIC ACIDS

The absorption spectra of nucleic acid constituents in the
near UV are characterized by several $\pi\pi^*$ transitions between
300 and 180 nm (4). In nucleic acids it is not possible to separate
the contributions of the individual bases. Stacking and hydrogen
bonding interactions between bases which are responsible for the
helical and double helical structures of nucleic acids lead to a
strong hypochromic effect of the first main UV absorption band
whose maximum is around 260 nm. Denaturation of the nucleic
acid for example upon heating, induces a hyperchromism. The
transition is cooperative for double helices (such as DNA) but
shows no cooperativity for single stranded molecules (5).

Transition moments are known in individual bases on the basis of absorption spectra of crystals. In DNA linear dichroism of
molecules oriented by application of an electric field or flow dichroism studies have shown that the main transition around 260 nm
is polarized in the plane of the base pairs (which are perpendicular
to the helix axis) as expected for $\pi\pi^*$ transitions.

EXCITED STATES OF NUCLEIC ACIDS AT ROOM
TEMPERATURE

The fluorescence of nucleic acid constituents is very weak in aqueous solutions at room temperature. The fluorescence quantum yields are in the range 10^{-5} - 10^{-3} (4). Assuming that the lowest excited singlet state which emits fluorescence corresponds to the first $\pi\pi^*$ transition it is possible to calculate the fluorescence lifetime from the experimentally determined fluorescence quantum yield and the radiative lifetime calculated from the area of the first absorption band. This leads to values around 10^{-12} s.

Information on the fluorescence lifetime of nucleic acid constituents also comes from studies of the sensitized Eu^{+++} fluorescence arising from energy transfer from the excited organic molecule (6). Transfer can take place from both the singlet and triplet states. However owing to the very different lifetimes of these two excited states, the two transfer processes will occur over quite different ranges of Eu^{3+} concentrations. This method allowed the determination of the excited state parameters of nucleotides assuming that singlet state energy transfer was controlled by diffusion.

However the determination of the singlet state lifetimes of nucleic acid constituents rests upon assumptions which are difficult to verify : i) in the first method it is assumed that there is no hidden state of lower energy ($n\pi^*$ state for example) which would be responsible for fluorescence emission even though it could not be detected under the long wavelength absorption band ; ii) in the energy transfer method it is assumed that energy transfer from the lowest singlet state to Eu^{3+} ions is controlled by diffusion.

If the singlet state lifetimes are as short as estimated from the two above methods then one should expect the fluorescence of nucleic acid constituents to be polarized because their lifetime would be shorter than or of the same order of magnitude as the rotational correlation time of these molecules in solution. This is what has been recently reported for different nucleic acid constituents in aqueous solutions (7). The polarization of the fluorescence of the molecules with the smallest fluorescence quantum yields (10^{-4} - 10^{-5}) is as high as in rigid medium at low temperature.

Depolarization occurs when the fluorescence quantum yield increases because the fluorescence lifetime reaches values of the same order of magnitude as (or higher than) the rotational correlation time. From these polarization measurements one may conclude that the fluorescence lifetimes of nucleic acid bases are really in the 1-10 ps range.

Triplet state parameters at room temperature in fluid aqueous solutions have been deduced both from energy transfer to Eu^{3+} ions (6) and from flash spectroscopy (8). These two methods gave similar results. The lifetimes of the pyrimidine triplet states are in the microsecond range. Intersystem crossing rate constants have also been determined. From the fluorescence and intersystem crossing quantum yields (see table 1) one can immediately see that more than 99 % of the excitation energy is dissipated from the first excited singlet state by a non radiative transition to the ground state. There is no explanation yet available for the very high rate of these non radiative processes. The nature of the solvent obviously plays a great role. For example triplet state quantum yields for uracil and thymine are about 10 times higher in acetonitrile than in water (8).

Table 1 : Luminescence characteristics of nucleotides in aqueous solutions at pH 7 at 77 K and 300 K

77 K

	$\phi_F^{(a)}$	τ_F (ns)	$\phi_{ISC}^{(b)}$	$\phi_P^{(b)}$	$\tau_P^{(b)}$ (s)
TMP	0.30	3	$<3 \times 10^{-3}$	8×10^{-3}	0.33
CMP	5×10^{-2}	0.3	0.03	0.01	0.34
AMP	5×10^{-3}	3	0.02	0.015	2.4
GMP	0.17	5	0.15	0.07	1.3

300 K

	$\phi_F^{(c)} (\times 10^5)$	$\tau_F^{(d)}$ (p sec)	$\phi_{ISC}^{(d)}$
TMP	4	18	8×10^{-3}
CMP	3	3.6	1.5×10^{-3}
AMP	0.3	1.3	4×10^{-4}
GMP	0.2	4.4	4.6×10^{-4}

(a) P.I. Honnas & H.B. Steen (1970) Photochem. Photobiol. 11, 67
(b) references 12 & 4 ; (c) references 4 & 9 ; (d) reference 6 assuming that the rate constant for energy transfer to Eu^{3+} is diffusion controlled : $k_t = 5 \times 10^9$ $M^{-1}.s^{-1}$)

Much less is known about excited states in DNA itself at room temperature. The fluorescence quantum yield is low ($\sim 10^{-5}$) and the fluorescence spectrum is shifted to longer wavelengths as compared to the corresponding mixture of mononucleotides. This fluorescence seems to arise from excimer (or exciplex) states formed between adjacent bases or base pairs (9).

To determine whether energy transfers can occur in DNA attempts have been made to use intercalated dyes as energy acceptors (traps). A sensitized fluorescence of acridine dyes or ethidium bromide has been observed in such complexes. This sensitization occurs over 5-10 base pairs which is probably the result of a direct base-dye transfer rather than a base-to-base transfer followed by trapping at the site of the intercalated dye (10). It was also observed that ethidium bromide inhibits the formation of photoinduced thymine dimers (10). In contrast to the sensitized fluorescence which does not depend on the excitation wavelength within the DNA absorption band, the inhibition of photodimer formation is markedly wavelength dependent. This was ascribed to a triplet transfer mechanism, one dye molecule being able to quench about 17 base pairs.

Using picosecond pulses recent results show that the rise-time of the sensitized fluorescence of dyes intercalated in DNA is less than 20 ps (11). But this result does not provide evidence for base-to-base energy transfer in DNA preceding the base-to-dye transfer step. As shown above the singlet state lifetime of DNA bases is of the order of 1-10 picoseconds so that direct energy transfer from this singlet state to a dye should occur in the same time range.

Before concluding this section it is worth mentioning that in some nucleic acids, e.g., tRNAs some unusual bases can be found which can have much higher fluorescence quantum yields than usual bases and therefore can be used as intrinsic fluorescent labels.

EXCITED STATES OF NUCLEIC ACIDS AT LOW TEMPERATURE

This subject has been reviewed several times and the reader is referred to these recent reviews (12-15). The results can be summarized briefly as follows :

i) in frozen water-ethylene (propylene) glycol mixtures at 77 K all nucleic acid constituents emit fluorescence and phosphorescence.

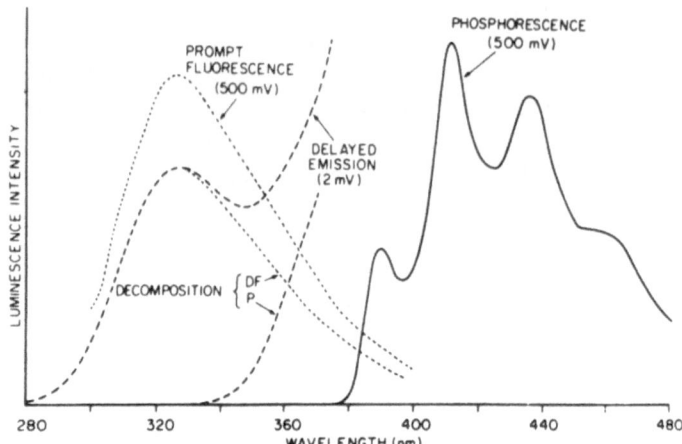

Figure 1 : Luminescence of poly(A) in a water-ethylene-
glycol glass (1v/1v) at 77 K. The dotted line is the fluorescence spectrum and the full line the phosphorescence
spectrum both recorded with the same sensitivity. The
broken line represents the luminescence spectrum recorded in the phosphorescence mode with an increased sensitivity (2 mV instead of 500 mV). A delayed fluorescence
is clearly observed around 330 nm whose spectrum is identical with that of prompt fluorescence (from reference 16).

The fluorescence quantum yields are much higher than at room
temperature (up to 1000 times higher for TMP which has the highest fluorescence quantum yield of 0.30 at 77 K)
ii) the fluorescence of polynucleotides and DNA is usually characterized by a red shifted excimer fluorescence
iii) a delayed fluorescence is observed in poly(A) a polynucleotide
in which extensive delocalization of the triplet state energy does
occur (figure 1). Originally attributed to triplet-triplet anhili-
tation (16) this delayed fluorescence is more likely to result from
ejection of electrons in a biphotonic process followed by electron-
cation recombination (17).
iv) the triplet state energy is always localized on the lowest lying
triplet level. The triplet energy levels decrease in the order
C > G > A > T. In DNA the triplet energy is localized on T. This

triplet level can be populated via triplet-triplet energy transfer from acetophenone.

v) energy transfer does occur in DNA at the singlet and triplet levels but these energy transfer processes are rather limited (18, 22). Rapid triplet transfer to quenchers and acceptors occurs in DNA over 5 to 30 bases depending on the A-T content. At the singlet level energy migration occurs over 5-10 bases at most but there is no evidence for long range migration.

PHOTOCHEMICAL REACTIONS IN NUCLEIC ACIDS

This subject has been reviewed recently in detail (19). Many different types of photoproducts have been identified in UV irra-

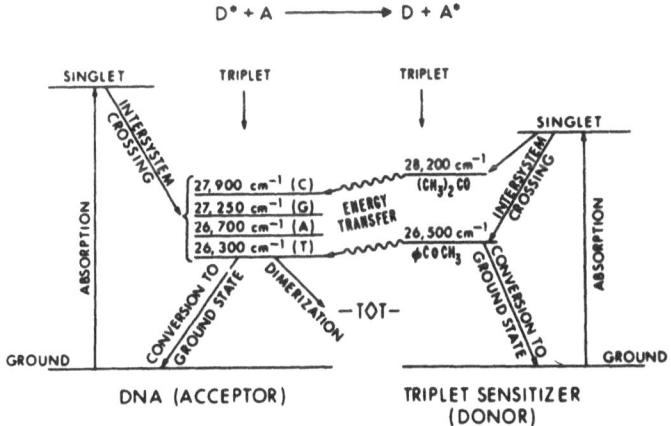

Figure 2 : Energy levels (lowest singlet and triplet states) of nucleotides, acetone and acetophenone. The two ketones can be excited at wavelengths where nucleic acid bases do not absorb but can transfer their energy at the triplet level thus populating the triplet states of the bases (from R. O. Rahn & M. H. Patrick, in reference 1, vol. 2, pp. 97-145).

diated DNAs or RNAs. To appreciate the biological importance of any given photoproduct one might contemplate the possibility of using two methods : i) one can try to eliminate the photoproducts of interest and to compare the biological behavior of the irradiated DNA before and after removal of this photoproduct ; ii) one may use special irradiation conditions to try and produce only the photoproduct of interest.

The two methods have been used to investigate the role played by pyrimidine dimers in survival of bacteria, in the induction of mutations, in the induction of malignant transformation... Method i) makes use of photoreactivating enzymes which specifically remove pyrimidine dimers in a light dependent enzymatic reaction (20). Method ii) makes use of what has been learnt from excited state studies of nucleic acid constituents. As summarized above the triplet state of DNA is localized on thymine. It is possible to populate directly this triplet state without exciting the singlet state(s) of DNA by using acetophenone as a photosensitizer (21). Acetophenone can be excited above 310 nm. It has a triplet state energy above that of thymine but below that of other bases and can therefore transfer its energy to thymine by a triplet-triplet transfer process (figure 2). This leads to the selective formation of thymine-containing pyrimidine dimers in DNA whose biological role can then be investigated.

REFERENCES

(1) J. Jagger (1976) in "Photochemistry & Photobiology of Nucleic Acids, S. Y. Wang Ed. Academic Press, vol. 2, pp. 147-186

(2) R. W. Hart, R. B. Setlow and A. D. Woodhead (1977) Proc. Nat. Acad. Sci. USA, 74, 5574-5578

(3) R. B. Setlow (1977) Nature, 271, 713-717

(4) M. Daniels (1976) in reference (1), vol. 1, pp. 23-108

(5) M. H. Patrick (1976) in reference (1), vol. 2, pp. 1-34

(6) J. Eisinger and A. A. Lamola (1971) Biochim. Biophys. Acta, 240, 299-312 and 313-325

(7) J. P. Morgand and M. Daniels (1978) Photochem. Photobiol. 27, 73-75

(8) C. Salet and R. Bensasson (1975) Photochem. Photobiol., 22, 231-235

(9) a. P. Vigny (1973) C. R. Acad. Sci. Paris, Série D, 277, 1941-1944

   b. P. Vigny and M. Duquesne (1976) in "Excited States of Biological Molecules", J. B. Birks Ed., Wiley, pp. 167-177

(10) B. M. Sutherland and J. C. Sutherland (1969) Biophys. J., <u>8</u>, 490 and <u>9</u>, 1045

(11) S. L. Shapiro, A. J. Campillo, V. H. Kellman and W. B. Goad (1975) Optics Comm., <u>15</u>, 308-310

(12) J. Eisinger and R. G. Shulman (1968) Science, <u>161</u>, 1311-1319

(13) J. Eisinger and A. A. Lamola (1971) in "Excited States of Proteins and Nucleic Acids", R. F. Steiner and I. Weinryb Eds., Mac Millan, pp. 107-198

(14) C. Hélène (1973) in "Physicochemical properties of Nucleic Acids", J. Duchesne Ed., Academic Press, vol. 1, pp. 119-142

(15) W. W. Hauswirth and M. Daniels (1976) in reference (1), vol. 1, pp. 109-167

(16) C. Hélène and J. W. Longworth (1972) J. Chem. Phys., <u>57</u>, 399-408

(17) M. Bazin, R. Santus and C. Hélène (1973) Chem. Phys., <u>2</u>, 119-128

(18) M. Guéron, J. Eisinger and A. A. Lamola (1973) in "Basic Principles of Nucleic Acid Chemistry, P. O. P. Ts'o Ed., Academic Press, vol. 1, Ch. 4

(19) M. H. Patrick and R. O. Rahn (1976) in reference (1), vol. 2, pp. 35-95

(20) H. Harm (1976) in reference (1), vol. 2, pp. 219-263

(21) A. A. Lamola (1969) Photochem. Photobiol., <u>9</u>, 291

(22) W. C. Galley (1968) Biopolymers, <u>6</u>, 1279-1296

EXCITED STATE INTERACTIONS AND PHOTOCHEMICAL

REACTIONS IN PROTEIN-NUCLEIC ACID COMPLEXES

Claude HELENE

Centre de Biophysique Moléculaire

45045 Orléans Cedex (France)

INTRODUCTION

Protein-nucleic acid complexes are of central importance in molecular biology. The expression of genetic information and its regulation require the formation of such highly specific complexes. An investigation of the excited states and of the photochemical behavior of these complexes has different purposes : i) excited state properties may be used to obtain information on the mechanism of complex formation and on the nature of molecular interactions involved in these complexes ; ii) the study of photochemical reactions in protein-nucleic acid complexes should help us understand the action of UV radiations on biological systems ; iii) the formation of photochemical cross-links between a protein and a nucleic acid in a specific complex should provide information on the regions of the two macromolecules which are in close contact in the complex. The use of photochemical cross-linking reactions should be comparable to that of bifunctional chemical reagents with the advantage that the photochemical reaction directly links two chemical groups one on each macromolecule.

EXCITED STATE PROPERTIES OF PROTEIN - NUCLEIC ACID COMPLEXES

The investigation of excited-state interactions and energy transfer processes in protein-nucleic acid complexes relies mainly upon the observation of light emission from the excited states.

At room temperature only aromatic amino acid side chains emit fluorescence (1). The fluorescence of usual nucleic acid bases is very weak (quantum yields around 10^{-4}) and will be difficult to detect in the presence of a protein containing aromatic amino acids (2). In some cases (e. g. , tRNAs) the fluorescence of unusual bases may be observed. At low temperature (77 K), fluorescence and phosphorescence are emitted by both aromatic amino acids and nucleic acid bases (1, 3, 4).

1 - Stacking Interactions of Aromatic Amino Acids and Nucleic Acid Bases in Frozen Aqueous Solutions

Interactions between nucleic acid bases and aromatic amino acids were first observed in frozen aqueous solutions at 77 K (5, 6, 7). Under these experimental conditions the formation of molecular aggregates induces interactions between solute molecules. The results can be summarized as follows :

i) tryptophan and nucleic acid bases form electron donor-acceptor (EDA) complexes in which the indole ring of Trp is the electron donor and the base is the electron acceptor. These complexes are characterized by a quenching of the fluorescence of both components and by the appearance of a new fluorescence at longer wavelengths with a much lower quantum yield (5, 6).

ii) tyrosine and pyrimidine bases also form EDA complexes. This is accompanied by a quenching of the fluorescence of both molecules. In the complexes formed by tyrosine and purine bases, only tyrosine fluorescence is quenched but the purine fluorescence is not markedly affected (7).

iii) the fluorescence of nucleic acid bases is not affected by the presence of an equimolar concentration of phenylalanine. An excited state interaction has been observed only with protonated cytidine under acidic conditions (8).

2 - Energy Transfer Processes in Frozen Solutions

The formation of stacked arrays of molecules in frozen aqueous solutions made it possible to study energy transfer processes between nucleic acid bases and aromatic amino acids (9). Mixed aggregates can be formed in which the concentration of one of the components is small as compared to the other one. Very efficient energy transfer processes at the triplet level were observed from bases to tryptophan (10) and from tyrosine to bases (11), in agreement with triplet state energies which decrease in the or-

der Tyr $>$ C $>$ G $>$ A $>$ T $>$ Trp (12). At the singlet level, an efficient
transfer occured from tyrosine to bases. In all cases the excitation energy migrated step-by-step from one excited molecule to
its nearest neighbors in the aggregate until it was trapped by the
molecule which had a lower singlet or triplet state energy. For
example, one molecule of tryptophan was able to trap the triplet
state energy of about 200 adenosine molecules (10). About 70 tyrosine molecules were able to transfer their singlet energy to one
base (11).

An efficient energy transfer at the triplet level was also observed in poly(A)-(Lys-Trp-Lys) complexes (10). The results and
the mechanisms were similar to those described for adenosine-
tryptophan aggregates. The triplet excitation energy migrated
from base to base in poly(A) until it was trapped by a Trp residue
inserted between two adjacent bases in the polynucleotide.

3 - Quenching of Aromatic Amino Acid Fluorescence in Oligo-peptide-Nucleic Acid Complexes at Room Temperature

At room temperature, excited-state interactions between
nucleic acid bases and aromatic amino acids were investigated in
oligopeptide-nucleic acid complexes. Oligopeptides of the general
sequence Lys-X-Lys, where X is an aromatic residue, were
shown to interact with nucleic acids at low ionic strength due mainly to strong electrostatic interactions of lysyl residues with phos-
phates. Different physical techniques were used to investigate these interactions including nuclear magnetic resonance (13), circular dichroism (14) and fluorescence (15). The fluorescence of the
aromatic amino acid was always quenched in the complexes. A
comparison of the results obtained by the different methods allowed us to draw the following conclusions :

i) the tryptophyl residue of the oligopeptide Lys-Trp-Lys
form stacked complexes with bases and this stacking interaction
is strongly favored in single-stranded as compared to double-
stranded nucleic acids (15, 16). Fluorescence quenching is due to
the formation of stacked complexes. An electron donor-acceptor
interaction is responsible for this quenching as observed in frozen
aqueous solutions (see above).

ii) the tyrosyl residue of the oligopeptide Lys-Tyr-Lys (and
of other related oligopeptides such as Lys-Ala-Tyr-Ala-Lys) is
stacked with bases only in single-stranded polynucleotides and in
single-stranded regions of nucleic acids (e.g., in tRNAs). Howe-

ver a strong fluorescence quenching is observed with both double-
stranded and single-stranded polynucleotides or nucleic acids.
Different mechanisms can be proposed to explain this fluorescence quenching : a) stacking with bases does quench tyrosine fluorescence in aggregates (see paragraph 1 above) ; b) hydrogen bonding of the OH group of tyrosine to acceptor groups on the nucleic
acid (base, sugar or phosphate) might result in fluorescence
quenching due, e.g., to an excited-state proton transfer ;
c) singlet energy transfer from tyrosine to bases (which are not
fluorescent) should lead to tyrosine fluorescence quenching. This
transfer should be efficient as shown from the calculation of critical Förster distances (18) ; d) a conformational change of the
peptide due to complex formation might bring the tyrosyl residue
close to a quenching group inside the peptide.

Mechanism b) was eliminated by investigating the binding of
the peptide Lys-Tyr(OMe)-Lys in which the OH group of tyrosine
is methylated. Fluorescence quenching is observed upon complex
formation with nucleic acids and is as efficient as in the case of
Lys-Tyr-Lys, even though neither hydrogen bonding nor proton
transfer can occur. Although phosphate monoanions quench tyrosine fluorescence (17) we recently observed that phosphate dies-
ters do not (T. Alev, J. J. Toulmé and C. Hélène, to be published).
Hypothesis d) seems very unlikely since similar results were obtained with Lys-Tyr-Lys, Lys-Ala-Tyr-Ala-Lys and peptides in
which both the amino- and carboxyl groups were blocked by acetyl
and ethylamide substituents, respectively. The conclusion of these
studies is that both stacking and energy transfer are responsible
for tyrosine fluorescence quenching in the complexes formed between oligopeptides and single-stranded nucleic acids. Since no
stacking was observed with double-stranded DNA, tyrosine fluorescence quenching in this case must be entirely attributed to
energy transfer to base pairs (34).

iii) the fluorescence of several oligopeptides containing phenylalanine (including Lys-Phe-Lys) is quenched in complexes formed with both single-stranded and double-stranded nucleic acids.
Nuclear magnetic resonance studies clearly demonstrate that stacking interactions are much more important in single strands. As
in the case of tyrosine, energy transfer to nucleic acid bases is
very likely responsible for phenylalanine fluorescence quenching
(34).

4 - Interactions Between Nucleic Acid Bases and Disulfides

The possible effect to disulfide bridges of proteins on the excited state properties of nucleic acids has been investigated in the complexes formed by cystamine and polynucleotides (19). The phosphorescence of poly(A) is strongly quenched by small amounts of cystamine in low temperature glasses (19) indicating that cystamine acts as a trap for the triplet excitation energy migrating amongst adenine bases in poly(A). The most likely mechanism involves an electron transfer from the adenine triplets to the disulfide bridge as already observed with tryptophan and tyrosine (20, 21).

5 - Fluorescence of protein-nucleic acid complexes

In most cases where fluorescence has been used to investigate protein-nucleic acid complexes a quenching of the fluorescence of aromatic residues has been observed (36). This is particularly striking in the complexes formed by tRNAs with aminoacyl tRNA synthetases. This observation lends some support to the idea that a general mechanism is responsible for this fluorescence quenching. A conformational change in the protein would not be expected to lead in all cases to such a quenching. From the studies of model systems reported above it may be tentatively concluded that tryptophan fluorescence quenching is certainly due to stacking of tryptophan residues with nucleic acid bases whereas in the case of tyrosine both stacking and energy transfer could be responsible for quenching. It must be remembered that stacking of tyrosine with bases is very unlikely in double stranded nucleic acids.

PHOTOCHEMICAL REACTIONS IN PROTEIN-NUCLEIC ACID COMPLEXES

The photochemical behavior of protein-nucleic acid complexes is expected to be different from that of the separated macromolecules. The formation of the complex may alter the mutual interactions of reacting groups in each macromolecule and therefore modify the kinetics and quantum yields of photoproduct formation. For example a distortion of the phosphodiester backbone of a nucleic acid upon binding a protein may modify stacking interactions between bases and thus affect the photodimerization of adjacent pyrimidines. A conformational change in the protein induced by complex formation may modify the environment and the photochemical reactivity of, e.g., tryptophyl residues or disulfide bridges. However more interesting will be new photochemical reactions which are specific of the particular protein-nucleic acid complex under investigation. The formation of cross-links between the two macromolecules is a very attractive method to obtain information on the location of the interacting parts in the complex (although this method has obviously its own limitations). The photosensitized splitting of pyrimidine dimers using tryptophyl residues of proteins as photosensitizers may prove useful in providing evidence for the involvement of these residues in complex formation. It should also be kept in mind that primary photoproducts could act as photosensitizers for other reactions. This is the case, for example, of N'-formylkynurenine, a photooxidation product of tryptophan (22).

1 - Photochemical Cross-Linking of Proteins to Nucleic Acids

This subject has been reviewed recently (23, 24). The formation of cross-links between proteins and nucleic acids has made it possible to locate the regions of tRNA molecules which interact with aminoacyl-tRNA synthetases (23) and the peptides of ribosomal S 4 protein which are in close contact with ribosomal 16 S RNA (25). The method has also been recently applied to the identification of those thymine bases of the lac operator which interact with the lac repressor (26). Using DNA in which thymine has been replaced by 5-bromouracil, irradiation of the lac repressor-operator specific complex leads to a cross-linking of the two molecules (27). This cross-linking reaction is due to the photochemical cleavage of the C-Br bond and to the high reactivity of the uracilyl radicals thus produced. In most cases however, the chemical nature of the crosslinks has not yet been determined.

The photochemical formation of covalent bonds between amino acid side chains and nucleic acid bases has been demonstrated in simple model systems. For example, thymine and uracil react with cysteine to form several photoproducts (29). Thymine can also react with lysine amino groups (30). Purines have been shown to form adducts at the C(8) position with alcohols and amines (31). Recently we have shown that carboxylic acids (side chains of Glu and Asp residues) could photochemically react with nucleic acid bases, especially thymine and uracil (F. Toulmé and C. Hélène, to be published). All these reactions could be involved in the formation of photochemical cross-links between proteins and nucleic acids.

2 - Photosensitized Splitting of Pyrimidine Dimers by Tryptophan-Containing Oligopeptides and Proteins

The splitting of pyrimidine dimers can be photosensitized by several types of molecules. Indole derivatives can act as photosensitizers with respect to both isolated dimers and dimers in DNA. Oligopeptides containing tryptophan residues, such as Lys-Trp-Lys, have been shown to bind strongly to UV irradiated DNA and to photosensitize the splitting of thymine dimers (16, 32). An electron transfer from the indole ring to the dimer was proposed as a very likely mechanism for the photosensitized splitting. This electron transfer requires a close proximity (a stacking interaction) of the indole ring and the pyrimidine dimer.

This reaction was thought of as a possible method to provide evidence for stacking interactions of tryptophyl residues of proteins with nucleic acid bases in protein-nucleic acid complexes especially in complexes involving single-stranded DNA where the presence of pyrimidine dimers is not expected to affect markedly protein binding. If a photosensitized splitting of thymine dimers could be demonstrated in such complexes, this would lend strong support to the involvement of stacking interactions of tryptophyl residue(s) with bases. We chose to study the protein coded by gene 32 of phage T 4 because this protein was known to bind selectively and cooperatively to single-stranded DNA. The fluorescence of the tryptophyl residues of gene 32 protein is quenched upon binding to single-stranded DNA or polynucleotides (33, 35). The presence of pyrimidine dimers does not affect this binding. Upon UV irradiation of the complex formed by gene 32 protein with DNA containing pyrimidine dimers a photosensitized splitting of dimers is observed (33, 35). These two results (fluorescence quenching, photosensitized splitting) are consistent with the involvement of tryptophyl residues of gene 32 protein in stacking interactions with bases.

CONCLUSION

This review shows that many different processes are expected to take place when protein-nucleic acid complexes are excited by UV radiations. Excited-state interactions between the chromophores of the two macromolecules may lead to properties which are characteristic of the complexes. These properties may be used to get some insight into the mechanism of complex formation. They may also contribute to our understanding of the photochemical reactions which take place under UV excitation. The photochemistry of nucleic acid-protein complexes certainly plays an important role in the behavior of cells and organisms submitted to UV radiations. Research in this area will also contribute to a better knowledge of the rules which govern the selective recognition of nucleic acids by proteins.

REFERENCES

(1) J. W. Longworth (1971) in "Excited States of Proteins and Nucleic Acids", R. F. Steiner and I. Weinryb Ed., Plenum Press, pp. 319-484

(2) M. Daniels (1976) in "Photochemistry and Photobiology of Nucleic Acids", S. Y. Wang Ed., Academic Press, vol. I, pp. 23-108

(3) J. Eisinger and A. A. Lamola (1971) in reference 1 pp. 107-198

(4) C. Hélène (1973) in "Physico-chemical Properties of Nucleic Acids", J. Duchesne Ed., Academic Press, vol. I, pp. 119-142

(5) T. Montenay-Garestier and C. Hélène (1968) Nature, 217, 844-845

(6) T. Montenay-Garestier and C. Hélène (1971) Biochemistry, 10, 300-306

(7) C. Hélène, T. Montenay-Garestier and J. L. Dimicoli (1971) Biochim. Biophys. Acta, 254, 349-365

(8) T. Montenay-Garestier and C. Hélène (1973) J. Chim. Phys. 70, 1385-1390

(9) T. Montenay-Garestier and C. Hélène (1973) J. Chim. Phys. 70, 1391-1399

(10) C. Hélène (1973) Photochem. Photobiol., 18, 255-262

(11) T. Montenay-Garestier (1976) in "Excited States of Biological Molecules", J. B. Birks Ed., Academic Press, pp. 207-216

(12) C. Hélène (1976) in reference 11, pp. 151-166

(13) J. L. Dimicoli and C. Hélène (1974) Biochemistry, 13, 714-723 and 724-730

(14) M. Durand, J. C. Maurizot, H. N. Borazan and C. Hélène (1975) Biochemistry, 14, 563-570

(15) F. Brun, J. J. Toulmé and C. Hélène (1975) Biochemistry, 14, 558-563

(16) J. J. Toulmé, M. Charlier and C. Hélène (1974) Proc. Nat. Acad. Sci. USA, 71, 3185-3188

(17) J. Feitelson (1964) J. Phys. Chem., 68, 391-397

(18) T. Montenay-Garestier (1975) Photochem. Photobiol., 22, 3-6

(19) T. Montenay-Garestier, F. Brun and C. Hélène (1976) Photochem. Photobiol., 23, 87-91

(20) J. Feitelson and E. Hayon (1973) Photochem. Photobiol., 17, 265-274

(21) D. V. Bent and E. Hayon (1975) J. Am. Chem. Soc., 97, 2612-2619

(22) P. Walrant, R. Santus and M. Charlier (1976) Photochem. Photobiol., 24, 13-19

(23) P. R. Schimmel, G. P. Budzik, S. S. M. Lam and H. J. P. Schoemaker (1976) in "Aging, Carcinogenesis and Radiation Biology", K. C. Smith Ed., Plenum Press, pp. 123-148

(24) K. C. Smith (1976) in "Photochemistry and Photobiology of Nucleic Acids", S. Y. Wang Ed., Acad. Press, vol. 2, 187-218

(25) B. Ehresmann, J. Reinbolt and J. P. Ebel (1975) FEBS Letters, 58, 106-111

(26) R. Ogata and W. Gilbert (1977) Proc. Nat. Acad. Sci. USA, 74, 4973-4976

(27) S. Y. Liu and A. D. Riggs (1974) Proc. Nat. Acad. Sci. USA, 71, 947-951

(28) C. Hélène (1976) in reference 23, pp. 149-163

(29) A. J. Varghese (1976) in reference 23, pp. 207-223

(30) M. D. Shetlar, H. N. Schott, H. G. Martinson and E. T. Liu, (1975) Biochem. Biophys. Res. Comm., 66, 88-93

(31) D. Elad (1976) in reference 23, pp. 243-260

(32) M. Charlier and C. Hélène (1975) Photochem. Photobiol., 21, 31-37

(33) C. Hélène, F. Toulmé, M. Charlier and M. Yaniv (1976) Biochem. Biophys. Res. Comm., 71, 91-98

(34) R. Mayer, F. Toulmé, T. Montenay-Garestier and C. Hélène (1978) J. Biol. Chem. (in press)

(35) C. Hélène and M. Charlier (1977) Photochem. Photobiol., 25, 429-434

(36) C. Hélène (1977) in "Excited States in Organic Chemistry & Biochemistry", B. Pullman and N. Goldblum Eds, Reidel, pp. 65-78.

PRIMARY PROCESSES IN RADIATION CHEMISTRY

Fabio Busi

Laboratorio di Fotochimica e Radiazioni d'Alta Energia

C.N.R. - Via Castagnoli 1 40126 Bologna - Italy

Radiation chemistry is that branch of chemistry which studies the chemical effects and the atomic and molecular phenomena involved with the interaction of high-energy, or ionizing radiation with matter.

The action of high-energy electromagnetic radiations occurs predominantly through the agency of fast-moving electrons, "photoelectrons", produced by photoelectric or Compton effect. The photoelectrons are produced with high kinetic energy which they transfer to the medium by exciting or ionizing the molecules in the vicinity of their trajectories. The primary products, excited or ionized molecules and the ejected electrons undergo secondary reactions which lead ultimately to stable products. The events following the absorption of ionizing radiations takes place on a large time interval. Table 1 summarized the time scale of the various processes induced by ionizing radiation.

The description of the primary processes is based largely on theoretical principles, but new experimental techniques, such as the use of synchrotron radiation, picosecond pulse radiolysis are now offering the possibility of direct observation and measurements of very short living transient species.

TABLE 1

- lg t	Events
18	Fast electron traverses molecule
16	Time for energy loss to electronic states
14	Fast ion-molecule reactions
	Fast excited states dissociation
12	Electron thermalization
11	Spur reactions in dielectric liquids
10	Spur reactions in water
9	
6	Homogeneous neutralizations
5	radical-radical reactions
4	radiative lifetime of triplets
3	Biological transformations.
2	

- Primary ionization and excitation yields

High-energy electrons transfer their energy by soft collisions
and hard collisions. A soft collision is a glancing coulombic
interaction of the high-energy electron with the electrons of the
atoms or the molecules of the target. In a soft collision the
electron transfers a small fraction of its kinetic energy, T. In
a hard collision the electron loses a large fraction of its energy.
The theoretical treatment for energy transfer in soft and hard
collisions are different. However hard collisions occur relatively
infrequently and extension of the soft collision treatment up to
an energy loss of ΔT_{max} = T/2 leads to an error of only a few
percent in the estimated yields of ionization. A simplified method
for the calculation of the interaction of a high-energy electron
with atoms or molecules is that proposed by Weizsäker-Williams and
Fermi which yields results quite easily, at some cost of accuracy.
Consider, following Platzman[1], the simpliest case of the interaction of a high-energy electron with an isolated molecule. When the
fast electron, in its rectilinear path, passes close to the molecule, the electronic molecular system is under the influence of
the impulsive electric field generated by the radiation. The force
exerted on individual electrons of the molecule varies with time

both in direction and intensity, and it can be resolved into two
time dependent components, one parallel, E_p, and the other perpendicular, E_t, to the direction of the radiation, figure 1.

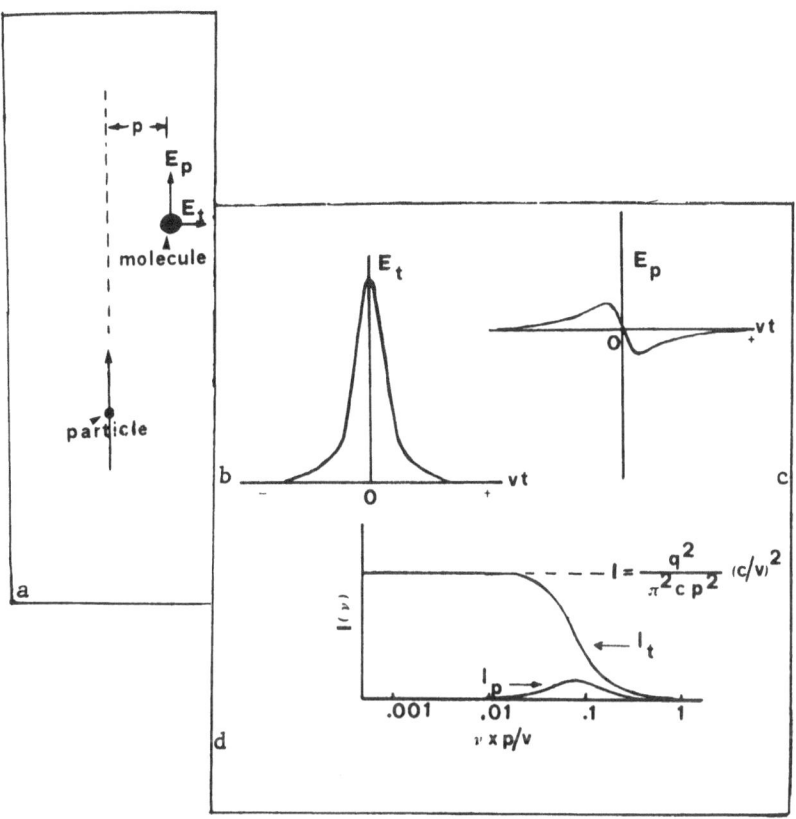

Fig. 1 - a) Soft collision of a fast moving charged particle and an
isolated molecule; b) variation of the perpendicular component of the electric field intensity, E_t, during the
impact; c) variation of the parallel component, E_p, during
the impact; d) frequency spectrum of the virtual photons.

The time dependence of the two components is given by:

$$E_p = - \frac{q \gamma v t}{(p^2 + \gamma^2 v^2 t^2)^{3/2}} \qquad (1)$$

$$E_t = \frac{\gamma q \, p}{(p^2 + \gamma^2 v^2 t^2)^{3/2}} \qquad (2)$$

where q is the electron charge, $\gamma^2 = 1/1 - \beta^2$ and $\beta = v/c$, v is the velocity of the electron, c the velocity of light, p is the closest distance of approach of the electron to the molecule. The equations (1) and (2) can be separated into Fourier components each of which is a purely harmonic function of time

$$E_{p(i)} = \text{constant} \times \sin(2\pi v_i t) \qquad (3)$$

$$E_{t(i)} = \text{constant} \times \cos(2\pi v_i t) \qquad (4)$$

The sum of all the individual $E_{p(i)}$ and $E_{t(i)}$ gives the function E_p and E_t. The functions (3) and (4) are completely equivalent to two pulses of plane polarized light of frequency v_i incident on the molecule in the direction parallel, $P_{p(i)}$, or perpendicular $P_{t(i)}$ to the propagation of the incident electron. The frequency spectra of the so-called "virtual photon" are reported in figures 2 d. The shapes of the spectra can be easily understood if one con siders the behaviour of the fields of the pulses P_p and P_t. The fields of pulses P_t are bell-shaped in time with a width $\Delta t \sim b/v\gamma$. The frequency spectrum therefore contains all frequencies up to $v_{max} \sim 1/\Delta t$. The fields of pulses P_p are similar to one cycle of a sine wave of frequency $v \sim \gamma v/b$ and its spectrum contains a limited range of frequencies in the interval around $\gamma v/b$. Each virtual photon is absorbed or not by the medium according to the formula for the absorption of light:

$$\sigma = \frac{\pi e^2}{mc} \, N \, \frac{df}{dv} \qquad (5)$$

which applies for transition to the continuum or

$$\sigma = \frac{\pi e^2}{mc} \, N \, f_s \qquad (6)$$

which applies for the discrete transition to the state of energy E_s. In the equations (5) and (6) σ is the linear absorption coefficient e and m are the electric charge and mass of the electron, c is the speed of light N is the number of molecules per unit volume, f_s is the oscillator strength and df/dv is the differential oscillator strength. To understand the significance of f_s one has to associate, according to the quantum theory, at each electron

in the molecule a large number of electronic oscillators with different characteristic resonant frequencies, ν_s. The number of oscillators with characteristic frequency ν_s is represented by the
oscillator strength f_s and is non integral and may be much smaller
than unity. The sum of all oscillator strengths in discrete
transitions and the integral oscillator strength over regions of
continuous distribution of frequencies is equal to the total number
of electrons, the atomic number, of the absorbing unit:

$$\sum_n f + \int (df/d\nu)\ d\nu = Z \tag{7}$$

The model of the virtual photon does not consider the exchange
effect between the incident electron and the electrons of the
molecule and is therefore valid, as the Born approximation for the
collision cross section, at energies of the incident electron much
higher than the energies of the electrons on the scattering molecule. Under the conditions of validity of the present method the
energy of the radiation may be assumed constant since it greatly
exceeds the energies of the absorbing spectrum of the molecule.
The number of virtual photons per unit energy interval can be shown
to be:

$$n(\nu) = constant/h\nu \tag{8}$$

Therefore it is possible to calculate the number of primary products in the states S that are formed by the action of ionizing radiation:

$$N_s = constant \times \frac{1}{E_s} \times f_s \tag{9}$$

and the number of primary ionizations

$$N_I = constant \times \frac{1}{E} \times \frac{df}{d\nu} \tag{10}$$

These equations are the "optical approximation" to the number of
primary products of radiation chemistry[2]. The yields in terms of
the absorbed energy are measured in radiation chemistry as the
number of molecules formed per 100 eV of energy absorbed and are
reported as G-values. The optical approximation hence gives, for
an absorbed energy D the yield of molecules in the state S

$$G_S = \frac{100 \times constant}{D} \times f_s/E_s \tag{11}$$

The optical approximation shows that any specificity is impossible

among the primary products in radiation chemistry. All possible
photochemical processes in the complete absorption spectrum for
light may occur with yields which are simply proportional to f_s/E_s
or $(1/E)$ $(df/d\nu)$. The total oscillator strength for valence-shell
electron transitions is approximately equal to the number of the
electrons in the shells whereas the value of the energy for a
transition to the state S is roughly proportional to the square of
the effective nuclear charge acting on the electrons which undergo
the transition. Therefore the probability of excitation or ionization of the valence-shell electrons, which occur at lower energy
E_s, greatly exceedes that of inner-shell electrons.

When the transition induced by the incident electron is the
ionization of the molecule, an electron is ejected with kinetic
energy T given by:

$$T = E - I - E'_{ion} \tag{12}$$

where E is the energy transferred from the radiation to the molecule, I its ionization potential and E'_{ion} is the internal excitation energy of the positive ion formed. The secondary electrons
with energy T, dissipate their energy in a number of interactions
that are described by the virtual photon model only if T is in the
region where Born approximation applies. Low-energy electrons,
i.e. the electrons ejected with an energy T lower than the Born
approximation limit, $T < \sim 100$ eV, may excite optical forbidden
transitions and this process may in fact have appreciable cross
section for low-energy electrons. A comprehensive treatment of
the radiation chemistry will, therefore, necessarily involve consideration of a very broad array of electron energies. The present
knowledge of the radiation induced phenomena do not permit the
description of the processes induced by low energy electrons. A
rigorous treatment requires the determination of the so called
degradation spectra.

When electrons, with initial energy T_0, are generated in a
homogeneous medium they lose their energy progressively. The medium
is therefore crossed by a flux of electrons with energy reduced to
various values $T < T_0$ and of electrons generated in the ionization
processes at different energy $T' \leqslant T_0/2$.

The spectral distribution of these electrons may be represented
by a function, $y(T)$ which gives the number of electrons of energy
between T and T + dT crossing an area of 1 cm^2 per unit time, and
has the dimension of length per unit energy. The differential

y(T)dT is the total distance travelled by all electrons in the medium as they transfer the energy dT at T. The function is the degradation spectrum and, for one particle, is the inverse of its stopping power or rate of energy lose $(\frac{dE}{dx})^{-1}$.

The number of molecules of primary product S formed by electrons of initial energy T_0 is thus given by the product of y(T) and the probability $N\ \sigma_s(T)$ of producing S per unit track length traversed at the energy T. σ_s is the cross-section for excitation to the state E_s and N is the density of atoms or molecules in the medium. The G-value of primary product S is

$$G_S = \left/ N \int_E^{T_o} y(T)\ \sigma_s(T)\ dT \right/ (100/T_o) \tag{13}$$

To solve equation (13) knowledge is required of the two functions y(T) and $\sigma_s(T)$. In order to calculate values of y(T) is necessary to know the differential ionization cross section, $\sigma_i(T,T')$, i.e. the energy distribution of secondary electrons of energy T' ejected in collisions by electrons of energy T, and the total cross section

$$\sigma_{tot} = \sigma_i(T) + \sum_s \sigma_s(T) \tag{14}$$

where $\sigma_i(T)$ is the ionization cross section and σ_s the cross-section for discrete excitation to the state $E_s < T$. The data available on the cross-sections are limited in scope and reliability; only recently Electron Energy Loss spectroscopy and Vacuum Ultra-Violet Spectroscopy using synchrotron radiation have successfully been applied to this problem but the data obtained are as now still limited in number. This is an open field for research which can give a fundamental contribution to the understanding of primary processes in irradiated systems.

In the collision theory it has been found that the cross-section has an asymptotic behaviour that prevails at high values of T and therefore σ_s should behave as

$$\sigma_s = A_s\ T^{-1}\ \ln T + B_s/T + C_s/T^2 + \dots \tag{15}$$

where A_s, B_s, C_s are constants. If we consider (Born approximation) only the first term of the right-hand side of equation 15, it can be shown that

$$\sigma_s = \frac{8 \pi z^2 R a_o^2}{T} \times (f_s R/E_s) \ln c_s T \tag{16}$$

where R is the Rydberg constant and a_o is the Bohr radius; in the continuum the expression for the cross-section is

$$\sigma_i = \frac{8 \pi z^2 R a_o^2}{T} \times (R/E) \, df/dE \ln c_s T \tag{17}$$

Only very recently using Electron Energy Loss spectroscopy experiments have been done for the determination of the constant c_s and the data available are still very limited. The equations   16 and 17 involve properties of a particular molecular transition through the constant c_s, located within the logarithm, and through the factor f_s/E_s. Because of the lack of knowledge of the constant it is necessary, in order to have at least an approximate solution of equation 16 and 17, to assume that the values of c_s are practically constant and thus the cross-section is insensitive to the value of c_s due to its location within the logarithm. We obtain, in this way, the formula of the optical approximation, in which the cross-section depends upon the molecular transition only through the factor f_s/E_s.

The optical spectrum of a molecule which is the complete set of oscillator strengths, is indicated by $f(E)$ and is, apart from a constant factor, the absorption spectrum for electromagnetic radiation extending from the visible region to and throughout the X-ray region. In the optical approximation the probability of producing a particular excited state by ionizing radiation is proportional to f_s/E_s. The complete function $(R/E)f(E)$ is usually called excitation spectrum. This function differs but is closely related to the optical spectrum. The excitation spectra have been obtained only for very simple molecules, He, Ne, Ar, H_2, CH_4 and H_2O vapour[2]. The excitation spectrum of a large number of molecules in the I.R., visible and U.V. range of the spectrum are available from detailed spectroscopic studies. Very little is known about the higher energy range of the spectrum. This is mainly due to the experimental difficulties of VUV spectroscopic experiments and to the low resolution of electron energy loss spectroscopy. The use of synchrotron radiation has recently solved some of these problems[3]. As an example we report the excitation spectrum of methane, figure 2.

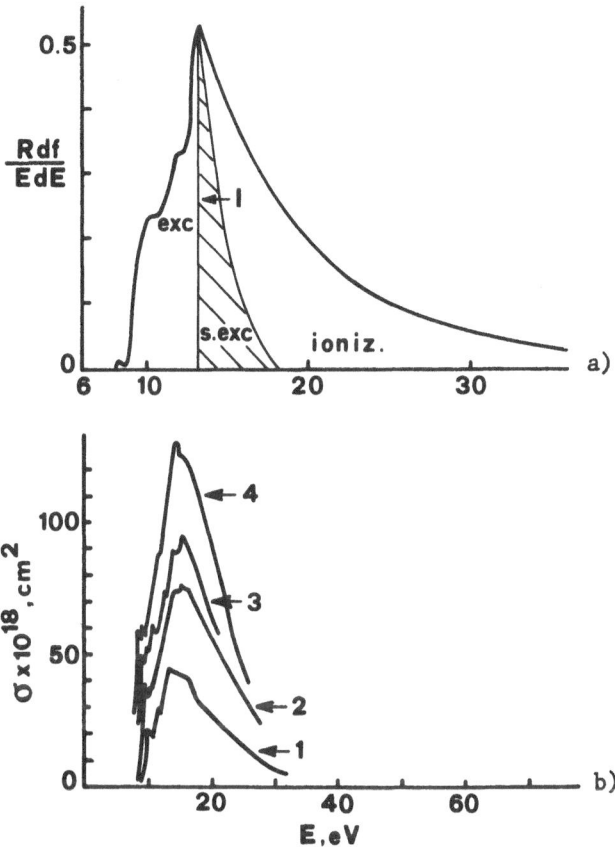

Fig. 2 - a) Excitation spectrum of methane, calculated by R.Platz-
man[2].

b) Absorption spectrum of methane (1), ethane (2), propa-
ne (3) and butane (4), measured with synchrotron
radiation[3].

The total oscillator strength corresponds to the area under the
curve and is, for the spectrum reported, 8.12. Methane has, of
course, a total oscillator strength of 10, but the 1.88 oscillator
strength belongs to K electrons are not included since they absorb
at higher energies than those reported. From the excitation spec-

trum of methane we can calculate the ratio of excitation to
ionization assuming all states below the ionization potential are
excited and all those above are ionized. The ratio so determined is

$$N_{ex}/N_i = 0.27 \tag{18}$$

However using the mean energy per ion pair formed by ionizing
radiation in methane the ratio obtained is

$$N_{ex}/N_i = 0.8 \tag{19}$$

The resolution of this serious disagreement has given interesting
results. When a charged particle interacts with a molecule, there
is an appreciable probability that the latter will be left tempo-
rarily in an excited state (superexcited states[1]) the energy of
which exceeds that required for ionization. The molecule may dissipate its energy by ejecting an electron and becoming ionized,
but the "super-excited states" may find other degradation ways as
for example dissociation into radical fragments. Experimental
evidence has been obtained on the existence of super-excited
states[4] and recently the application of synchrotron radiation to
this problem has been very successful[3]. The first experiment that
proved super-excited state formation in the radiolysis of gaseous
system was that by Jusse[5]. He found that the number of ion pairs
in pure helium or in pure neon is greatly increased by the addition
of a small amount of almost any impurity. Only 0.01 per cent of
the contaminant is required to give a pronounced effect. It has
been shown that the addition ionization arises from reaction of
the type

(1) $He^* + X \rightarrow He + X^+ + e^-$

If the experiments are repeated using Ar it is found that different
impurities give different total ionization. This is due to the
fact that Ar has a much smaller excitation energy of the excited
states and this energy is not much greater than the ionization
potential of the impurities. In this case the energy transfer from
the Ar to the impurity may then produce

(2) $Ar^* + AB \rightarrow Ar + AB^+ + e^-$

or

                    $Ar^* + AB \rightarrow Ar + A^\bullet + B^\bullet$

This behaviour is confirmed by the experiments on deuterated and
nondeuterated compounds. In any experiments, the ionization in the
deuterated compounds was found to be higher[6]. This is due to the
smaller probability of dissociation in deuterated compounds.

    We have seen that the theoretical calculation of the yields of
primary products in a low pressure gas system irradiated with high
energy radiation is possible only using the approximate method
derived from the Bethe formula for the cross-section for excitation
and ionization of molecules by high-energy electrons and normally
referred to as the optical approximation. In condensed media the
solution of the problem is even more complicated. The ionizing
particles interact in a dense medium more or less with the
electromagnetic field of a plasma of molecular electrons not necessarily belonging to a single molecule. Consequently, the resulting
activation may be delocalized, and involves the collective excitation of a few or many molecules and thus the optical spectrum can
be appreciably modified in comparison with that of isolated molecules. It has been shown that the alteration depends upon the
value of the dimensionless parameter

$$u_f = \underline{/}\, (h\,\nu\,p)^2/2RZ\,\underline{/}\,\underline{/}\,(R/E)\,(df/dE)\,\underline{/}\tag{20}$$

where νp is the so called plasma frequency and is the natural
frequency of oscillation that the electrons in matter would have
if they were not bound in atoms. If the parameter u_f is much less
than unity the alteration of the optical spectrum is negligible;
if it is not there is a modification of the spectrum, a shift to
higher frequencies, that increases with u_f. The parameter u_f, apart
from a constant factor $\underline{/}\,(h\,\nu\,p)^2/2\;RZ\,\underline{/}$, which depends upon the
nature of the medium and its density, has the same energy dependen-
ce function as the excitation spectrum. It is then possible to
represent in the same plot the optical spectrum and the parameter
u_f (figure 3). From the figure it can be seen that this parameter
in the case of ethanol is small and the same situation has been
found for many organic compounds, as liquid saturated hydrocarbons.
We may thus conclude that collective effects on the excitation
spectrum, although not negligible, are likely to be small for many
compounds and the calculations of the yield of primary products
for condensed matter may be approximately obtained   using the
optical approximation. Another substantial difference between the
radiolysis of low pressure gas systems and condensed systems is
that the final products may not be the same since radicals formed

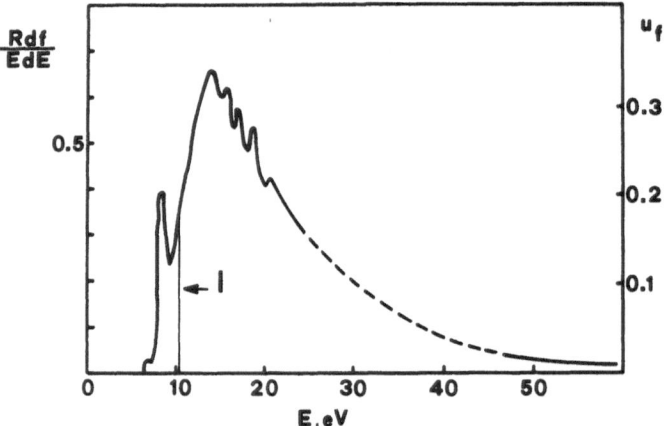

Fig. 3 - Excitation spectrum of ethanol.

by dissociation of excited and superexcited states remain localized
for an appreciable period of time in the cage formed by the surrounding molecules. In this cage they are likely to make many
collisions with the possibility of recombination. Ultimately there
is a very important difference in the fate of ejected electrons
in the gas phase and in the condensed media. In the gas phase the
electron loses its energy until it gets thermalized and then reacts
with a positive ion and yields neutral molecule generally formed
in an excited state that may dissociate or degrade to the ground
state.

The electron may also react with a molecule ánd form a negative
ion which will dissociate or neutralize with a positive ion. In
condensed systems the situation is more complex. We may represent
schematically the path of an energetic electron as in figure 4.

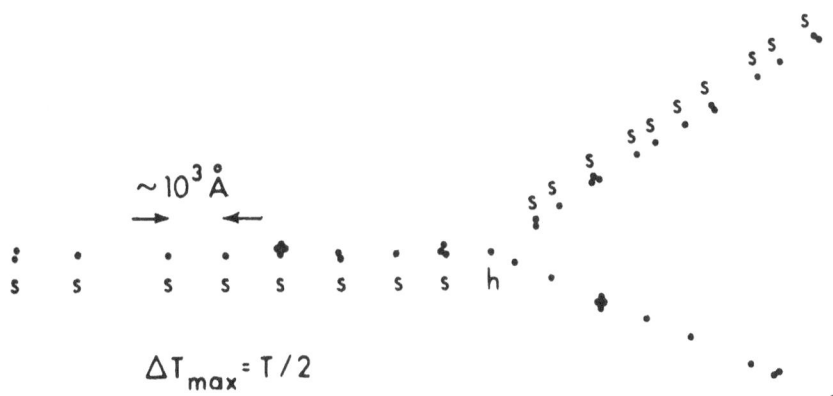

Fig. 4 - Schematic representation of part of the path of a high-
         energy electron through a liquid. Each dot represents
         a collision[7].

The dots represent a collision and correspond to regions of
the system where high concentrations of unstable product are
formed, these region are called "spurs". A schematic representation
of the ejected electron is shown in figure 5.

The circles represent the electron shells of the molecules. As
the electron moves through or near the electron clouds it passes
over a series of potential energy barriers. When the kinetic energy
of the electron is lower than the next potential barrier, it gets
trapped by short range repulsion forces. The energy of an electron
just before it gets trapped is on the average $\sim$ 1 eV and its
velocity $\sim$ 6×10^7 cm/sec. At this velocity an electron takes $\sim 10^{15}$

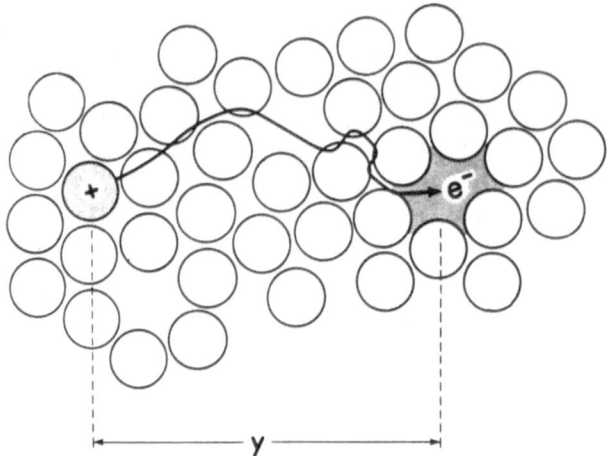

Fig. 5 - Schematic representation of the formation of a solvated
electron-ion pair[8].

sec. to traverse a molecule. Since electronic polarization of
molecules occurs in less than 10^{-15} sec., the electron immediately
finds itself in a electronically polarized hole. The potential well
of the hole increases as the various types of polarization of the
medium occur. The trapped electron is transformed into a solvated
electron over a period equal to the dielectric relaxation time of
the liquid. A solvated electron has definite energy levels and
shows characteristic absorption bands, which may be associated with
an excited level or with the ejection of the electron into the
continuum.

REFERENCES

1) R.L. PLATZMAN, The Vortex 23, 372 (1962).
2) R.L. PLATZMAN, Radiation Research, 1966: Proceedings of the
        Third International Congress of Radiation Research,
        Cortina d'Ampezzo, Italy, Ed. G. Silini (North-Holland
        Publishing Company, Amsterdam, 1967), pg. 20.
3) E.E. KOCH and A. OTTO, Int. J. Radiat. Phys. Chem. 8, 113(1976).

4) J.W. BOAG, Radiation Research, 1966: Proceedings of the Third
   International Congress of Radiation Research, Cortina
   d'Ampezzo, Italy, Ed. G. Silini (North-Holland Publishing
   Company, Amsterdam, 1967), pg. 43.

5) W.P. JESSE and J. SADAUSKIS, "Alpha-Particle Ionization in
   Mixtures of the Noble Gases", Phys. Rev. 88, 417 (1952)
   and Phys. Rev. 100, 1755 (1955).

6) W.P. JESSE and R.L. PLATZMAN, Nature 195, 790 (1962).

7) G.R. FREEMAN, Quaderni dell'Area di Ricerca dell'Emilia Romagna,
   ed. by Badiello and Busi, 2, 55 (1972).

8) G.R. FREEMAN, Quaderni de La Ricerca Scientifica, ed. by Busi,
   67, 41 (1970).

X-RAY RADIOLYSIS OF CONDENSED SYSTEMS: SOLID DNA AND DNA SOLUTIONS

Fabio Busi

Laboratorio di Fotochimica e Radiazioni d'Alta Energia

C.N.R. - Via Castagnoli 1, 40126 Bologna - Italy

We have seen in the previous lectures that the effect of ionizing electromagnetic radiation on a system is, at the first stage, the production of high energy electrons by photoelectric or Compton effect. The fast moving electrons transfer their energy to the molecules of the system in the vicinity of their trajectory by ionization or excitation processes. These primary products, excited molecules, positive ions and ejected electrons undergo secondary reactions which ultimately lead to stable products. The yields of primary ionization and excitation depend on molecular properties that are at the present largely unknown. The most important of these properties is the excitation spectrum, which we have seen is closely related to the absorption spectrum of the molecule extending from the visible region to the X-ray region. The synchrotron radiation has been used for experiments in this field[1,2]. The absorption spectroscopy and the experimental set up for experiment using synchrotron radiation are topics of other lectures so we will simply show, as an example, the result obtained on benzene, in the region up to 35 eV (figure 1)[2]. The upper spectrum is a generalized excitation spectrum with the three regions of excitation, super-excitation and ionization, the bottom one is the absorption spectrum of benzene. This spectrum covers the whole interesting region, the one of superexcited state formation.

Nevertheless from the absorption spectrum only one cannot obtain the yield of the superexcited state formed. An experiment can be easily designed to measure the absorption spectrum and the ioni-

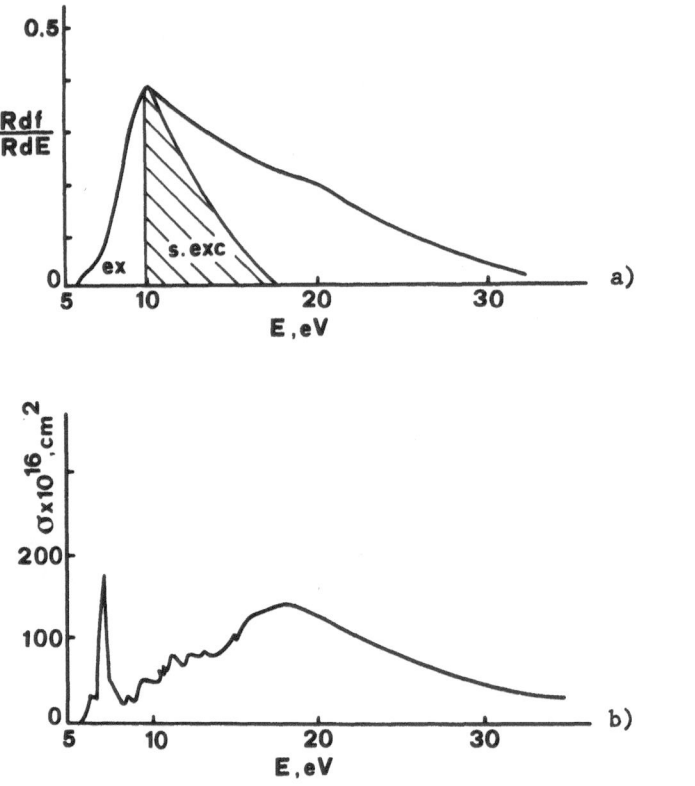

Fig. 1 - a) Excitation spectrum, b) Absorption spectrum of Benzene[2].

zation yield as a function of the photon energy. These two data
make it possible to calculate the yields of primary ionization and
thus the yield of super excitation at least within the limits of
the optical approximation. The more energetic part of the synchrotron radiation, the X ray region, may be utilized for studies of
the chemical or biological changes induced by ionizing radiation.
The complete study of the radiolysis of a system requires the
analysis of the final products and the identification of the
transient species that participate in the reaction mechanism. To

this end the pulse radiolysis technique is employed.

In pulse radiolysis the ionizing radiation is delivered in a pulse whose duration is normally shorter than the lifetime of the intermediates which one wants to investigate. Optical absorption and emission spectroscopy are generally employed as a fast method of characterizing the intermediates. One example of radiolysis equipment is shown in figure 2.

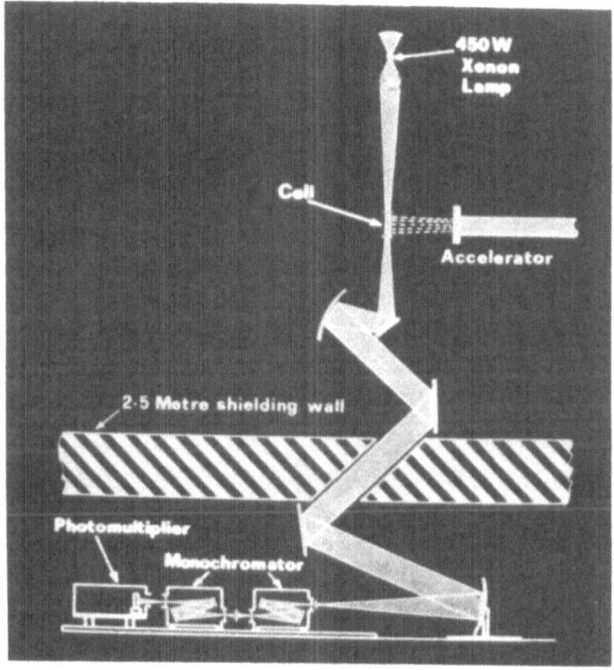

Fig. 2 - Pulse radiolysis equipment at Bologna, Laboratorio FRAE, CNR.

To illustrate the information it is possible to obtain with this technique we will describe one system that has been widely investigated. It is well known that high energy irradiation of DNA causes strand breaks, single or double. We will refer to the total effect of the irradiation as damage on the DNA.

The results obtained from DNA irradiated in the solid state, in spite of the complexity of the molecule and the different form in which it can be prepared, are very reproducible and can be explained on a very simple model.

Immediately after the radiation pulse a sample of DNA at 93°K shows an emission spectrum which consists of a fluorescence peak at 330 nm and a phosphorescence peak at 480 nm[4], figure 3.

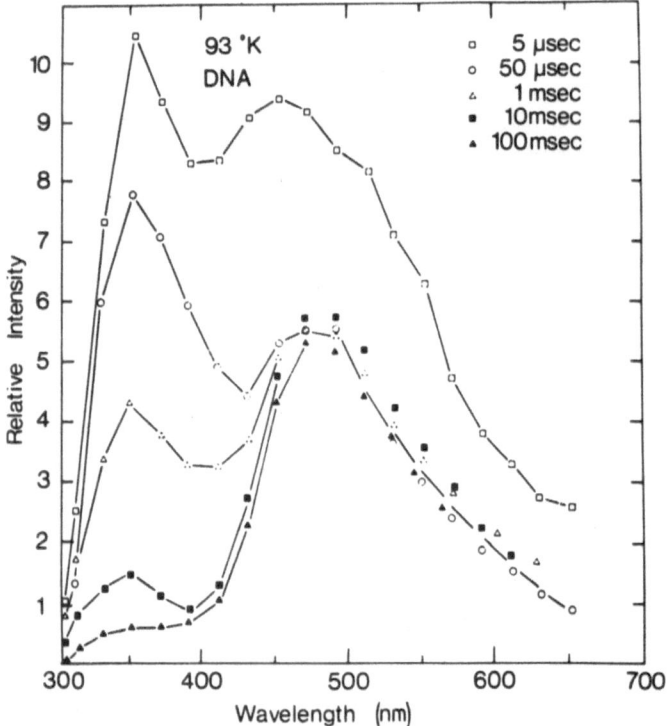

Fig. 3 - Emission spectra taken at 5 usecs, 50 μsecs, 1 msec., 10 msec., and 100 msec following pulse irradiation of DNA at 93°K.

Experiments on the emission spectra obtained from the irradiation of the four separate bases, adenine, guanine, thymine and cytosine have not given sufficient information for the identifications of

the species responsible for the emission from DNA itself. Adenine
and thymine present emission spectra similar to that of DNA.
Excitation ultraviolet experiments have produced evidence that the
emission from irradiated DNA occurs from the thymine bases only.
The analysis of the decay of luminescence at a specific wavelength,
and at different temperatures gives a possible mechanism for the
luminescence in irradiated DNA. From the decay curves of emission
intensities, at a specific wavelength, values of the intensity at
selected times were measured for a large range of temperatures,
and these data were then plotted for each selected time with
temperature as the abscissa, figure 4. This plot shows that above
170°K the curves exhibit maxima which occur at lower temperature

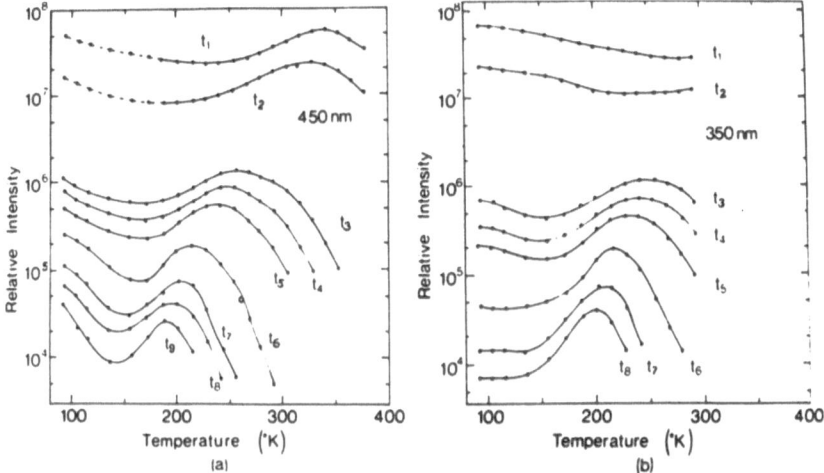

Fig. 4 — Derived plots of temperature dependence of the luminescence
         intensity at various times after irradiation. The data on
         this plot are taken from the decay curves measured at
         various temperatures between 93°K and 380°K. t_1 - t_9 are
         15 µsec., 40 µsec, 1msec and 2 msec, 4 msec, 15 msec ,
         50 msec, 100 msec and 150 msec respectively.
         a   recorder at 450 nm.
         b   recorded at 350 nm.

with increase in time after irradiation. This is the behaviour of
a temperature controlled release of electrons from traps leading
to luminescence. When an ionization event occurs in the molecule,
charge recombination yields excited molecules which give the
emission. If in high energy irradiation, where the ionization

events may occur in any site of the molecule, the emission occurs
only from the thymine, the energy deposited at random throughout
the molecule gets ultimately localised on the thymine bases before
emission occurs. Experiments on DNA in which 5-bromouracil is
incorporated show a dramatic effect on the luminescence of DNA[3];
the yield of luminescence increases and the emission spectrum, as
the 5-bromouracil incorporated increases, approaches that observed
with pure 5-bromodeoxyuridine. The effect may be interpreted in
terms of the transfer of energy, which is initially deposited
randomly on a DNA molecule, to a specific site, i.e. the 5-bromo-
deoxyuridine bases and the enhanced emission indicates that a
5-bromodeoxyuridine is more efficient at localizing excitation
energy than the thymine base. The proposed mechanism is based on
DNA excited states formed through an ionic mechanism and does not
include the possible contribution of DNA excited states formed by
direct action of high-energy radiation. For a complete description
of the luminescence mechanism of irradiated DNA it is necessary
to investigate the irradiation in the energy range from VUV to low
energy X-ray, which synchrotron radiation now makes possible.

The amount of energy emitted from the DNA is only a small fraction of the energy absorbed. This is due to the competition between
charge recombination and separation.

When electrons escape from the positive charge, the positive ion
decomposes and free radical centers will be established in the
molecule. This has been proved by EPR experiments on irradiated
DNA. The EPR spectra of irradiated solid DNA containing a small
amount of methylnaphthaquinone show that the radical yield of the
DNA mixture was greater than that found for pure DNA. Methylnaphthaquinone can oxidize rapidly and quantitatively the bases of DNA.
This has been demonstrated by pulse radiolysis experiments in
solution. Therefore charge transfer from the ionized molecule to
the oxidizing agent reduces the probability of recombination and
thus increases the number of positive ions which decompose forming
radicals. The presence of these compounds increases the yield of
the damage on DNA. This indicates that the yield of radical formed
is proportional to the damage on DNA. This can be easily confirmed
by experiments in aqueous solutions.

If one irradiates an aqueous solution containing DNA in the presence or absence of cysteamine (RSH) it is found that cysteamine
has a protective action on DNA. This action is due to the ability
of cysteamine to repair the radical damage by transfer of a hydrogen atom from the SH group. This reaction can be directly demon-

strated using pulse radiolysis. The radical RS formed from cysteamine in the hydrogen transfer reaction

$$X^{\cdot} + RSH \rightarrow RS^{\cdot} + XH$$

combines with another molecule of cysteamine and produces a radical

$$RS^{\cdot} + RS^{-} \rightarrow RSSR^{-}$$

which has an intense absorption band in the visible[5], figure 5.

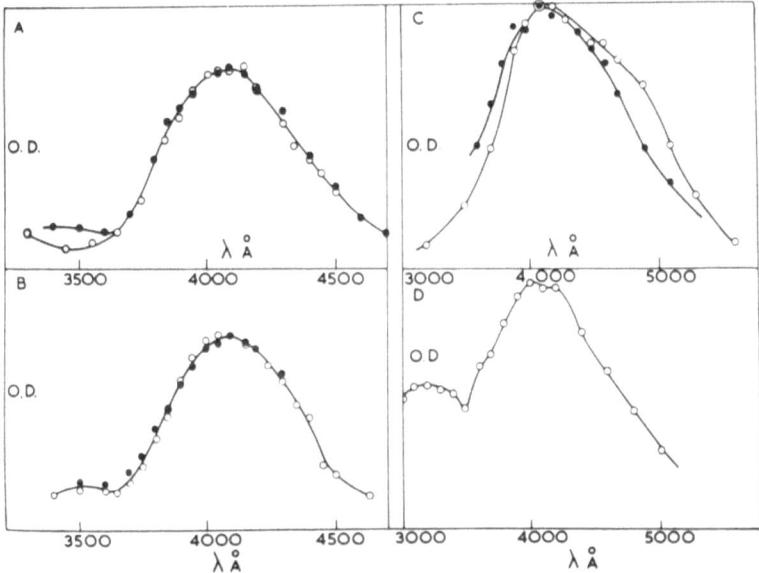

Fig. 5 - Transient absorption spectra in disulphides and SH com-
         pounds.
         A) ○-Cysteamine hydrochloride (pH8); ●-Cystamina (pH8).
         B) ○-Cysteine hydrochloride (pH8); ●-Cystine (pH8).
         C) ○-Glutathione (oxidized) (pH6.6); ●-Glutathione (re-
            duced) (pH8.4).
         D) Lysozyme, 1.5×10^{-4} M.

The radiation induced damage on DNA in solution may be due to
secondary reaction of the DNA with transient species formed by

energy absorbed by the solvent or to direct action of the incident radiation. The determination of the fraction and the effect of energy absorbed by the solute in high concentrated solutions of complex molecules is still an unsolved problem connected with the determination of the degradation spectra of low energy electrons.

REFERENCES

1) M.B. ROBIN, Higher Electronic States of Polyatomic Molecules, Academic Press, New York 1974, Vol. 1, 1975, Vol. 2.
2) E.E. KOCH and A. OTTO, Int. J. Radiat. Phys. Chem. 8, 113(1976).
3) E.M. FIELDEN, Quaderni de La Ricerca Scientifica, 68, 63 (1970).
4) C. NICOLAU, O. KORNER and A. CRISTEA, Studia. Biophys.,Berlin, 1, 59 (1966).
5) G.E. ADAMS, Quaderni de La Ricerca Scientifica, 68, 27 (1970).

UV PHOTOBIOLOGY: DNA DAMAGE AND REPAIR*

Betsy M. Sutherland

Biology Department
Brookhaven National Laboratory
Upton, New York 11973

INTRODUCTION

Photobiology deals with the interaction of light and living systems. However, we will talk about only one kind of photobiology: the interaction of light in the wavelength range 200-300 nm (ultraviolet light) and its interaction with biological systems, in specific, living cells (1).

Why ultraviolet light? The first and most important reason is that this wavelength range is strongly absorbed by many biologically-important molecules.

Why biological molecules? Since we want to deal with photobiology, we must work in the range of doses in which the cell is still alive. This poses both a problem of measuring photoproducts in small numbers and provides a challenge to us to develop methods which enable us to do this.

When ultraviolet light interacts with a cell, what are the important targets which determine the fate of the cell? Cells are mostly water by weight. However, the important molecules in the cell which absorb in the ultraviolet range are a) proteins, be they enzymes or structural proteins - perhaps membrane components, and b) nucleic acids, the DNA which contains the genetic information and the RNA's which are structural molecules, mediators in

*Research supported by U. S. Department of Energy, Research Cancer Development Award CA00466, and the National Cancer Institute Grant CA23096.

263

protein synthesis and also take part in the transmission of
genetic information.

How could we decide which of these targets is a critical one
for the cell? A popular method of doing this is action spectroscopy, measurement of the efficiency of different wavelengths in
producing a given biological effect. If the action spectrum is
straightforward, and many of them are, one can match the action
spectrum for the given biological effect with the absorption
spectrum of different biological macromolecules to get an idea of
which molecule is doing the absorption, and thus which is the
chromophore or "target" for the resulting biological damage. As
an example, in the 1930's action spectroscopy provided important
information that DNA was the target for the action ultraviolet
light in the case of the killing of bacteria. Very recently
Rothman and Setlow have shown that DNA is the important target
for the killing of mammalian cells by ultraviolet light.

What are the biological effects of damage to DNA? In the
first place, as we have said, a cell can die. The second
possibility is: the cell does not die, but is affected by mutation,
a change in the DNA leading to altered cell properties. The third
possibility is: no effect on the cell. Now how can this happen?
Let us consider the biological properties of DNA. It contains
information in the genetic code which is read in groups of three
called triplets. Each triplet specifies a tRNA intermediate,
which in turn specifies an amino acid which is assembled into a
growing protein chain. Since there are four bases and $4^3 = 64$,
there are 64 possible triplets vs. about 20 amino acids. The
cell's solution to this is degeneracy (more than one triplet
specifying the same amino acid). So even if a triplet, for
example UUU, is mutated to UUC, both triplets specify the amino
acid serine and thus the protein coded for by that DNA segment
will not be changed. Second, related triplets frequently code
for amino acids with similar size and charge. So even if the DNA
triplet GUA were mutated to GCA resulting in substitution of the
amino acid alanine for valine, the protein would likely show
little structural or functional differences since alanine and
valine are closely related amino acids. The third possibility
is an alteration of a nonessential part of the protein. Most
proteins are present in excess anyway, so the cell can limp along
even with a true change in its DNA and in the protein for which
that DNA coded.

How does one measure a mutation? Bacteria are frequently
used and we will talk about them here in some detail as an example.
Many bacteria can be grown on defined chemical media consisting
only of salts, minerals, and a carbon source. Upon mutation to
auxotrophy, cells must be supplemented with an exogenous material,

for example, amino acids, vitamins, a nucleic acid base and so
forth. In the case of bacterial viruses, mutations can be
detected by their lack of growth on certain host, their slow
growth or the appearance of the plaques which they make on a plate
of bacteria. If the cell does not grow on a defined medium, the
problem of finding mutations is somewhat harder. In mammalian
cells, for example, in which growth on defined media is rather
difficult to achieve, it is just in the past few years that
systems have been developed for examining mutants. Here again one
generally screens for the lack of some metabolic enzyme by
specially developed selection procedures.

Chemical Damage to DNA

     We have discussed the biological consequences of the detri-
mental effects of UV on cells. What is some of the chemistry
leading to these biological changes? When ultraviolet light
impinges upon DNA, although there are many photoproducts formed,
one of these has been implicated in much of the resulting
biological damage (2). Among the multiplicity of products formed
in DNA, are pyrimidine dimers (formed between two adjacent
pyrimidines on the same DNA strand), hydrates of pyrimidines,
adducts within a DNA molecule (mainly pyrimidine to pyrimidine),
DNA-DNA cross-links, and a product for mainly in dry and frozen
DNA originally termed the "spore photoproduct." Of these, the
dimer has been shown to be largely responsible for death and
mutation in the bacteria and in simple eucaryotes, and now
evidence is accumulating that the same is true in mammalian cells.

     What happens to a cell which receives damage to its DNA? Is
the cell fated to death or mutation? Evidence now indicates this
is not the case. Cells can repair damage to their DNA. There
are at least three major modes of repair of DNA in biological
systems. You might just imagine the possibilities. First is the
reversal of the damage; the second mode is the cutting out of
damage; and the third mode is one in which the cell procrastinates
and tries to ignore the damage. We will talk about these mechanisms in more detail in this and in subsequent lectures. Much of
what we know comes from E. coli. We will discuss not only what we
know from bacterial cells, but also what we know and perhaps what
we wish to extrapolate to higher cells.

     What are the consequences of failure of repair? In bacterial
cells, as we have said, cells can be mutated or die but what about
people? Are there any implications that failure of DNA repair in
people can lead to harmful effects? There are now a number of
human diseases which have been implicated as resulting from lack

of DNA repair: one of these is xeroderma pigmentosum, which is
the propensity to sunlight-induced skin cancer (3). Most of
these individuals die before the age of about 20 from invasive
tumors. Other such diseases are Fanconi's anemia and progeria.

PHOTOREACTIVATION

We have noted that there are three kinds of DNA repair. We
will discuss the one which, at least in concept, is the simplest -
simple reversion of the lesion called photoreactivation (4,5).

Photoreactivation was first found in 1949 by Dulbecco and by
Kelner. They showed that UV-induced killing could be reduced by
exposure to visible light after the ultraviolet. A decade later
Rupert, in both the E. coli and in yeast, found that the active
agent was a photoreactivating enzyme (PRE) and the substrate was
ultraviolet-irradiated DNA. Rupert showed that the steps in
this reaction were several: first, complex formation between PRE
and UV'd DNA which did not require light, and then a repair step
that was dependent on the presence of light. (Note that this
reaction, unlike some which are known in photosynthesis, is not the
activation of an enzyme by the light and in its subsequent action
in an active state on a substrate. This enzyme must be complexed
to the substrate when the light energy is supplied.) The Setlows
showed that pyrimidine dimers in DNA were the photoreactivable
lesion. They also provided evidence that the pyrimidine dimer was
the only substrate. They showed that the product of the action
of the photoreactivating enzyme on DNA were pyrimidine monomers.
Photoreactivating enzymes are interesting not only because they
catalyze a biological important repair step, but as unique proteins
which require light for photolysis (6). Photoreactivating enzymes
have been studied from a variety of procaryotic and eucaryotic
sources. Most of photoreactivating enzymes have the following
functions in common: first they have all been shown to be proteins,
they require lignt for their function, they monomerize pyrimidine
dimers, they have a molecular weight in the range of 35-40,000 and
most are active in the wavelength range of 300-500 nm. It is the
last of these properties which provides both interest and also a
useful analytical tool.

The peaks of the action spectra of various enzymes range
from about 360 nm to 380, 405 to even 435 nm, for the enzyme from
Streptomyces griseus. The short-wavelength extent of the action
spectrum is somewhat limited by the wavelengths at which one
starts making dimers as well as photoreactivating them, that is
about 300 nm. The long wavelength extent varies with the enzyme.

Many photoreactivating enzymes do not use light wavelength longer than about 500 nm. However, we have found that the human enzyme can use light which is between 500-600 nm and Giese has also found that the photoreactivating enzyme from Blepharisma can use light in this wavelength range.

One of the primary laws of photochemistry states that photochemical action occurs only when there is light absorption, so we would expect photoreactivating enzymes which act in the region 300 to say 500 or 600 nm to absorb in this region. Eker has found that the enzyme from Streptomyces griseus seems to contain an intrinsic chromophore which absorbs in the region where the Streptomyces enzyme is active. However, other photoreactivating enzymes, that from E. coli, that from the silverfish Thermobia domestica and perhaps the one from yeast, do not show any visible absorption in the region of the photochemical action. The solution to this dilemma was found by J. C. Sutherland and his postdoctoral fellow K. L. Wun (7). They found that the complex of E. coli with UV-irradiated DNA generated a new absorption which agreed in magnitude (molar extinction coefficient) and wavelength range with that predicted by the action spectrum of the E. coli enzyme acting in purified form in vitro or acting in the E. coli cell (in vivo). The new absorption was dependent on the presence of dimers in the DNA, and on complex formation, as it did not appear when PRE was added to irradiated DNA in the presence of so much salt that no complex was formed. The absorption also disappeared the same kinetics as dimers were monomerized.

For many years it was thought that photoreactivating enzyme was absent in mammalian cells, even though there were a few reports of biological photoreactivation of tumor formation in mice and of human erythema. In 1974 a photoreactivation enzyme was first isolated from human leukocytes (6,8,9). Since then Dr. Helga Harm has shown photoreactivating activity from rabbit, cow, cat and human cells. Photoreactivating enzyme from human cells has been characterized as a protein of a molecular weight of about 40,000 with action spectrum extending from about 300 to at least 600 nm.

If photoreactivating enzyme is present in these cells, do they use the enzyme to photoreactivate dimer lesions in their own DNA, or perhaps does the enzyme not have access to the DNA? A case of particular interest is that of human skin. Sunlight forms pyrimidine dimers in the DNA of human skin cells. If we wish to measure photoreactivation of dimers in skin, we would be faced with a problem: most measurements of repair, dimer excision, photoreactivation or whatever, depend on the use of radioactive DNA. In order to have enough radioactivity to test, one would have to have a very radioactive human! A solution to this problem has been worked out in the laboratory of R. B. Setlow, using the technique developed in the laboratory of F. W. Studier at Brookhaven. The technique is electrophoresis in alkaline agarose. DNA extracted from cells is treated with an enzyme which specifically makes a nick beside every pyrimidine dimer. The DNA is then denatured and electrophoresed on the alkaline gels. DNA which contains dimers and therefore is nicked by the "UV-specific endonuclease" will be smaller and therefore migrate further into the gel. The absent of dimers, however, will be seen by the lack of nicking by the enzyme and by the presence of larger molecular weight DNA. Using such a technique we have been able to show that leukocytes (white blood cells) taken directly from humans do have dimers formed in their DNA and they do photoreactivate these dimers in a reasonably short exposure (about 20 to 40 minutes) to visible light.

What is the role of pyrimidine dimers in the induction of skin cancer by UV? We have developed a model system in our laboratory which we believe will allow us to examine the role of photoreactivation and of DNA repair in the prevention of the induction of skin cancer by ultraviolet light. This technique involves a conversion of normal human cells by ultraviolet light to cells which are "transformed" that is, with many properties of cancerous cells. We find that ultraviolet light, at rather small doses, produces this conversion of normal cells to transformed cells. Furthermore, an exposure of the cells to photoreactivating light immediately after the UV decreases the rate of conversion of the cells from normal to transformed (10).

Although much progress has been made in understanding the photobiology and photobiochemistry of photoreactivation, many outstanding problems remain. The first is the structural similarities and differences in photoreactivating enzymes. Do they reflect evolutionary similarities, divergence, or do they reflect perhaps true differences in function? For example, does the enzyme from Streptomyces, which seems to have its own intrinsic chromophore, have a different method of function from the enzymes from E. coli and yeast, which do not contain intrinsic chromophores?

Finally, we can ask some questions about photoreactivation in vivo. Since the reactivation seems to be specific for pyrimidine dimers, we can use its action as an analytical tool; that is, if a reaction is reversible by a true photoenzymatic reaction, the chemical lesion which leads to the biological damage

is likely to have been a pyrimidine dimer. Finally, we can
explore the role of the photoreactivating enzyme in repair in
human cells, determine the biological consequences of the failure
of the enzyme to repair DNA.

REFERENCES

1. Jagger, J., Introduction to Research in Ultraviolet Photo-
    biology, Prentice-Hall, Englewood Cliffs, New Jersey, 1967
2. Setlow, R. B., Science 153, 379-386 (1966).
3. Robbins, J. H., Karemer, K. H., Lutyner, M. A., Festoff, B. W.
    and Coon, H. G., Ann. Int. Med. 80, 221-248 (1974).
4. Symposium on Molecular Mechanisms in Photoreactivation,
    Photochem. and Photobio. 25 (1975).
5. Setlow, J. K., Curr. Top. Radiat. Res. 2, 195-248 (1966).
6. Sutherland, B. M., Int. J. Cytol., in press (1978).
7, Wun, K. L., Gih, A., and Sutherland, J. C., Biochemistry 16,
    921-924 (1977).
8. Sutherland, B. M., Nature 248, 109-112.
9. Sutherland, B. M., Runge, P. and Sutherland, J. C., Biochem.
    13, 4710-4715 (1974).
10. Sutherland, B. M., Proceedings of the International Conference
    on Solar Carcinogenesis, Airlie House, in press (1978).

UV PHOTOBIOLOGY: POSTREPLICATION REPAIR

Betsy M. Sutherland

Biology Department
Brookhaven National Laboratory
Upton, New York 11973

We have discussed two repair modes: 1) photoreactivation,
in which dimers are reversed in DNA; and 2) excision repair, in
which lesions are removed from the DNA. What happens if it is
time for the cell to synthesize a whole new copy of DNA before
the lesions are reversed or removed? One might image a number
of scenarios. 1) The DNA repair enzymes may stop at the damage
and never start again. 2) The DNA repair enzymes might go on
inserting a few bases, not necessarily the right ones, and then
copy as usual. 3) Finally, the DNA repair enzymes might fall
off the DNA and resume synthesis further downstream.

What does happen in biological systems? There is evidence
that our second possibility, insertion of a few bases at random,
can happen in vitro. This might also happen in vivo. There is
also good evidence from a number of systems, both bacterial and
mammalian, that the third possibility can happen in cells; that
is, that the DNA repair enzymes fall off the DNA, restart later
downstream and leave a gap.

The evidence for the second possibility, the insertion of
incorrect bases is as follows. Setlow and Bollum tested the
ability of DNA polymerase to copy ultraviolet irradiated DNA
(1). They found that the product DNA had a decreased number of
A's incorporated into the new DNA. Thus the original parental
strand of DNA which contained thymine dimers could not code

*Research supported by U. S. Department of Energy, Research
Cancer Development Award CA00466, and the National Cancer
Institute Grant CA23096.

correctly for the AA sequence which should have been inserted
opposite. They also found that there was a decrease in the
number of AA nearest neighbor sequences, thus implying that the
incorporation by the DNA polymerase across from a pyrimidine
dimer was inaccurate. They also found that there was a decrease
in the total amount of synthesis as if the enzyme got to the
damaged region and spent much longer than usual than "deciding"
what to do before it could proceed down the DNA.

From E. coli comes major evidence that the third possibility
can happen; that is, that the replicating enzymes might fall off
the DNA and restart leaving a gap. Much of this evidence comes
from the laboratory of Howard-Flanders and his colleagues (2).
They found first that gaps were made in DNA synthesized just
after UV. This was done by taking an excision-defective E. coli
which was prelabeled with ^{14}C thymidine. After ultraviolet
irradiation the cells were exposed to ^{3}H thymidine for 60 minutes
and then sedimented in alkali to separate the DNA strands. They
found that the newly synthesized DNA resulting from the UV
irradiated cells had a smaller molecular weight and that from
un-UV'd cells. They also found distance between the gaps was
approximately equal to the distance between dimers in the DNA.
(They knew the average molecular weight of the pieces, the
UV exposure and thus the number of dimers in the genome.) They
also have shown that pyrimidine dimers are opposite gaps in the
newly synthesized DNA. To do this, they used bacterial episomes,
small pieces of DNA which can exist in the bacterial cells
independent of the cellular chromosomes. They UV'd a bacterial
cell which was excision⁻ and which contained an episome with
a genetic marker. They transferred the episome to excision-
proficient or excision-deficient hosts, and found that the
capacity of the recipient for excision repair made no difference.
This implies that the structure which was being transferred into
the cells was not able to be repaired by excision repair. However,
when the episome was transferred to a cell and then photoreactivated (and the control kept in the dark), it was found that
the photoreactivated samples gave much more biological activity
as recognized by the presence of the marker of the episome. This
implies the structure which was being transferred was susceptible
to photoreactivation. As you can tell, a dimer in a DNA opposite
a gap can be acted on by photoreactivating enzyme since no
incision and thus breakage of the DNA would be involved. Finally,
they have estimated the size of the gaps by chromatography to be
about 1,000 nucleotides.

What happens after the gap is created in the cell?
Howard-Flanders has shown in bacterial cells there is slow gap
filling after ultraviolet irradiation. If the E. coli, which

we used in the experiment just discussed above (that is, excision minus, ^{14}C thymine labeled, then ultraviolet-irradiated, allowed to incorporate ^{3}H thymidine and then sedimented in alkali) had been allowed to grow longer and had been sampled at different times of incubation after the UV, it would be found that immediately after the UV molecular weight of the DNA is large. (Remember that these bacteria are excision deficient.) After a few minutes the molecular weight is smaller; that is, during DNA synthesis the synthesis on the damaged template gives rise to DNA of lower molecular weight than if the synthesis were occurring on undamaged template. After a few hours, the molecular weight of the DNA became large again implying that somehow the gaps were filled. The exact mechanism by which the gap filling occurs is still not well understood. It is thought that somehow the E. coli can use information from the multiple copies of the bacterial chromosome present in the cell to provide the correct information for synthesis of the DNA. Again we should point out that there are no ideas as to exactly which enzymes can participate in these repair processes, even though E. coli would seem the ideal system since the genetics are reasonably well worked out.

What happens in mammalian cells? The problems of examining postreplication repair in mammalian cells are many. First the molecular weight of the intact DNA is so large that the sedimentation of molecules is not as independent DNA molecules but aggregated, tangled masses. Thus in order to be able to examine the DNA at all, it must be fragmented into molecular weights less than about 5×10^{8}. Using this procedure, Alan Lehmann in England has found that there are gaps in newly replicated material formed after UV (3). The distance between the gaps is approximately equal to the distance between dimers. In marsupial cells it has been shown that after UV the newly synthesized DNA is smaller than the DNA synthesized on undamaged templates. If the cells are UV'd and then photoreactivated, the DNA is larger. This would imply that the gaps which made the DNA smaller were indeed opposite dimers. There is also gap filling after ultraviolet irradiation. However, there is disagreement about the size of the DNA made after ultraviolet irradiation, the interpretation of size changes, and even whether postreplication repair is an independent repair process were some aberrant excision.

XP cells are also deficient in postreplication repair. Lehmann has shown that if cells are exposed to ultraviolet light and then grown, the molecular weight of the DNA can be followed by fragmenting the DNA into pieces at least smaller than 5×10^{8}.

After UV-irradiation, normal cells show a reduction in the
molecular weight of the newly synthesized DNA, then rapidly
convert this DNA to the 5×10^8 maximum. However, XP cells reach
this level much more slowly than do normal cells. The striking
thing, however, about Lehmann's data is that the slopes of the
lines (log molecular weight vs. time), are the same for XP and
normal cells. However, XP cells begin the DNA synthesis period
at a much smaller molecular weight than do the normal. This
implies that perhaps the defect in the XP's is too much nicking
of the DNA rather than not enough gap filling.

Another group of patients is diagnosed by clinical signs as
xeroderma pigmentosum; however, cells from these XP "variants"
excise dimers as well as do cells from normal individuals. They
also have virtually as much unscheduled DNA synthesis as do
normal cells. Lehmann finds that the XP variants are even slower
than XP's to reach the limit molecular weight, again because the
molecular weight is initially smaller but the rates of increase
are the same as normal. XP variants are also deficient in photoreactivation. It is not yet clear exactly what is the molecular
cause of the XP in these cells which are normal with regard to
excision.

MAJOR PROBLEMS IN DNA REPAIR:
PROSPECT OF SYNCHROTRON RADIATION STUDIES

During these lectures, I have pointed out a number of
important areas in DNA repair which must be solved before we can
correlate DNA repair deficiencies and human diseases. For example,
what is the molecular cause of XP? Another important problem
is the target for the induction of human skin cancer by sunlight.
What is the molecular lesion which leads to skin cancer?

Synchrotron radiation offers advantages for all of these
studies. For example, it can provide important information on
the molecular target for skin cancer induction. One way for
studying this is by determining the action spectrum. Are the
wavelengths which inactivate DNA more important or are the
wavelengths which inactivate protein more important? These
action spectra could be performed on experimental animals or on
cultured cells by looking at UV transformation.

Examining the molecular lesion would be more difficult.
However, we can examine the role of pyrimidine dimers in
producing these effects if we can show that a biological damage
caused by ultraviolet light can be reversed by longer wavelength

light in a photoenzymatic reversal. Then we would suspect that dimers were an important lesion in the production of skin cancer. However, the proper controls must be done to show that photoreactivation and not a nonspecific light effect (for example, on cell growth), was reversing the lesions. Again here action spectroscopy is crucial. If we could show that the action spectrum of the light reversal event was the same as that for the action spectrum of the purified photoreactivating enzyme, we would have good evidence that the light mediated amelioration of damage was due to true photoreactivation.

Synchrotron offers two major advantages for such biological studies. The first is light intensity. Most biological studies have been limited to mercury lines in the ultraviolet, In most sources, for adequate wavelength purity, the irradiation times have been many minutes to hours. During these irradiation times, the cells' metabolic processes are proceeding; these may complicate the experiment or even lead to possible errors. The second is the advantage of a continuum. Most ultraviolet action spectra have been limited to the mercury lines. In many cases there are not enough for good resolution of action spectrum. (However, we should notice that the absorption spectrum of biologically important molecules; that is, molecules essentially in a water solution, is very broad so it would not be fruitful to do action spectra, say at 1 Å spacing.)

In addition to discussing the advantages that the synchrotron now offer to biological studies, we must consider some possible problems. The first is the timing of the synchrotron beam. The pulsed photons from the synchrotron may give entirely different photochemistry, and this must be examined with regard to the possibility of multi-photon effects. This might lead to entirely different photobiology - which, while of possible interest in itself, might tell us nothing about cells and their responses to solar insults. Finally, just as in any biological studies, we must have careful and informed biology, as well as sophisticated light sources, so they both can be used to the maximum advantage.

REFERENCES

1. Setlow, R. B., J. Cell, Comp. Physiol. <u>64</u>(1), 51-68 (1967).
2. Howard-Flanders, P., Rupp, W. D., Wilkins, B. M., and Cole, R. S., Cold Spring Harbor Symposia on Quantitative Biology <u>33</u>, 195-205 (1968).
3. Lehmann, A. R., Life Sci. <u>15</u>, 2005-2015 (1975).
4. Lehmann, A. R., Kirk-Bell, S., Arlett, C. F. Paterson, M. C., Lohmann, P. H. M., de Weerd-Kastelein, E. A., and Bootsma, D., Proc. Nat. Acad. Sci. USA <u>72</u>, 219-223.

UV PHOTOBIOLOGY: EXCISION REPAIR*

Betsy M. Sutherland
Biology Department
Brookhaven National Laboratory
Upton, New York   11973

In addition to photoreactivation, the reversal of damage
produced in DNA by ultraviolet light, there is another repair
process, excision, in which the damaged regions are cut out of
the DNA. Although we will deal mostly with damage inflicted on
DNA by ultraviolet light, there are also excision systems for
chemical damage as well. Most of what we know about excision
comes from E. coli, but we can extrapolate to mammalian systems;
how well the extrapolation fits the actual case is not yet clear.

There are two and perhaps three types of excision repair.
The first is nucleotide excision in which a few to many bases,
including the damaged area, are removed. There is base excision
in which only the damaged base is severed from the DNA backbone,
then nucleotide excision proceeds just as usual. Finally, in a
very recent study, there is a hint that damage may be removed
directly from a base leaving the DNA intact without any need
for incision into the DNA backbone or any new synthesis. (Note
that both nucleotide excision and base excision do require new
synthesis into the DNA.)

I. Nucleotide Excision

We are gradually developing a good idea of the major steps
in nucleotide excision. First is the recognition of damage by

---

*Research supported by U. S. Department of Energy, Research
Cancer Development Award CA00446, and the National Cancer
Institute Grant CA23096

an enzyme called a UV-endonuclease. The next is incision by the
nuclease into the phosphodiester backbone, generally at a site
near the damage of damaged base. Then there is polymerization
by covalent extention of the damaged strand using the complementary undamaged strand as template. Virtually simultaneously
with polymerization there is excision of the damaged region.
Finally, there is ligation of the newly synthesized region to
the parental repaired strand.

    Even in E. coli exactly which enzyme participates in which
step is still largely unknown. For the endonuclease step there
have been several reports of isolation and characterization of
enzymes which seem to have had some properties expected of the
enzyme carrying out the incision step. However, none of these
enzymes has turned out to have definitive evidence in its favor.
DNA polymerase I seems to participate in both polymerization
and excision although it is possible there are other enzymes
in the cell which can serve as backup enzymes should polymerase I
be defective. Finally, DNA ligase is responsible for joining
the newly-synthesized strand to the parental repaired strand.

    An important tool in studying any multi-step pathway is
complementation. First let us see how we could use the complementation to study DNA repair in E. coli and then later we will
see how this can be used in mammalian cells. We know that in
excision repair, damaged DNA is acted upon by a endonuclease, then
by a polymerase and finally by a ligase. If we had a cell which
was a mutant in the endonuclease, the repair pathway could not
proceed. If we had another cell which was defective in the
polymerase, the repair pathway still could not proceed. However,
if we could combine the properties of these cells, we would
have at least one good copy of the endonuclease, the polymerase,
and the ligase, and thus the two mutants could be said to
complement each other. This complementation approach has been
used both in vivo and in vitro. Most recently it has been used
by Erling Seeberg to isolate a UV-endonuclease from E. coli.
We shall see later in this lecture how this approach can be used
also in mammalian cells.

    What happens in higher organisms with regard to nucleotide
excision repair? They do seem to remove damage from their DNA
by cutting out; however, the number of enzymes, their identity
and the possible alternative pathways, are not yet known. The
bacterial model provides just that: a good working model,
although not necessarily reality.

    How would we measure excision in higher organisms? First,
we could look for a damaged piece which would be cut out and
might appear in the smaller molecular weight portions of DNA.

We could look for the insertion of new bases. Two major
procedures have been used to look for the insertion of new bases
into DNA. The first of these is called unscheduled DNA synthesis.
Ordinarily cells synthesize new DNA only during a limited portion
of the cell cycle called S or synthesis. However, if the cells
have been exposed to UV, one gets DNA synthesis even though the
cell is not in the normally scheduled period of the cell cycle
for synthesis. Thus this incorporation of new DNA is called
unscheduled DNA synthesis. It seems to represent the insertion
of the new bases corresponding to the portion of the DNA which
were replaced due to the removal of damaged bases.

The second method of looking for the insertion of new DNA
bases is by bromodeoxyuridine photolysis. In this method cells
which are undergoing repair synthesis are supplied with bromodeoxyuridine (which is an analog of thymidine). It is
incorporated into the DNA; when the cells are exposed to long-
wavelength UV (e.g. 313 nm) the bromodeoxyuridine absorbs the
light producing free radicals which make breaks in the DNA. So
whenever there is new synthesis in the presence of bromodeoxyuridine, by the long wavelength photolysis one can induce DNA
breakage and thus cells which have undergone DNA repair can be
recognized by smaller size of the DNA.

Nucleotide excision repair provided the first correlation
of DNA repair defect and possible human disease. Cleaver first
noticed that xeroderma pigmentosum cells underwent unscheduled
DNA synthesis at a lower level than did normal cells (1). These
cells also show decreased repair synthesis as detected by bromodeoxyuridine photolysis.

Complementation analysis has also been carried out on
xeroderma cells in culture. Cells from two individuals are
fused using heat inactivated sendai virus. The fused product
of the two cells is called a heterokaryon. Just as in complementation in E. coli discussed above, there is complementation of
unscheduled DNA synthesis if the two xeroderma cells are from
individuals with different defects in the same repair pathway.

There are at least five and perhaps seven complementation
groups in XP, as defined by the fusion method (2). However,
there are no data on which function is missing in which XP cells.
Furthermore, our understanding of the proteins which function
in DNA repair in normal humans is fragmentary at best.

In the past few years Tanaka and his associates and Hanawalt
and his associates have been able to put a dimer-specific endonuclease into XP cells (3,4). When this is done, the cells then
undergo unscheduled DNA synthesis at roughly the normal level.

Surprisingly enough, all XP cells tested, no matter what the
complementation groups are complemented by the exogenous dimer-
specific endonuclease! One possibility is that an endonuclease is
missing in all known XP's. Perhaps defects in other repair
enzymes would be lethal to cells. On the other hand, cells which
show complementation by the UV-endonuclease are might be undergoing an alternate excision pathway initiated by the UV-endonuclease and thus the process might not relate to normal excision
at all.

II. Base Excision Repair

The second kind of excision repair is called "base excision"
in which the bond between the sugar and the damaged base is
severed (5). Since this bond is the N-glycosyl bond, the enzyme
releasing the damage base is called an N-glycosylase. The result
of the action of the enzyme is just the release of the damaged
base from the DNA. After the release of the damaged base, there
is a nick inserted into the sugar-phosphate backbone at the site
of the sugar which lacks the base. The enzyme performing this
step has been termed an "apurinic endonuclease" for historical
reasons: the first DNA which was generated with missing bases was
prepared by removing purines from the DNA, thus the origin of
the terms apurinic DNA and "apurinic endonuclease" (6). After
the nicking of the sugar-phosphate backbone, there is excision
and resynthesis just as before. Both the endonuclease and the
N-glycosylase have been found in bacterial and in human cells.
For example, for the uracil N-glycosylase the substrate is DNA
damage in which thymine in DNA is demethylated to produce uracil.
The N-glycosylase breaks the N-glycosyl bond in between the base
uracil and the rest of the DNA, leaving the sugar-phosphate backbone intact and releasing the uracil from the DNA. Then the
endonuclease makes a nick at the damaged site but it would
not make release the damaged base, for example, uracil, from the
DNA.

Are there defects in this mode of excision repair in humans?
Indeed, some xeroderma cells show deficiencies in the level of
the apurinic endonuclease. Linn and his group have been able
to purify partially such an enzyme from human cells and thus can
determine the source of this deficiency: is it due to a decrease
in the number of enzyme molecule or to defective enzymes (6)?
Ways of distinguishing these possibilities involve the characterization of the physical, chemical and kinetic properties of the
enzyme; in fact, Linn and his associates have found that the
apurinic endonuclease present in some XP cells shows altered
kinetic properties. This implies a change in enzyme structure
rather than a control mutation. This poses a problem: XP cells

have been shown to be deficient in photoreactivating enzyme, in excision repair, and in apurinic endonuclease. Now one might think that these deficiencies might be due to a common control mechanism which would just decrease the levels of several repair enzymes. But the evidence from Linn's lab indicates that the apurinic endonuclease, at least, is not merely present in lower numbers but is actually an altered enzyme. If these kinetic studies on partially purified enzymes are valid, this means that there are multiple defects in XP. However, the level of XP in the population (1/200,000) is too high for XP to result from a requisite three or four separate mutations. (If this were the case, one would expect to have almost no XP's in the human population!) Thus this poses one of the important problems in DNA repair and human disease today: exactly what is the molecular origin of XP?

III. Damage Removal

The third kind of excision repair is simple removal of damaged region of a base in DNA without excision of a stretch of DNA and without excision of the base. An example of this seems to be the removal of methyl groups which have been added to DNA bases by methylating agents. The cells could excise the entire region (nucleotide excision), they could just remove the base (base excision), but both these repair pathways are costly in terms of energy to resynthesize, and to ligate the new strand back to the parental strands. Pegg has some evidence that rat liver contains an activity which can merely remove the methyl groups from the damaged base (7). This enzyme has not been purified and it remains to be seen if the activity can be isolated and characterized. It is also not known if its activity exists in human cells, nor is it known if XP's are deficient in this enzyme. Problems in determining these points will include the insensitivity of the assay (which require many grams of substrate of cellular material for a few assays, each assay requiring many milligrams of DNA).

Thus although excision repair has been one of the most studied of the repair mechanisms, there are still many important questions which remain to be answered. First, it is not really known exactly which enzymes in bacteria, for which we may take E. coli to be a prime example, participate in excision repair. If there are deficiencies in our knowledge in bacteria, one might say that our knowledge in human cells is in chaos. There is really no complete or partial idea as to exactly which enzymes participate in excision repair in normal cells, much less which enzymes are deficient in xeroderma cells.

There are really no good explanations at this point as to the molecular cause of such apparent DNA repair deficiency diseases, such as XP. Indeed we are suffering an embarrass of riches, with too many molecular defects. Finally the possibility of a new and intriguing repair system in which only the damaged portion of base is removed without any new synthesis is an intriguing one which deserves further study.

REFERENCES

1. Cleaver, J. E., Nature 218, 652-656 (1968).
2. Tanaka, K., Sekiguchi, M. and Okada, Y., Proc. Nat. Acad. Sci. USA 72, 4071 (1975).
3. Setlow, R. B., Science 153, 379-386 (1966).
4. Smith, C. A. and Hanawalt, P. C., Proc. Nat. Acad. Sci. USA 75, 2578-2602 (1978).
5. Lindahl, T., Nature 259, 64-66 (1976).
6. Kuhnlein, U., Penhoet, S. and Linn, S., Proc. Nat. Acad. Sci. USA 73, 1169-1173 (1976).
7. Pegg, A., Nature 274, 182-184 (1978).

FAR AND NEAR ULTRAVIOLET RADIATION PRODUCTS AND THEIR REPAIR

Emanuel Riklis

Israel Atomic Energy Commission, Radiobiology Department

Nuclear Research Center-Negev, Beer-Sheva, Israel

INTRODUCTION

The chronological history of our knowledge of the molecular mechanisms and the biochemical events involved in the processes of repair of radiation damage is short, but the thinking that led to these recent exciting developments is as old as the fields of radiobiology and photobiology. Ever since different survival curves were obtained for mutant strains of the same microbial cells, the search has begun for the reasons which could explain this intrinsic difference in behaviour towards radiation. When thymidine dimers were identified as the major ultraviolet radiation product responsible for cell death, a breakthrough was reached by the discovery of the excision repair system (1,2,3) which gave a logical explanation to the differences in radio and photosensitivity of wild type "resistant" cells vs mutant "sensitive" cells, followed by developments with mammalian cells which led to our better understanding of many phenomena and diseases such as autoimmune diseases in which genetic damage is involved and DNA repair is impaired.

The fidelity of repair determines the ultimate fate of the cell and organism – excision repair is error-free, but other types of repair are error-prone, thus increasing the chance for mutagenicity and carcinogenicity. Evidence for the relation between DNA damage, repair and carcinogenicity has been accumulated during the present decade.

The discovery of thymine dimers (4,5) and the mounting evidence that small molecule, when formed by UV irradiation, is responsible for cellular damage but can be biochemically repaired by either of

283

the various mechanisms of repair of DNA, has led to much photochemical research on the various photoproducts which may be formed
in DNA in accordance with wavelength, concentration and other mo-
difying factors. Most, if not all of the research work on the
repair of photoproducts has been carried out on thymine dimers as
the substrate and the focal point of damage and repair, be it the
classical discoveries of excision repair (1,2,3) through those of
photoreactivation, post replication repair, and more recently SOS
repair.

PHOTOPRODUCTS

Although thymine dimers are responsible for most of the damage
and cell death from 254 nm UV light, it is not the only photoproduct
nor the only type of lesion, and other "minor" photoproducts,
crosslinks and interactions may be responsible for damage which
may lead to cell death or mutagenicity. Chemically induced damage
to DNA often resembles these products in both damage and repair.

The Far Ultraviolet Photoproducts which are formed in DNA and
are of biological importance include: Purine and Pyrimidine
heteroadducts: Pyrimidine hydrates; dihydro reductions; dimers
- all involving the unsaturated 5,6 double bond.
Crosslinks and adducts; chain breaks: single and double strand;
DNA-Protein crosslinks;
The near ultraviolet light photoproducts include also pyrimidine
dimers and single strand breaks, but the more important lesions
are those caused by chemicals, such as psoralen, which form cross-
links and adducts under the influence of NUV. Of all those, some
are known to be repaired, while others are either non-reparable or
still under investigation.

The relative number, not only of the specific photoproducts,
but also of the relative amounts of each type of damage, is highly
dependent on the nature of the radiation and its wavelength. For
example, the ratio of thymine dimers which are formed by far UV
radiation (254 nm) to those produced by 365 nm near UV light is
7.1×10^5, similar to the inactivation efficiency of E. coli mutant
cells by these radiations. Strand breaks however are more abundant
with NUV light than with FUV light, and the ratio of dimers to
breaks for 254:313:405 nm is 300:21:0.1 respectively.

Gamma radiation also produces more strand breaks than specific
radiation products. These are now being identified actively and
their importance as probable causes of damage also gains more
weight (6,7).

During this search, radiochromatographic techniques have been developed and proven as sensitive and reliable for separation of hydrolyzed irradiated DNA. If proper procedures are employed – it becomes evident that thymine dimers are not the only photoproducts formed by UV irradiation of DNA and that thymine hydroperoxide is not the only radiation product formed by gamma or x-radiation. The nature and relative amounts of products formed depend not only on the type of radiation but also on dose, concentration of irradiated solute, temperature and the presence of sensitizers. Only a few examples, from my own experience, will be mentioned here. Thus, the "spore product" was identified (8) following my observation that another photoproduct rather than thymine dimer is formed in dry DNA (9). Another photoproduct, the "trimer" formed in UV-irradiated frozen thymine solution is not normally seen, although formed in DNA, since it is heat-acid labile (10). With certain separation methods, different chromatography solvent systems and two dimensional chromatography, various photoproducts have been identified, including dihydrothymine which in most works developed hidden under the thymine peak (11). A complete review on the photoproducts can be read in books edited by Wang (12) and Smith (13). See Fig. 1.

Methods

Various methods have been employed in the research of DNA repair mechanisms; when excision repair was studied one tried to follow the fate of photoproducts in the DNA as well as in the acid soluble supernatant in which the excised products appeared. This gives direct chemical evidence that excision indeed took place, when the ratio of products to thymine (T<>T/T) is decreased in the DNA and increased in the acid soluble fraction. A popular method which provides evidence of repair synthesis is done by following the uptake of labelled thymidine into irradiated cells. The direct biological evidence of a complete repair cycle from incision to ligation may be obtained by studying sedimentation patterns. This will show also the repair of strand breaks by rejoining of the smaller DNA molecules to give the original size. If one wishes to follow the fate of photo- or radiation products and determine which of them are repaired, the only method possible is by separation of the DNA from the other cell constituents and performing a chromatographic analysis. This can be done on paper, but it is important not to simply cut the paper and transfer it for liquid scintillation counting, but rather to count the whole paper on a high sensitivity radiochromatogram scanner, then redevelop in second dimension with another solvent system, and scan-count again. This, when done in a simple excision repair experiment, revealed that two additional products which appeared with an Rf lower and higher than thymine, are also excised during the process of repair (14). One of these is probably dihydrothymine. Some of the photoproducts formed by

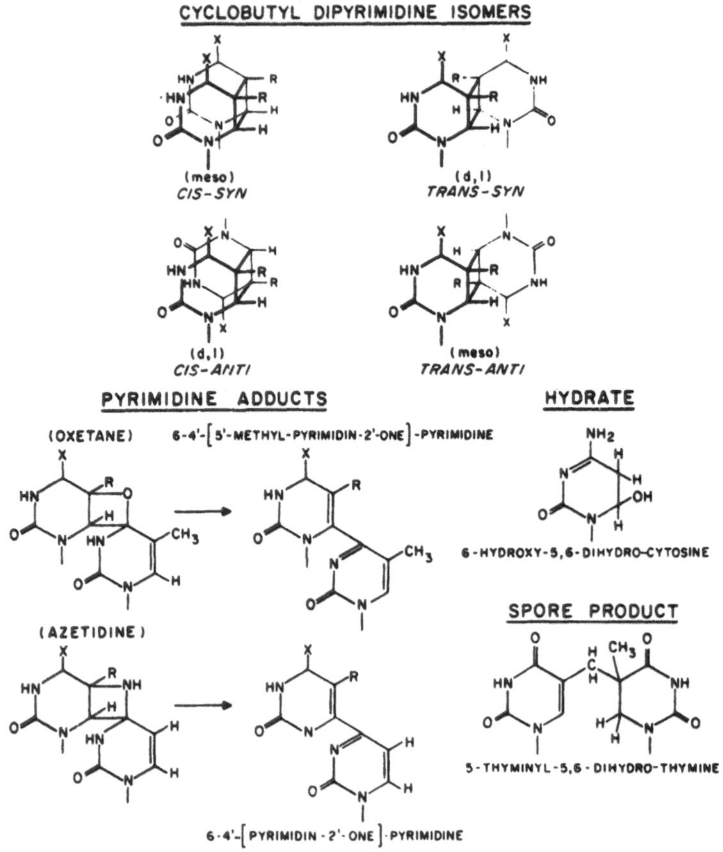

Fig. 1. PYRIMIDINE PHOTOPRODUCTS ISOLATED FROM UV
IRRADIATED DNA

The arbitrary group X is: C = O when R is CH₃
C – NH₂ when R is H.

Table I. SEPARATION OF PHOTOPRODUCTS BY RADIOCHROMATOGRAPHY

Photoproduct	Material	Solvent				
		A	B	C	D	E
PI	Trimer	0.13	0.07	0.02	0.22	-
PII	Cis-syn dimer	0.26*	0.13	0.12	0.35	0.49
PIII	Adduct	0.28	0.13	0.12	0.38	-
PIV	Cis-anti dimer	0.33	0.17	0.12	0.40	0.56
PV	Urea	0.46	0.27	0.24	0.42	0.41
PVI	Dihydrothymine	0.58	0.54	0.44	0.66	0.63
PVII	Thymine	0.61	0.54	0.31	0.68	0.63
PVIII	n-Propyl urea	0.75	0.62	0.55	-	-

Rf values: ±0.02
Whatman No. 3MM. paper chromatography, (ascending)

Solvents:
A: n-Butanol, acetic acid, water. (80:12:30)
B: n-Butanol, water (86:14)
C: Isoamyl alcohol, MEK, isopropanol, ammonia, water. (40:30:13:7:10)
D: 2-Butanol saturated with water.
E: Isopropanol, ammonia, water (80:10:20)
*: Rf of 0.23 with Whatman paper No. 3.

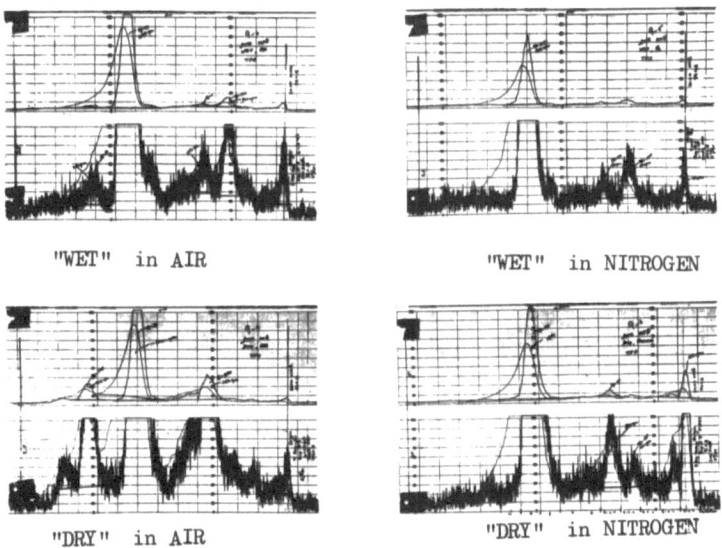

"WET" in AIR   "WET" in NITROGEN

"DRY" in AIR   "DRY" in NITROGEN

Fig. 2. TYPICAL RADIOCHROMATOGRAMS OF UV-IRRADIATED DNA

far UV radiation of frozen thymine or DNA and the solvent system
used for their separation are indicated in Table I.

A typical radiochromatogram of photoproducts from DNA which
has been UV-irradiated under different conditions is shown in
Fig. 2.

A relatively new approach to the problem of photoproducts and
their repair utilizes the fact that antibodies are formed against
UV-irradiated DNA (15,16) and x-irradiated DNA (Riklis and Slor,
unpublished). Irradiated DNA, when injected into rabbits produces
a range of antibodies. By separating them on columns which contain
the specific photo or radioproducts one can thus obtain immunolo-
gical evidence as to which of the products is recognized as damage
and induces production of specific antibodies against it.

                              REPAIR

Most types of damage can be repaired in normal cells. These
repair processes may be classified as: strand rejoining - single
strand breaks, double strand-breaks; photoreactivation; base excision repair; post-replication repair; SOS repair. Specific products
and base damage are recognized and repaired either by excision, in
cells which have the excision repair system, or by post replication
repair. If these systems do not function, the cell can still
employ recombination repair and call to its aid the SOS repair, the
latter being however error-prone which may lead to more genetic
damage. Single strand breaks are quickly repaired by joining and
it has been suggested that also double strand breaks are repairable.

Crosslinks, such as formed by psoralen and near UV light are
efficiently repaired. Even BUdR which sensitizes cells to UV radiation is released as uracil when the DNA is not too heavily
labelled with it (17).

Only few types of damage are not repaired by normal cells:
[125]Iodine which irradiates the cell with Auger electrons causes
damage in the form of 4 to 5 breaks per single strand per disin-
tegration. Only 50% of these breaks are repaired in DNA of Chinese
Hamster cells, as compared to 70-100% repair of breaks caused by
the same dose of X-rays (18). In T_1 and T_4 phage, Iodine-125 causes
double strand breaks, which are less prone to repair than single
strand breaks. Our own results showed (19) that the effect of
hyperthermia is smaller in cells grown with [125]IUdR than on ^{3}H-TdR
labelled cells, meaning that there is a priori more non-repairable
damage, leaving less possibility for the repair system to be
affected by hyperthermia. DNA to protein crosslinks are apparently
not repaired (20). This type of damage is of great importance in
mammalian cells.

Finally, the combined lethal effects of far UV and near UV, sunlight-like exposures are believed not to be entirely additive (21) suggesting that the cell killing photolesions induced by these radiation are only partly the same. When near UV light is employed in combination with photosensitizer it certainly causes the production of a different set of photoproducts which turn out to be not only responsible for cell killing but also for increased mutagenicity. Thus, and only as an example, we have been able to show (22) by using a modified Ames system, the very high mutagenic property of 4,5,'8 trimethylpsoralen plus NUV light, as well as of chlorproma in - both drugs used in phototherapy or general medicine, and the complete lack of mutagenic activity of hematoporphyrin which has been used in combination with 600 nm light for tumor phototherapy.

CONCLUSIONS

The intensity of UV radiation, its wavelength and the conditions of irradiation, were shown to affect the relative production of photoproducts in DNA. Wave length dependence is obvious, as some products which are formed by one wavelength are split by another. The spectrum of products will determine the fate of the organism, as some products are and others are not repairable. Synchrotron radiation may therefore open a completely new spectrum of photoproducts in the near UV, far UV and especially in vacuum UV region which has not yet been investigated, and thus widen the scope of understanding damage and its repair.

Studying the photochemistry and photobiology of DNA, recognizing the products, and understanding the processes of DNA repair had already clarified our understanding of the causes and development of certain diseases and indicated the possibility of employing these scientific disciplines for their cure by phototherapy or radiotherapy.

REFERENCES

1. Setlow R.B. and Carrier W.I., Proc. Nat. Acad. Sci. USA 51, 226 (1964).

2. Riklis E., Canad. J. Biochem. 43, 1207 (1965).

3. Howard-Flanders P. and Boyce R.P., Proc. Nat. Acad. Sci. USA 51, 239 (1964).

4. Beukers R. and Berends W., Biochim. Biophys. Acta 41, 550 (1960).

5. Wang S.Y., Nature (London) 188, 244 (1960).

6. Cerutti P.A., in "Photochemistry and Photobiology of Nucleic Acids" vol. 2 p. 375(1976) (edit. S.Y. Wang) Academic Press New York.

7. Bonicel A., Mariaggi N, and Teoule R., in "DNA Repair and Late Effects" Proc. IGEGM Intl Symp. Tel-Aviv (1978) (edit. E. Riklis, H. Slor and H. Altmann).

8. Donnellan J.E. and Setlow R.B., Science 149, 308 (1965).

9. Riklis E. and Simson E., Abstr. Biophys. Soc. 8th Ann. Meet. Chicago (1964).

10. Kabantchick Y. and Riklis E., Isr. J. Chem. 6, 102p (1968).

11. Riklis E., in "New Trends in Photobiology", Anais Acad. Bras. Cienc. 45, 221 (1975).

12. Patrick M.H. and Rahn R.O., in "Photochemistry and Photobiology of Nucleic Acids" vol. I p. 296 (1976) (edit. S.Y. Wang) Academic Press, New York.

13. Smith K.C. (editor) "Photochemical and Photobiological Reviews" vol. II (1977) Plenum Press, New York.

14. Riklis E., in "DNA Repair and Late Effects" Proc. IGEGM Intl Symp. p. 133 (1976) Vienna, (edit. H. Altmann) Rotzer Druck Eisenstadt.

15. Seaman E. Van Vunakis M. and Levin L., J. Biol. Chem. 18, 5709 (1972).

16. Slor H., Nivy S., Lev-Sobe T. and Friedberg E.C., in "DNA Repair and Late Effects" Proc. IGEGM Intl Symp. p. 27 (1976) Vienna, (edit. H. Altmann) Rotzer Druck Eisenstadt.

17. Ben-Hur E., Prager A. and Riklis E., Photochem. Photobiol. 27, 559 (1978).

18. Painter R.B., Young B.R. and Burki H.J., Proc. Nat. Acad.Sci. USA 71, 4836 (1974).

19. Riklis E. and Ben-Hur E., in "Modification of Radiosensitivity of Biological Systems," P. 141 (1976), IAEA, Vienna.

20. Todd P. and Han A. in "Aging, Carcinogenesis and Radiation Biology" p. 83 (1975) (edit. K.C. Smith) Plenum Press, New York.

21. Elkind M.M., Han A. and Chang-Liu C.M., Photochem. Photobiol. 27, 709 (1978).

22. Riklis E., Green M., Ben-Hur E. and Prager A., Abstr. ICN-UCLA Intl Symp. DNA Repair Mechanisms, Keystone 1978.

STRUCTURE DETERMINATION by X-RAY ABSORPTION SPECTROSCOPY -

Including Applications from the Study of Molybdenum Proteins

Stephen P. Cramer

Exxon Research and Engineering Company

Linden, New Jersey U.S.A. 07036

Lecture 1 - Curve-Fitting EXAFS Analysis

Since Dr. Eisenberger has already discussed in some detail the properties of synchrotron radiation and its utility for x-ray absorption spectroscopy, in my lectures I will concentrate on a description of the curve-fitting method of EXAFS analysis and the applications of this method to determination of the structures of molybdenum proteins. Today, I will begin by reviewing the nature of EXAFS, and then I plan to go into some detail on how one actually obtains structural information from an EXAFS spectrum. I hope this will give you an appreciation of what EXAFS analysis can and cannot do, and that you will eventually be able to read the literature and distinguish between some of the very good and very bad work that has been done in this field.

1. - What EXAFS Is -   The extended x-ray absorption fine structure, EXAFS, is simply the modulation of the absorption coefficient μ compared to the smooth average absorption μ_s, and normalized by the absorption coefficient of the free atom, μ_0. Thus, representing the EXAFS by χ, and as illustrated in figure 1,

$$\chi \equiv \frac{\mu - \mu_s}{\mu_0} \qquad (1)$$

Since the smooth absorption of the atom in a particular sample μ_s and the free atom absorption μ_0 are nearly equal, for the purpose of normalization the EXAFS may alternatively be defined as $\mu - \mu_0/\mu_0$ or $\mu - \mu_s/\mu_s$.

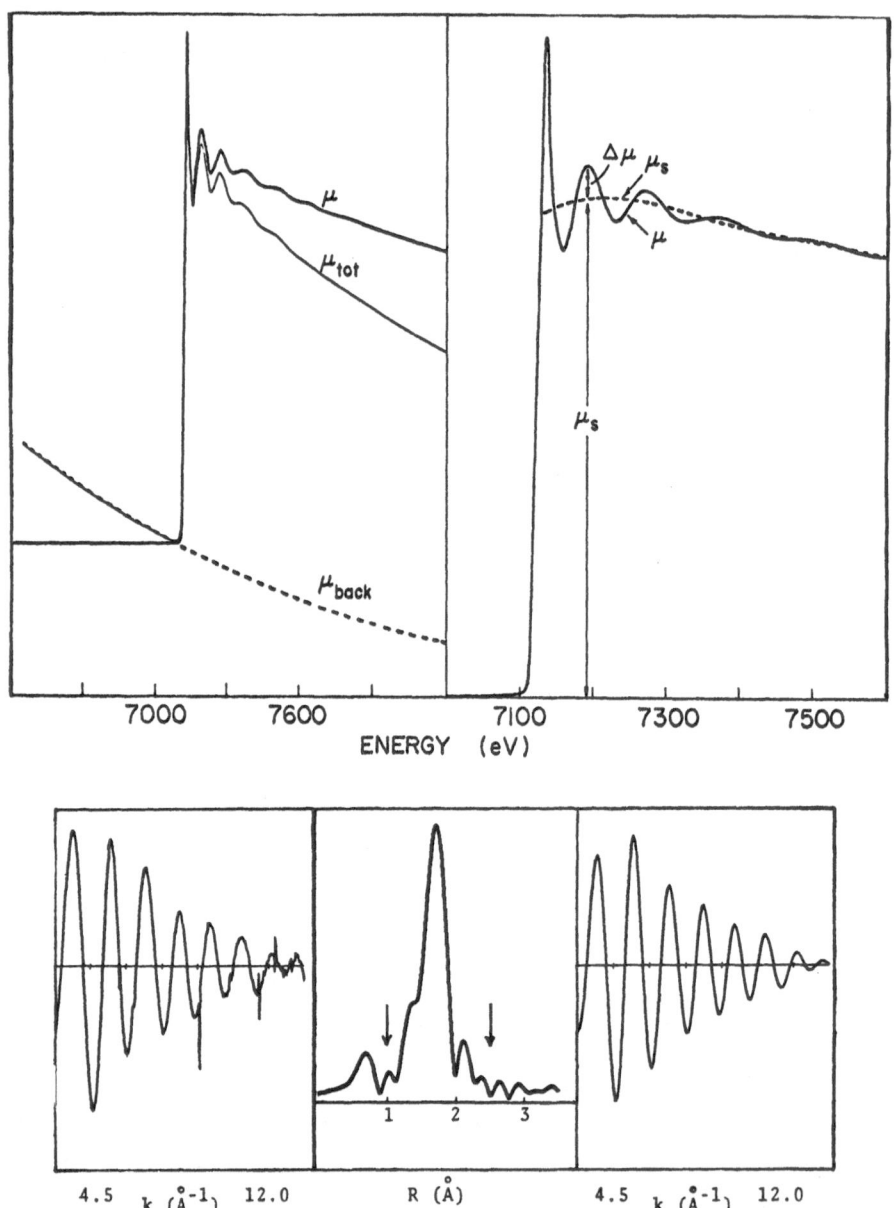

c) conversion to d) Fourier trans- e) final Fourier-
   k scale   form and window   filtered EXAFS

Figure 1 - EXAFS Data Processing - Illustrated with Fe(C₅H₅)₂ EXAFS

Extracting the EXAFS from an absorption or fluorescence excitation spectrum is a non-trivial part of the data analysis procedure. The extraction process typically involves a) isolating the absorption of the atom and edge of interest from the background absorption by other atoms and other edges of the same element, b) extracting the oscillatory part, $\Delta\mu$, from the smooth background μ_s, and normalizing to μ_0, c) defining E_0 and converting from an energy scale to a k-scale through:

$$k = [(2m/\hbar^2)(E-E_0)]^{1/2} \tag{2}$$

d) Fourier-transforming the data to frequency space, and isolating the EXAFS component(s) of interest from the EXAFS of other shells and low frequency baseline drift, and e) back-transforming to k-space, thereby obtaining the final, Fourier-filtered EXAFS spectrum. The detailed numerical procedures involved in each of these steps have been described previously (1), and a typical operation is illustrated in figure 1. In some cases, the Fourier-filtering procedure is omitted, and the data after step (c) is either used directly or smoothed and then used. The main point I would like to make is that unlike many spectra which can be interpreted and presented virtually on the chart paper on which they were recorded, EXAFS spectra are subject to a great deal of numerical manipulation before the analysis has even begun. From reading the literature it is often difficult to tell exactly how the data were processed, and one should always be on guard against overmanipulated data.

2. Why EXAFS Is - EXAFS results from the interference between the outgoing photoelectron wave from an x-ray absorbing atom and the backscattered waves from surrounding atoms. As illustrated in figure 2, if the outgoing and backscattered waves interfere constructively there will be an enhancement in the x-ray absorption, while destructive interference results in a decreased absorption rate. By increasing the photon energy, we decrease the photoelectron wavelength, and thereby produce a periodic succession of constructive (absorption maxima) and destructive (absorption minima) interferences.

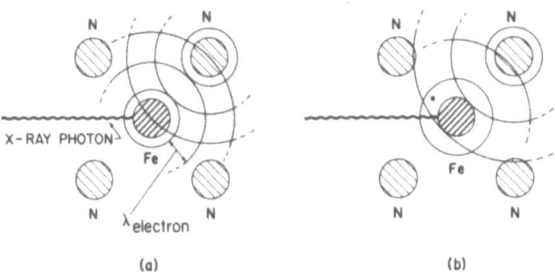

(a)   (b)

Figure 2 - The Mechanism Leading to EXAFS

The following set of equations can be used to illustrate the mathematical relationship between the final state interference and the absorption coefficient.

$$I = I_0 e^{-\mu x} \tag{3}$$

$$\mu = n_c \sigma_a \tag{4}$$

$$\sigma_a(E) = 4/3 \ \pi^2 \ \alpha(h\nu) \ |M|^2 \ N(E) \tag{5}$$

$$M = <f|r|i> \tag{6}$$

$$<f| = <f_0| + <f_s| \tag{7}$$

$$<f_s| = i \ \frac{e^{ikr_1 + i\delta_1}}{kr_1} \ f(\pi) \ e^{ikr_1} \ e^{-ikr_1 \cdot \hat{r}} \qquad r > r_0 \tag{8}$$

$$<f_s| = [i \frac{e^{2ikr_1 + 2i\delta_1}}{2kr_1^2} f(\pi) \ ] \ \psi_0 \qquad r < r_0 \tag{9}$$

$$\chi = \Delta \ |M|^2 / \ |M|^2 = - \ \frac{|f(\pi)|}{kr_1^2} \ \sin(2kr_1 + 2\delta_1 + \alpha_s) \tag{10}$$

Equations (3) and (4) merely relate the absorption coefficient μ to the incident and transmitted fluxes I and I_0 and the cross section σ, while equation (5) is essentially the Golden Rule which relates the cross section to a matrix element. Within the electric dipole approximation, the appropriate matrix element is given by equation (6), and equation (7) breaks the final state up into the sum of an outgoing photoelectron wave $<f_0|$ and a scattered wave $<f_s|$. Outside the atomic radius r_0 of the absorbing atom the scattered wave is described by equation (8), where δ_1 is the absorbing atom phase shift and $f(\pi)$ is the scattering amplitude of the scatterer at distance r_1. Inside r_0 the scattered wave is described by equation (9), where ψ_0 was the wave function in the absence of a scatterer. Finally, in equation (10) we arrive at the EXAFS effect in terms of scattering amplitudes, phase shifts, and distances, where $f(\pi)$ has been rewritten as a magnitude $|f(\pi)|$ and a phase shift α_s.

The reason why both phase shifts and distances enter into the EXAFS effect is illustrated in figure 3. Here we have introduced a new nomenclature in which quantities dependent on the absorber or scatterer get (a) or (s) subscripts respectively, while quantities dependent on both get subsripted with the absorber first. If the relative phases between outgoing and backscattered waves were only governed by the distance travelled by the scattered wave, then the oscillatory part of the EXAFS would be simply $\sin(2kR_{as})$. However, additional k-dependent phase shifts arise because of the

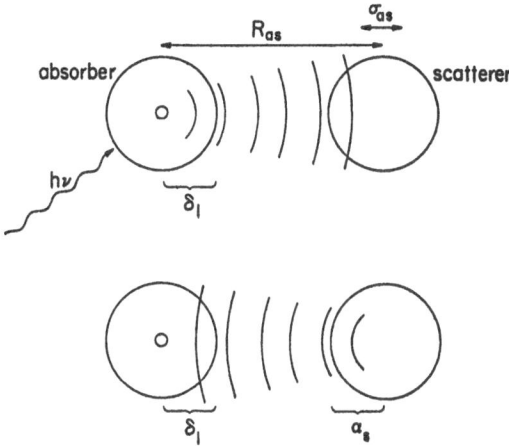

Figure 3 - Schematic Representation of the Factors Involved in the
EXAFS Equation. The interference between the outgoing
and backscattered waves in a small region near the
absorber nucleus determines the EXAFS.

potentials of the absorbing and scattering atoms. The combined
effect of these additional phase shifts is to change the EXAFS
oscillations to sin $(2kr + \alpha_{as}(k))$, where the total phase shift
α_{as} has contributions from both absorber and scatterer:

$$\alpha_{as}(k) = 2\delta_1(k) + \alpha_s(k) - \pi. \tag{11}$$

The additional π is added as a matter of convention to make the
EXAFS amplitude positive when the phase shifts are defined as
above.

While the phase of the EXAFS depends only on the distance
R_{as} and a total phase shift, the EXAFS amplitudes are sensitive to
many factors: the backscattering amplitude of the scatterer
$f_s(\pi,k)$, the distance R_{as}, the mean square deviation of that distance σ_{as}^2, and the number of equivalent scatterers N_s. When these
effects are summed over all neighboring atoms, we have a modification of equation (10) which becomes in our new nomenclature:

$$\chi(k) = \sum_s N_s |f_s(\pi,k)| e^{-2\sigma_{as}^2 k^2} \frac{\sin(2kR_{as} + \alpha_{as}(k))}{kR_{as}^2}. \tag{12}$$

For scattering atoms beyond the first coordination sphere, additional phase shifts and inelastic loss factors may be required, but we will ignore them for the moment. From equation (12), it is apparent that EXAFS contains information about the types, numbers, and distances of scatterers in the immediate environment of the particular x-ray absorber under study. The goal of curve-fitting EXAFS analysis is to extract this information from experimentally recorded spectra.

3. The Empirical Curve-Fitting Approach - Examination of equation (12) shows that in order to obtain distance information from EXAFS, knowledge of the total phase shift $\alpha(k)$ is required, while calculation of scatterer numbers requires knowing the distance, scattering amplitude, and Debye-Waller factor as well. Although both $\alpha_{as}(k)$ and $|f_s(\pi,k)|$ can now be calculated from first principles (2,3,4), phase shifts and total amplitudes can also be obtained experimentally from the spectra of suitable model compounds. In the preceding lecture Dr. Eisenberger presented both theoretical and experimental arguments for the transferability of EXAFS phase shifts. Thus, once we have obtained the phase shift for a particular a-s pair, we can use it to obtain distances in other compounds with the same pair of atoms. I will show you that the use of empirical phase shifts and amplitudes actually involves fewer adjustable parameters than does application of the theoretical values, and therefore the empirical technique has certain advantages for the EXAFS analysis of complicated structures.

A suitable model compound for obtaining empirical phase shifts and amplitudes has a set of equivalent scatterers at equal distances from the absorber, with a known distance which is resolvable from any other absorber-scatterer distances in the sample. Once the EXAFS of the a-s component is isolated (see figure 1), it is curve-fit as the product of an amplitude function $A(k)$ and a phase-shifted sine wave $\sin(2kR_{as}+\alpha(k))$. If one normalizes for N and R, then the amplitude function will depend on the Debye-Waller factor, k, and the scattering amplitude:

$$A(k) = |f_s(\pi,k)|e^{-2\sigma_{as}^2 k^2}/k \tag{13}$$

while the total phase shift $\alpha_{as}(k)$ will reflect absorber and scatterer contributions as given in equation (11). The particular form of the parameterized functions used to represent $A(k)$ and $\alpha_{as}(k)$ is not really important, one should use as complicated a function as is necessary to obtain a good fit to the data.

3.1 Phase Shifts - Empirically derived total phase shifts all have the same basic appearance; they are nearly linear functions of k with a negative slope and slightly positive curvature (see figure 4). In most cases the phase shift can be adequately modelled by a quadratic expression:

$$\alpha(k) = a_0 + a_1 k + a_2 k^2,$$ (14)

although in some cases expansion using inverse powers of k gives a better fit. Since all of these parameters are fixed after fitting the model compound spectrum, there is of course no reason not to use more complicated expressions, and this will probably be necessary for work over very wide ranges in k-space or with very high Z scatterers. Theory predicts that the phase shifts will become more positive as the scatterer Z increases, and such a trend is observable in the empirical phase shifts of figure 4. The net effect of the phase shift on the EXAFS oscillations is to lower their average frequency and to slightly compress the oscillations at higher k values.

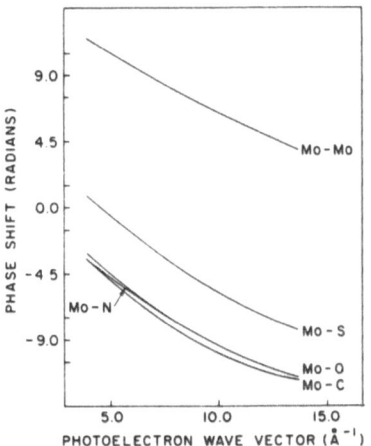

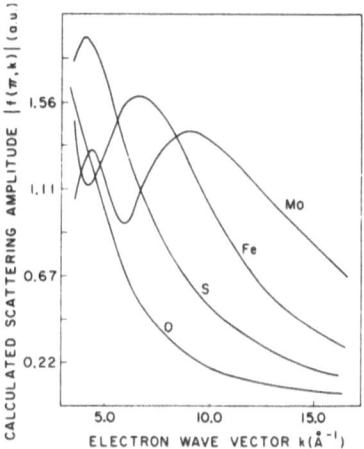

Figure 4 - Empirical Phase Shifts   Figure 5 - Backscattering
                                                  Amplitude Trends

3.2 Amplitudes - The shape of the EXAFS amplitude envelope is dependent on both the electron-atom scattering amplitude $|f_s(\pi,k)|$ and the Debye-Waller factor $\exp(-2\sigma_{as}^2 k^2)$. If one knows the appropriate Debye-Waller factor for the model being used, it can be removed from the observed total amplitude function to eventually yield $|f_s(\pi,k)|$. Approximate theoretical curves for these amplitudes are shown in figure 5. At high k values the amplitudes rise linearly with Z, while the low k behaviour is complicated and actually changes shape as the atomic number increases.

When only $|f_s(\pi,k)|$ is assumed transferable, then σ_{as}^2 must be included as a variable in the curve-fitting procedure. However, one can also assume that the Debye-Waller factor for model and unknown are nearly equal, and parameterize the total amplitude function of equation (13). As mentioned previously, the particular functional form of the parameterization is not really important, but one expression which proved useful in past studies was:

$$A(k) \overset{\sim}{=} c_0 e^{-c_1 k^2}/ k^{c_2}. \tag{15}$$

The combined effects of phase shift and amplitude functions on the EXAFS are illustrated in figure 6 for Mo-O, Mo-S, and Mo-Mo interactions. Figure 6a shows the single shell fits used to obtain phase shift and amplitude parameters. However, since the model compounds had different numbers of scatterers and different Mo-X distances, in figure 6b I have illustrated the combined effects of phase shift and amplitude for a single hypothetical Mo-X interaction at 2 Å. Note the substantial phase differences between the three waves, despite the fact that they all represent the same distance. In particular, Mo-O and Mo-S are nearly π out of phase, making O and S particularly easy to distinguish by EXAFS. The phase differences between S and Mo are less pronounced, but distinguishing between the latter two is aided by the dramatic differences in the amplitude envelope.

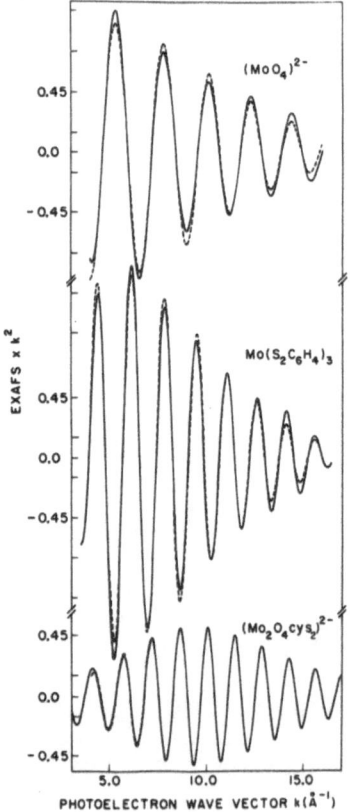

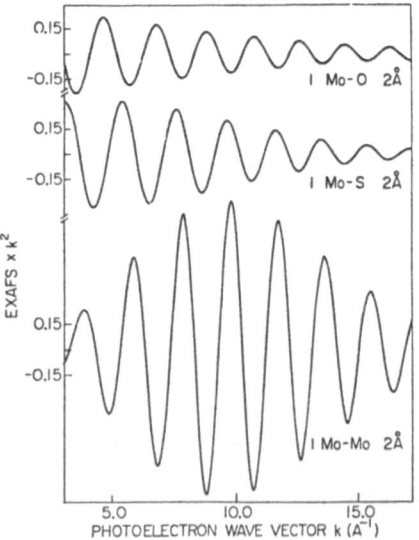

Figure 6a - Single-Shell Fits for
(left)   Parameterization

Figure 6b - Combined Effects of
            Phase Shifts and A(k)

3.3 Optimization - There are two distinct levels of refinement in curve-fitting EXAFS analysis. On the first level one postulates a set of scatterers at a distance R_{as} and varies both the number of scatterers N_s and the distance R_{as} in order to obtain the best fit. Numerically, this can be accomplished by minimizing the difference (f) between χ_{obs} and χ_{calc} in a non-linear least squares routine:

$$f = \sum_{n} k^m (\chi_{calc} - \chi_{obs})^2 / n, \tag{16}$$

where

$$\chi_{calc} = \sum_{s} N_s \frac{c_0 e^{-c_1 k^2}}{R_{as}^2 \, k^{c_2}} \sin (2kR_{as} + a_0 + a_1 k + a_2 k^2). \tag{17}$$

For each different elemental type of scatterer in the fits, a different set of phase shift and amplitude parameters is required.

The second level of curve-fitting refinement involves varying the elemental types of atoms postulated as scatterers and/or adding extra components to better fit the data. One usually has some chemical information about what types of atoms are reasonable candidates at various distances, and thus the entire periodic table need not be covered. Nevertheless, this part of the analysis requires the greatest caution in order to avoid false conclusions. Since every addition of an extra wave will produce some improvement in the fit, the most difficult part of the analysis is often deciding when to stop.

One procedure which has proven useful for the analysis of unknown structures is illustrated on the flow chart of figure 7. From examination of the Fourier transform of the EXAFS we can get an idea of what the major frequency components are, but we do not know the elemental type of scatterer involved. We then propose a structure, refine the atom numbers and distances, and see if what we get is chemically reasonable. If so, we check to see if we have neglected any minor components, while if not, we postulate another structure and begin the cycle again.

3.4 Pitfalls of Curve-Fitting - Perhaps the best cure to overenthusiasm about any curve-fitting data analysis is direct experience with the procedure. Initially, it appears quite miraculous to be able to reproduce the essential features of an EXAFS spectrum with a simple sum of damped sine waves. However, the successful reproduction of the EXAFS does not necessarily guarantee the physical significance of the calculated scatterer numbers and distances. There are several problems in the curve-fitting procedure which, although not insurmountable, could lead to false conclusions despite the presence of a very good fit.

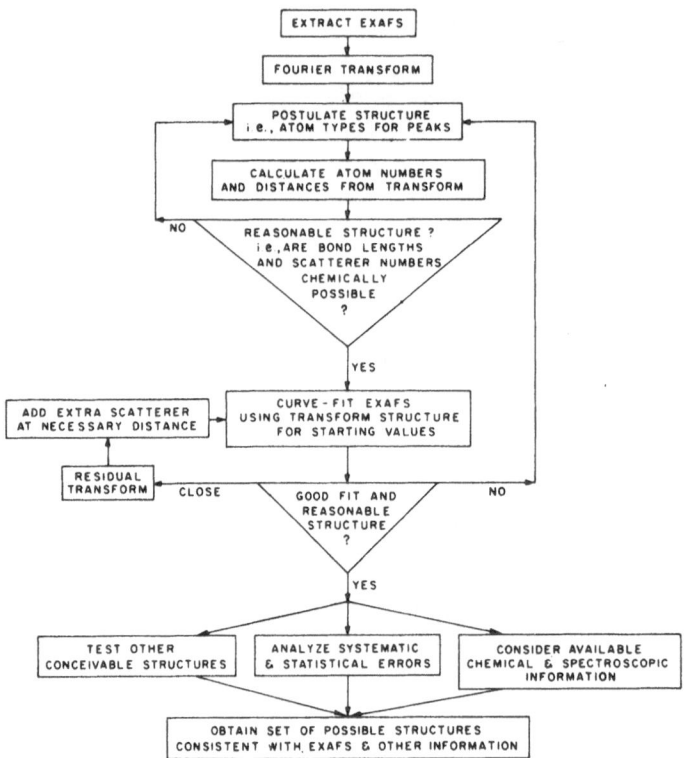

Figure 7 - A Systematic Procedure for Structure Determination

3.4.1 Parameter Correlation -   In previous lectures Dr. Phillips
warned repeatedly of the dangers of fitting fluorescence decay
data with more than two exponentials. This is because as one
introduces more unconstrained variables into the fits, they begin
to correlate with each other, and the significance of the individual values declines. Parameter correlation in the present case
has been minimized by using only one parameter that affects EXAFS
amplitudes - N_s, and one parameter which affects the phase - R_{as}.
With this procedure at least four EXAFS components can be included in the fits without severe correlation problems. Parameter
correlation is less of a problem in EXAFS than in fluorescence
decay data because sine waves are more orthogonal than are exponentials. Still, severe problems exist when two close but non-
equal distances are present, that is, when ΔR is on the order of
0.1 Å. In this case the resolution attainable depends on both the
length of the data set and on our knowledge of the amplitude en-
velopes of the components. A useful indicator of parameter correlation is the standard deviation calculated for the final values,
which gets quite large when correlation is a problem.

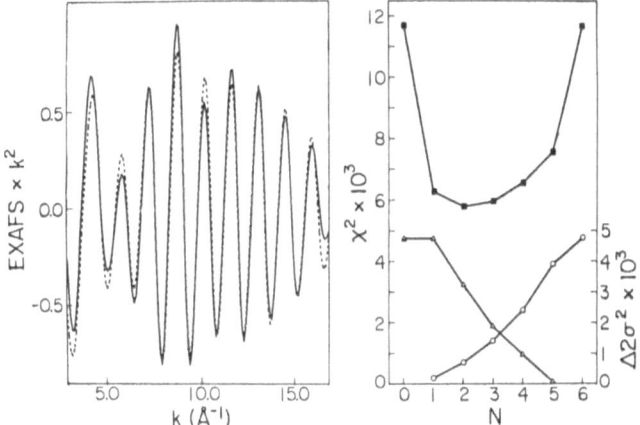

Figure 8 - Curve-Fitting Results for Mo(IV) (aq.). On the left is the Mo EXAFS (——) and the best 3-wave fit (----). The solid squares on the right plot show χ^2 as a function of N (the number of short Mo-O bonds), while Δ and O represent the Debye-Waller factors of the long and short Mo-O bonds respectively, relative to molybdate.

A nice example of parameter correlation effects is illustrated in figure 8. Here, in the curve-fitting analysis of aqueous Mo(IV), the fits were slightly different from those discussed previously, because we constrained the numbers of scatterers to integer values and varied the Debye-Waller factors. The number (N) of short Mo-O interactions was varied from 0 to 6, while the number of long Mo-O interactions was constrained to 6-N. As figure 8 clearly shows, there is a strong correlation between the postulated number of atoms and the Debye-Waller factor, so that fits with 1, 2, or 3 short Mo-O distances all gave similar values of χ^2, but quite different values for $\Delta 2\sigma^2$. (5)

3.4.2 Local Minima - In all curve-fitting procedures it is necessary to verify that the minimum in χ^2 found by variation of a set of variables is the absolute minimum obtainable. Often such a proof is impossible analytically, and one is forced to try a wide range of starting values instead. In many cases there are local minima at which the change in χ^2 with an incremental change in the variables is zero, despite the fact that if one jumped to a different region in variable space an even smaller χ^2 could be obtained. The simplest case of local minima relevant to EXAFS is that of fitting one sine wave with another over a finite range of data.

If the true function is sin (2kR), and the fitting function

is sin (2kR'), then the optimization program will try to minimize
the integrated square of the difference:

$$\int_{k_{min}}^{k_{max}} (\sin 2kR - \sin 2kR')^2 dk = \qquad (18)$$

$$k - \frac{\sin 4Rk}{8R} - \frac{\sin 4R'k}{8R'} - \frac{\sin 2(R-R')k}{2(R-R')} + \frac{\sin 2(R+R')k}{2(R+R')} \Big|_{k_{min}}^{k_{max}}.$$

$$\Delta R \neq 0, \ R \neq 0, \ R' \neq 0$$

Although the difference between the two sine waves will be zero
when R=R', inspection of equation (18) shows that the integrated
square of the difference is certainly not a monotonic function of
R'. Under typical conditions, when Δ is small, the most important term in equation (18) is the $\sin(2\Delta k)/2\Delta$. When evaluated
between k_{max} and k_{min}, this term yields:

$$\frac{\sin(2\Delta k)}{2\Delta k} \Big|_{k_{min}}^{k_{max}} = \frac{\sin(\Delta(k_{min}+k_{max}))\cos(\Delta(k_{max}-k_{min}))}{\Delta}. \qquad (19)$$

Thus, when Δ is small but nonzero, there will be subsidiary minima
in χ^2 at intervals characteristic of the fitting range. For a
typical fitting range of 4-12 Å^{-1}, these minima occur at about
0.4 Å to each side of the true value of R. Once this phenomenon
is appreciated, it becomes trivial to check on each side of an
optimized distance to see if one is in a local minimum. This
becomes especially important in multishell fits, where it is not
always possible to directly see the relationship between the individual EXAFS components and the sine waves used to fit them.

3.4.3 Wrong or Incomplete Postulated Coordination Spheres - With
a completely unknown structure, one must postulate the elemental
type(s) of scatterer(s) contributing to the EXAFS, in order to
apply the appropriate phase shift and amplitude functions. Often,
several different coordination spheres will be plausible from
chemical information and the Fourier transform, in which case each
possible structure must be tested by the optimization program. The
problems associated with this procedure are due to the facts that
1) every addition of another component adds two new degrees of
freedom and will produce some improvement in the fit, and 2) even
addition of the wrong scatterer type can result in a better fit.
Thus, although identification of the principal EXAFS components
of an unknown structure is often unambiguous, one is often faced
with the situation of wondering whether or not a minor component
which marginally improves the fits is real. In such cases it helps
to identify particular features of the spectrum which are improved
by the additional component, and then to consider the reliability

of those features. Ideally, one should remeasure the spectrum over
a wider range of k-space, or under different conditions, and then
see if the suspect component is still required.

To complete this section on curve-fitting analysis, I have
included representative examples of single-shell and multi-shell
fits for various molybdenum compounds. Notice that with single-
shell compounds the EXAFS corresponds to a simple damped sine wave,
while for multi-shell compounds a complicated beat pattern arises.
As the EXAFS becomes more complex, one loses the ability to assign
certain features to individual components, and it therefore becomes
more difficult to say whether or not a particular wave is correct.
In figure 9a, the top two waves were fit correctly, while the
bottom case shows an attempt to fit Mo-O EXAFS with Mo-S parameters.
Similarly, in figure 9b, the top two patterns were fit with the
correct parameters, while the bottom curves show an attempt to
fit the Mo-Mo component with Mo-O parameters.

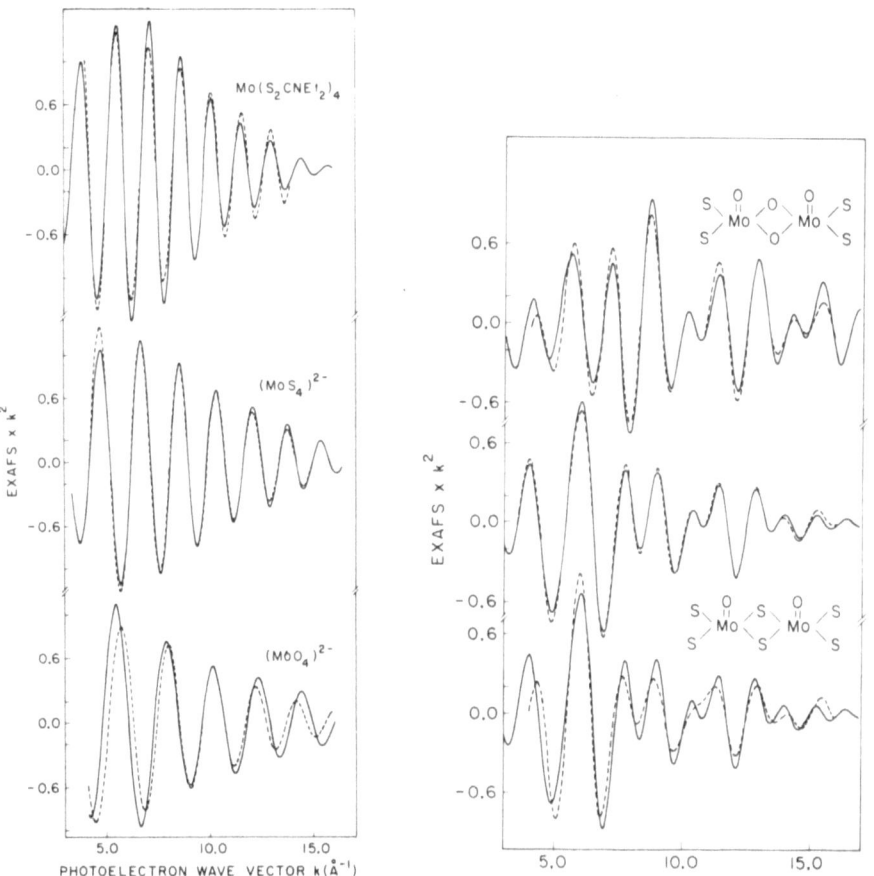

Figure 9a - Single Shell Fits   Figure 9b - Multi-Shell Fits

Lecture 2 - X-RAY ABSORPTION SPECTROSCOPY of NITROGENASE

In the next two lectures I would like to illustrate the application of x-ray absorption spectroscopy with examples from the study of molybdenum enzymes. The first lecture of this pair will cover already published work on the nitrogenase Mo-Fe protein, the Fe-Mo cofactor, and Fe-Mo-S model compounds. I will not attempt to present this material in chronological order, but will instead show how the totality of absorption edge and EXAFS data on nitrogenase and recent model compounds points toward the existence of a Mo-Fe-S cluster in this protein.

The second lecture will involve recent x-ray studies of the enzymes sulfite oxidase and xanthine oxidase. This body of work is in much less finished form, but perhaps for just this reason it will allow one to see more clearly how a structure is finally obtained from the combination of EXAFS fits, edge correlations, and crystallographic information. The goal of this work is to obtain a self-consistent picture for the molybdenum oxidases which will explain the chemical and structural differences between a number of proteins which contain a similar molybdenum cofactor. We would like to know which ligands belong to the cofactor and which belong to the protein (if any), and ultimately, how the protein controls the catalytic activity of molybdenum.

2.1 Introduction to Nitrogenase - The important reaction catalyzed by nitrogenase is the reduction of dinitrogen to ammonia (6). Substrates such as dinitrogen and other small unsaturated molecules are bound and reduced by the Mo-Fe protein, while electrons are shuttled to the catalytic site by the Fe protein from ferredoxin or flavodoxin electron donors. The reduction of dinitrogen is also coupled with the hydrolysis of ATP, and the overall process may be written:

$$\text{Fe protein}^{\text{red}} \qquad\qquad \text{Fe protein}^{\text{ox}}$$

$$\text{MgATP} \qquad\qquad\qquad \text{MgADP} + \text{P}_i \qquad\qquad (20)$$

$$\text{H}_3\text{O}^+ \qquad\qquad\qquad\qquad \text{H}_2\text{O}$$

$$\text{N}_2 \xrightarrow{\quad\text{Mo-Fe protein}\quad} 2\text{NH}_3$$

The Mo-Fe protein has a molecular weight of about 220,000, and contains 2 Mo, 24-32 Fe, and a nearly equal number of acid-labile sulfides, while the Fe protein which reduces it has a molecular weight of 50,000 and 4 Fe in a tetranuclear Fe-S cluster. A reaction which is always competing for electrons in this process is the reduction of H_3O^+ to H_2; since the relative numbers of electrons which go into hydrogen evolution and N_2 fixation depend

on the reaction conditions, I have not included any coefficients
for the Fe protein, MgATP, or protons in equation (20).

Because dinitrogen must be reacted or "fixed" before it
becomes available for incorporation into biological molecules,
nitrogen fixation is biochemically and ecologically critical. The
availability of fixed nitrogen is often the limiting factor in protein synthesis, and providing fixed nitrogen through fertilizer is
an energy intensive process whose cost has risen dramatically in
recent years. Understanding the detailed molecular mechanism of
nitrogen fixation, apart from its chemical and biochemical interest,
might eventually have significant technological implications.
Finally, apart from its biological significance and potential
applications, the low temperature, low pressure reduction of dinitrogen to ammonia is a supreme catalytic achievement which every
chemist should like to understand for its own sake.

Until the advent of x-ray absorption spectroscopy using synchrotron radiation, there was no spectroscopic way of examining the
molybdenum in nitrogenase. Therefore, the results obtained by this
method provided the first (and as yet only) structural information
about the Mo site in this protein. Besides the Mo-Fe protein, we
also obtained preliminary spectra on the iron-molybdenum cofactor
(FeMo-co), a low molecular weight fragment which can be isolated
from nitrogenase by acidification and NMF extraction of the intact
protein (7). Elemental analysis shows the cofactor to contain
$8Fe:6S^{2-}:1Mo$, and under certain conditions FeMo-co exhibits EPR
signals similar to nitrogenase (8). This cofactor may well contain
the active site of nitrogenase, since addition of FeMo-co to inac-
tive proteins of mutants which are lacking in cofactor can restore
nitrogenase activity.

2.2 The Nitrogenase Mo Absorption Edge Region   In figure 2.1, the
Mo absorption components of lyophilized *C. pasteurianum* Mo-Fe protein (Cp1), crystallized *A. vinelandii* Mo-Fe protein (Av1), and
NMF-extract Fe-Mo cofactor (FeMo-co) are compared. Following the
absorption edge peak at about 20030 eV, there are two distinct
maxima near 20090 and 20160 eV. Another weak peak at 20240 eV is
evident in the Cp1 and Av1 spectra, but this feature is either
absent or obscured by noise in the FeMo-co data. On a visual level
of comparison then, the spectra for all three samples are very
similar, and one can conclude that Mo has a similar coordination
sphere under lyophilized, crystallized, and NMF-extracted conditions. By comparison of the absorption edge region with model
compound spectra, and by curve-fitting the EXAFS, we have obtained
a good idea as to what this Mo site actually is.

Detailed pictures of the absorption edge region for Cp1, Av1,
and FeMo-co in their as-isolated (semi-reduced) states are

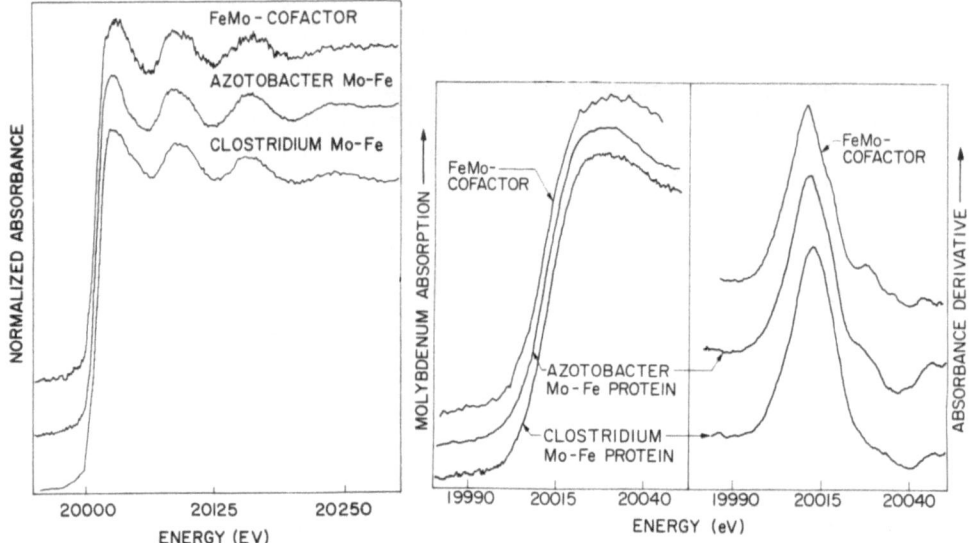

Figure 2.1 - Mo Absorption   Figure 2.2 - Mo Absorption Edge Region
            Components

presented in figure 2.2. Comparison with the model compound data
of table 2.1 shows that the nitrogenase and cofactor edge inflection points all fall in a region characteristic of sulfur ligation.
Furthermore, the shape of the absorption edge, a monotonic rise
with a single inflection point, permits us to rule out Mo=O bonds
for nitrogenase Mo. When such groups are present, a new low energy
is invariably observed, with sufficient intensity to produce a two-
inflection point absorption edge as illustrated in figure 2.3.

Table 2.1 Mo Complex Absorption Edges

Compound	Formal Mo Ox State	Inflection Point (eV)
MoS_2	IV	20 010.4
$Mo(S_2CNEt_2)_2(S_2C_6H_4)_2$	V	20 010.5
MoS_4^{2-}	VI	20 011.1
$MoCl_3$	III	20 010.7
$MoCl_4$	IV	20 012.7
$MoCl_5$	V	20 014.3
MoO_2	IV	20 015.4
$MoO(OH)_3$	V	20 017.7
MoO_3	VI	20 018.9
Nitrogenase or FeMo-co		20 011$\pm$1

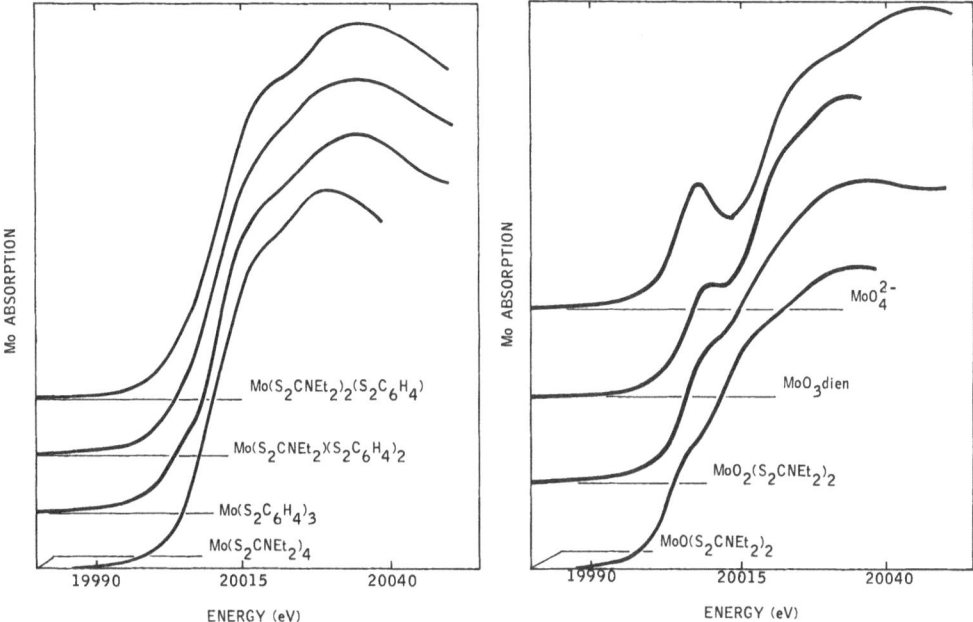

Figure 2.3a - Mo Edges of   Figure 2.3b - Mo Edges of
Compounds with all S Ligands Compounds with Mo=O Bonds

Although we can say from the edge position that it is
unlikely that the Mo has more than one oxygen or nitrogen ligand, nothing can be said about the Mo oxidation
state from this data. This is because when sulfur ligation is predominant, the absorption edge position is
insensitive to the changes in formal oxidation state.

> The molybdenum is presumably attached to a
> cysteinyl residue and may be expected to be
> present as an oxomolybdate.
>
> G. Schrauzer (9)

The fact that we can rule out oxo groups from a simple
absorption experiment is extremely useful in eliminating
a large number of possible models for nitrogenase molybdenum. Although the chemistry of the higher oxidation
states of molybdenum is dominated by Mo=O species, it
appears that a more unusual structure exists in nitrogenase.

2.3 The EXAFS Region -   The EXAFS of Cp1, Av1, and
FeMo-co are compared in figure 2.4, and as with the edge
spectra, there are strong similarities in all three
cases. Some discrepancy exists between FeMo-co and the
intact protein spectra near about 9 Å^{-1}, and further
work is necessary to see if these differences are real.
However, in all three cases we observed a strong beat
pattern in the EXAFS, indicating that at least two Mo-X
distances must be accounted for.

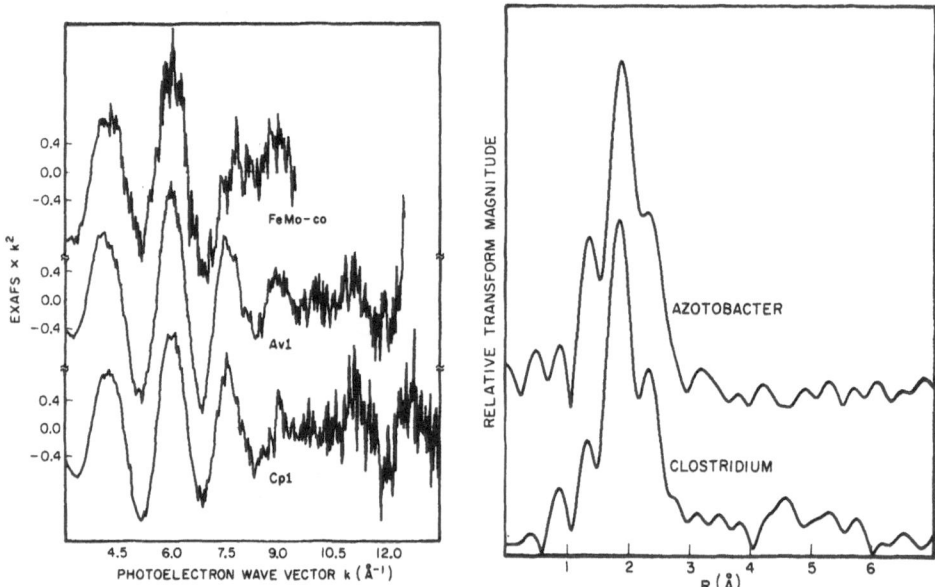

Figure 2.4 - EXAFS of N$_2$ase   Figure 2.5 - Fourier Trans-
                                  forms of N$_2$ase EXAFS

        As expected, the Fourier transforms of both Av1
and Cp1 EXAFS show two major peaks, one at about 2 Å and
one at 2.35 Å. When average phase shifts are applied to
these peaks, it turns out they represent Mo-X distances
of roughly 2.35 Å and 2.7 Å. However, from the transform alone we have no clue as to the nature of X. From
the absorption edge data it appeared that Mo was surrounded by sulfur ligands, and it was reasonable to
assign the major peak in the EXAFS transform as a Mo-S
interaction at about 2.35 Å. Still, the nature of the
longer distance component was not immediately obvious,
since it could represent a longer Mo-S interaction, or
Mo-Fe, or Mo-Mo. Thus, a major goal of the curve-fitting
analysis was to determine the elemental type of scatterer
at about 2.7 Å.

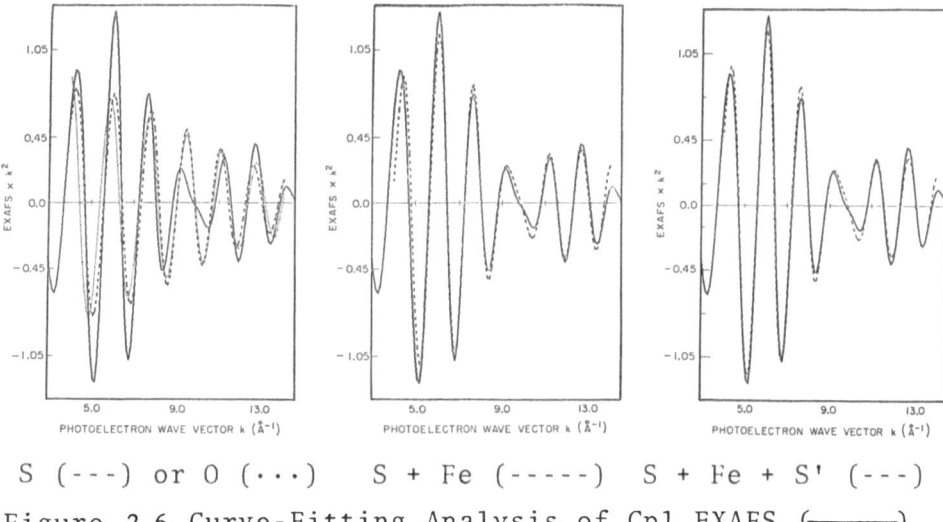

S (- - -) or O (· · ·)   S + Fe (- - - - -)   S + Fe + S' (- - -)

Figure 2.6 Curve-Fitting Analysis of Cp1 EXAFS (————)

In figure 2.5 is shown a series of fits which illustrates the logic we used in analyzing the Cp1 EXAFS. First, the data was Fourier-filtered, keeping only the components which contributed between about 1 and 3 Å in figure 2.5 The first fit with a single Mo-S wave was able to reproduce many of the principle peaks and troughs in the EXAFS, while using a Mo-O wave gave a worse fit, especially near the ends of the spectrum. Addition of a second wave for Mo-Fe permitted reproduction of the beat pattern of the EXAFS, and Mo-Fe parameters gave much better results than did Mo-S or Mo-Mo. Final addition of another longer Mo-S interaction gave a fit which reproduced the EXAFS almost perfectly. The values obtained from this fit corresponded to 3 or 4 S at 2.35 Å, 1 or 2 longer S at 2.49 Å, and 2 or 3 Fe at 2.72 Å. Results within 0.01 Å of these were obtained on the Av1 data. Although EXAFS provided types, numbers, and distances of atoms surrounding the molybdenum, it was still necessary to use crystallographic information and chemical intuition to obtain the structure of the Mo site.

When these x-ray absorption studies were begun, there were really no good Mo-Fe-S model compounds available, and two different types of sulfido-bridged Mo-Fe interactions seemed possible. In one case (structure I) the Mo is part of a cubane-like structure similar to previously studied $[Fe_4S_4(SR)_4]^{n-}$ complexes, while in an

alternate possibility (structure II), the Mo is part of
a more extended array of Fe and S.

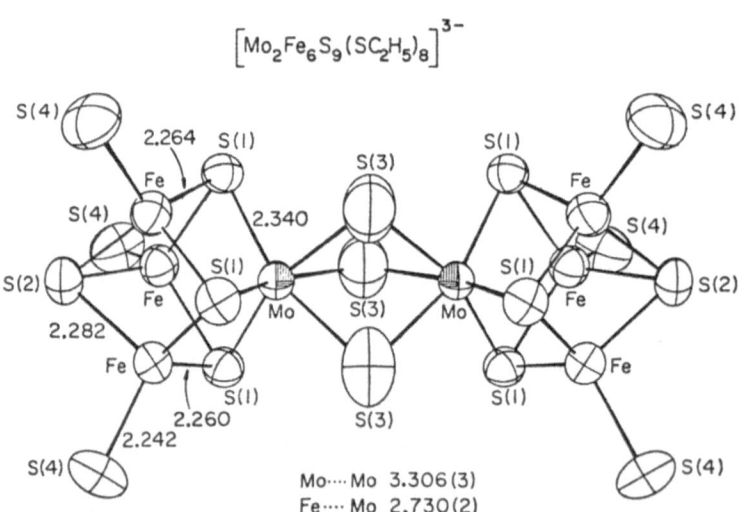

Recently, synthetic work in two different laboratories
(10) has yielded compounds with structures similar to (I).
We subsequently obtained Mo EXAFS data for one of these
compounds, $[Mo_2Fe_6S_9(SC_2H_5)_8]^{3-}$, and its EXAFS shows considerable similarity with nitrogenase EXAFS. The crystal
structure of the MoFe model structure (10a) is given in
figure 2.7, while the EXAFS analysis is shown in figure
2.8. Although the MoFe core appears to be the same in
both cases, it appears that the ligation of Mo outside
the cluster core is different in the model and in the
protein. These differences can be attributed to the
nature of the S-bridge in the synthetic analogue, and if
one reproduces just the core components of the EXAFS,
the correspondence between the model and N_2ase is quite
good. It seems at this point that the case for a structure similar to (I) in nitrogenase is virtually airtight.

Figure 2.7 - Crystal Structure of Holm's MoFeS Model

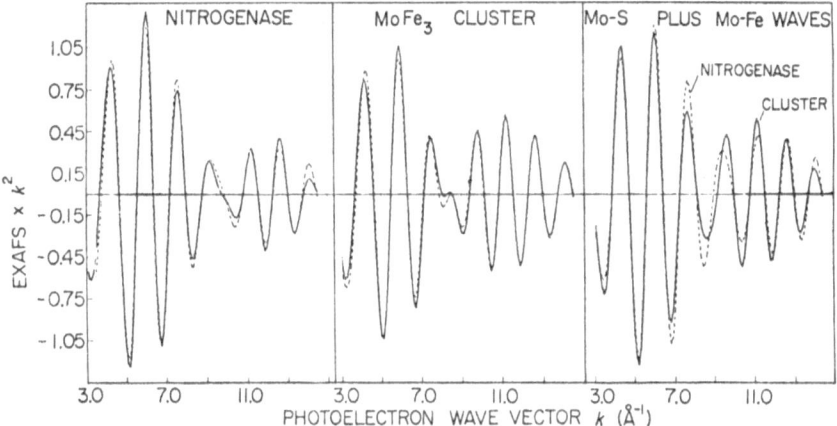

Figure 2.8 - Comparison of Cpl and $[Mo_2Fe_6S_9(SC_2H_5)_8]^{3-}$

2.4 - Chemistry - Until now, I have said little about studying the chemistry of nitrogenase. A large number of redox levels are available for this enzyme, and we have examined the Mo absorption edges of four of these states. As you can see in figure 2.9, for the reversibly dye-oxidized or reversibly fully-reduced states, no significant changes can be detected in the Mo edge. However, for the irreversibly air-oxidized enzyme, a second inflection point is now observed, corresponding to the formation of Mo=O bonds. A comparison of the air-oxidized nitrogenase data with that for two Mo compounds with Mo=O bonds is given in figure 2.10.

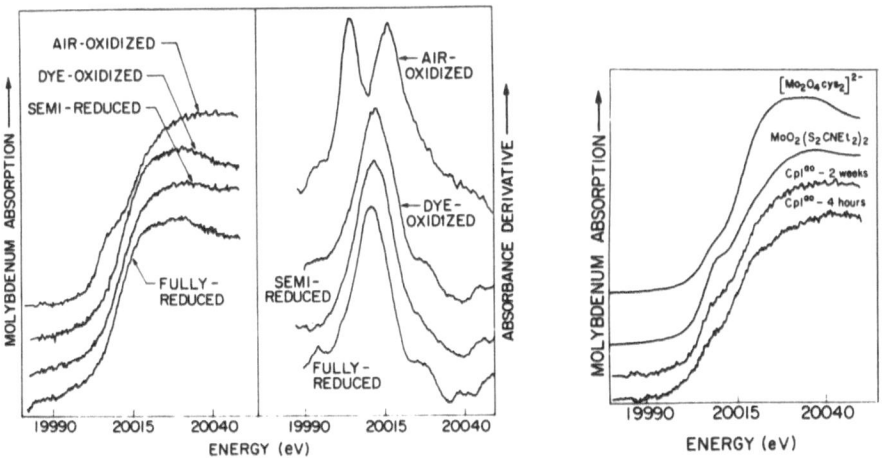

Figure 2.9 - Mo Edges of Different Protein Oxidation Levels

Figure 2.10 - Oxidized Mo Edges

Because the charge in a cluster such as $\underline{I}$ is highly de-localized, we would not expect significant edge shifts for the Mo upon reduction or oxidation. Thus, we cannot say whether or not a species such as this is involved in either the EPR or the over-all charges in protein oxidation level. However, we are extremely interested in determining whether or not Mo is involved in inter-actions with N_2ase substrates. Unsaturated adducts such as CO or N_2 might produce significant changes in edge shape and position, and studies along these lines are currently in progress.

At this point, it is useful to consider what were the important factors which enabled us to "solve", at least partially, the Mo site structure of N_2ase. Without synchrotron radiation, of course, none of the data could have been collected. I think it was also essential that we obtained a library of spectra on over two dozen Mo compounds, for by developing and testing the curve-fitting procedures on known structures the reliability of the method could be constantly tested. It also helped that we had in hand EXAFS spectra of 2Fe-2S and 4Fe-4S clusters; parallels between known Fe-S cluster chemistry and conceivable Mo-Fe-S chemistry were constantly suggesting themselves. It was fortunate that coincident with our work, other groups had shown that the Fe-Mo cofactor and the Mo cofactor were biochemically distinct, for it would be impossible to reconcile structure I with the Mo sites of sulfite oxidase and xanthine oxidase, where the iron is known to be tied up in hemes or 2Fe-2S clusters. Finally, one crucial piece of information was the theoretical Fe backscattering phase shift calculated by Lee *et al.*, for at the time no suitable Mo-Fe models were available. Thus, it was only through the combination of parallel work in chemistry, biochemistry, and physics that a plausible structure for N_2ase Mo site emerged.

Lecture 3 - XANTHINE OXIDASE AND SULFITE OXIDASE

It is an obvious extension of the previous nitrogenase work to use EXAFS to solve the structures of other molybdenum pro-teins. In this regard, I have been collaborating with Professor Harry Gray at Cal Tech and Professor K. V. Rajagopalan at Duke University on the enzyme sulfite oxidase, while Professor Keith Hodgson, Thomas Tullius, Donald Kurtz and Steve Conradson have been working on xanthine oxidase at Stanford University. Because the work on sulfite and xanthine oxidasesmakes such a nice story when presented together, I will discuss the results of both groups at the same time.

3.1 - Sulfite Oxidase, Xanthine Oxidase, and the Mo Cofactor -

Both sulfite oxidase and xanthine oxidase catalyze two-electron substrate oxidations which may also be viewed as oxo-transfer
reactions. For sulfite oxidase the physiological electron acceptor
is cytochrome c, while for xanthine oxidase O_2 serves as the electron sink. In table 3.1 we compare some of the physical and chemical properties of the two enzymes.

Table 3.1 Properties of Sulfite and Xanthine Oxidases

	M. W.	Prosthetic Groups	Mo Potentials
sulfite oxidase	110,000	$2x(Mo,b_5)$	~ 50 mv $(6{\rightarrow}5)$ ~ -50 mv $(5{\rightarrow}4)$
xanthine oxidase	275,000	$2x[Mo,FAD,(Fe_2S_2(SR)_2),$ $(Fe_2S_2(SR)_2)'^2]$	~ -355mV $(6{\rightarrow}5)$ ~ -355mV $(5{\rightarrow}4)$

$$SO_3^{2-} \xrightarrow{\text{sulfite oxidase}} SO_4^{2-}$$

Sulfite oxidase and xanthine oxidase are related by a common molybdenum cofactor ("Mo-co") which can be removed from the intact protein by acidification and extraction with N-methylformamide (11) .
The extracted cofactor can then be used to reconstitute other Mo
enzymes which are deficient in Mo-co, such as the nitrate reductase
manufactured by a mutant of *Neurospora crassa* in the nit-1 assay.

Much of the early cofactor biochemistry assumed that nitrogenase and other Mo enzymes contained the same Mo cofactor; this
unfortunate error was caused by working with impure nitrogenase
samples which probably contained residual nitrate reductase. We
now know that Mo-co and nitrogenase FeMo-co are completely different entities, and in fact, they can be nicely separated on a
G-100 column in NMF. From the previous work on nitrogenase we feel
reasonably certain that FeMo-co contains a MoFeS cluster, and now
by examining sulfite oxidase and xanthine oxidase we would like to
elucidate the nature of Mo-co.

We know sulfite oxidase contains only one other type of prosthetic group, a cytochrome-b_5-like unit, and it appears reasonable
to regard the enzyme as a 2-electron (oxo transfer) Mo catalytic
center which is reoxidized in 1-electron steps by the b_5 site.

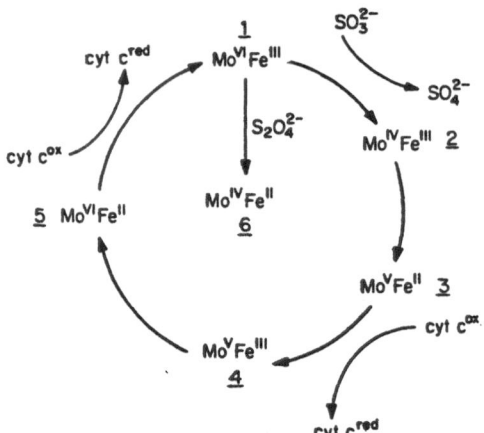

Figure 3.1 - Catalytic Cycle

Note that, as shown in figure 3.1, reduction of the
enzyme by sulfite results in the Mo(V) state, because
the initial Mo(IV) state is rapidly reoxidized by the b_5
site. We have preliminary data which suggests that the
Mo(VI)-Mo(V) potential is roughly 50 mV, the Mo(V)-Mo(IV)
potential is roughly -50mV, and the b_5^{ox}-b_5^{red} potential
is about 70mV, which agrees nicely with the above scheme.
Unpublished kinetic results indicate that the rate determining step is the reduction of Mo by sulfite, with subsequent electron transfer between Mo and heme being very
fast.

For xanthine oxidase, the presence of a large number of prosthetic groups with similar potentials results
in a more complicated reaction scheme, in which electrons are distributed throughout the enzyme according to
the relative potentials of the various sites. The active
site is still the Mo, however. Significantly, the Mo
potentials appear to be about 300 mV lower than the sulfite oxidase values. Just as with sulfite oxidase, however, the Mo is presumably Mo(VI) in the resting state
enzyme, and Mo(IV) in the presence of excess dithionite.

3.2 - Mo Absorption Edges of Sulfite and Xanthine Oxidases
As shown in figures 3.2 and 3.3, there are dramatic
differences between the Mo absorption edges of nitrogenase and xanthine oxidase, while xanthine oxidase and sulfite oxidase exhibit nearly identical edges. From the
previous discussion of nitrogenase edges it should be

evident that the shape of the oxidase edges indicates
the presence of doubly-bound oxo groups, and comparison
with the data in figure 2.3b suggests that one or two
Mo=O bonds are present in all of the oxidase states yet
examined. These edges are the most solid evidence yet
for Mo=O in the molybdenum oxidases, although such bonds
were hinted at as early as 1966 because of the strong
anisotropy in the EPR g values.

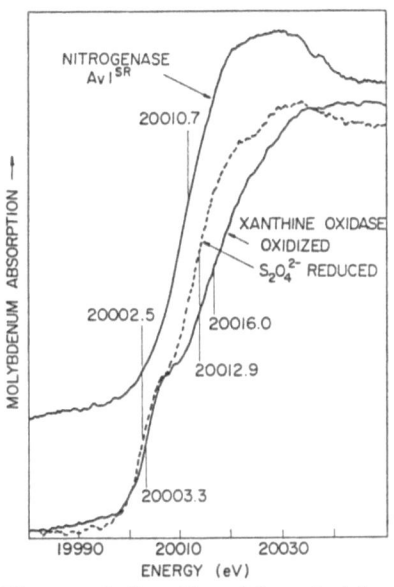

Figure 3.2 - Xanthine Oxidase
vs. N₂ase

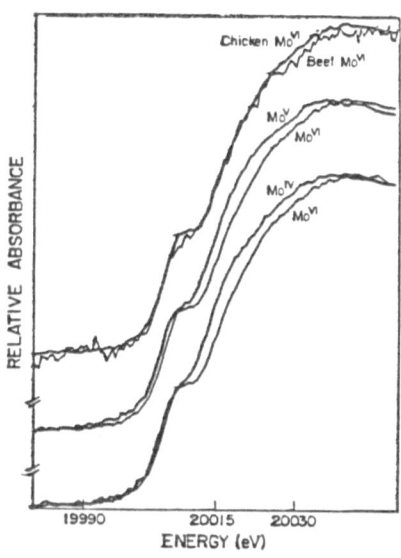

Figure 3.3 - Sulfite Oxidase

3.3 The EXAFS Region

As with nitrogenase, the EXAFS of sulfite and xanthine
oxidases shows a pronounced beat pattern, indicating the presence
of at least two components, hence two different Mo-scatter interactions. Early EPR work, chemical studies, and the position of
the absorption edge all hint at the presence of sulfur ligands,
and this is confirmed by the EXAFS analysis. For xanthine
oxidase, addition of a Mo-S wave to the short Mo-O wave results
in a fit which reproduces all of the essential features in the
EXAFS. The curve-fitting analysis by Tullius, Hodgson, Kurtz, and
Conradson also found significant improvement through addition of
another long Mo-S wave, while addition of a final Mo-N wave produced a negligible improvement in the fit. The scatterer numbers
and distances obtained by this group are given in Table 3.2, and
a representative three wave fit is shown in Figure 3.4.

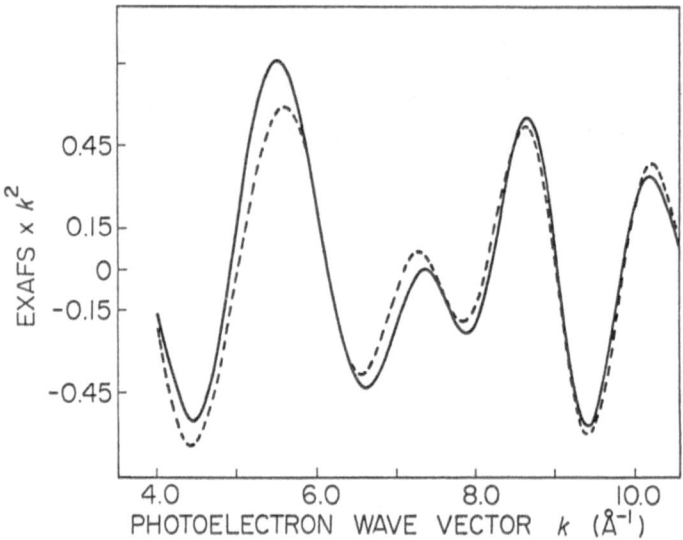

Figure 3.4 O,S,S´ Xanthine Oxidase Fit

Before I discuss the structural implications of the xanthine
oxidase results, I would like to present the EXAFS data for sulfite
oxidase as well. Here we are fortunate to have spectra for the
oxidized (Mo(VI)), sulfite-reduced (Mo(V)), and dithionite-reduced
(Mo(IV)) cases. We shall attempt to derive a structural model for
the oxidases which can explain both the changes in structure with
Mo oxidation state and the difference between xanthine oxidase
and sulfite oxidase. I hope you will appreciate that the Mo
oxidase EXAFS story is not yet as tidy as the previous N_2ase work.
However, I think it will be educational to see an EXAFS structure
analysis in progress, so that the logic of the method will be more
obvious.

Table 3.2 Xanthine Oxidase Fitting Results[a]

Shells	χ^2	O	S	S'
O	3.35	1.71[1.8]		
O,S	0.67	1.71[1.5]	2.52[2.0]	
O,S,S' short	0.64	1.72[1.5]	2.52[2.0]	2.16[0.2]
O,S,S' long	0.29	1.71[1.5]	2.54[2.1]	2.84[1.1]

[a]Fourier-filtered data

In Figure 3.5, I have presented the EXAFS and fits for the
various sulfite oxidase oxidation levels, as well as data on two

model compounds, $MoO_2(SCH_2CH_2)_2NCH_2CH_2SCH_3$ and $MoO(S_2CNEt_2)_2$. In all five cases, the EXAFS Fourier transform shows 2 peaks, corresponding to Mo=O and Mo-S interactions. Finally, I have illustrated the improvement in the quality of the fit, as measured by x^2, as various components are added. The scatterer numbers and distances obtained from the fits are in Table 3.3.

As you can see, EXAFS does not immediately give you the structure around the x-ray absorbing atom, it merely gives approximate numbers of atoms and distances. Furthermore, since every addition of a new component produces some improvement in the fit, one must decide which components are real. Finally, you always have to keep in mind the possibility of not seeing a particular Mo-X interaction.

For sulfite oxidase and xanthine oxidase, the most likely conclusion is that there are two oxo groups in the oxidized state, with Mo=O bond lengths of 1.71 Å in both cases. There also appear to be two medium-distance sulfurs at 2.42 Å in sulfite oxidase and at 2.54 Å in xanthine oxidase, and in both enzymes, a longer Mo-sulfur distance of 2.84 Å. Addition of a Mo-N wave produced no significant improvement in either fit, but this experience was also encountered in fitting the EXAFS of the above tripod ligand. Basically, a single Mo-N component is impossible to detect amidst strong Mo=O and Mo-S EXAFS.

Although on the basis of the EXAFS alone we might be forced to stop here, there are two facts from oxomolybdenum chemistry which allow us to go quite a bit further. First, oxomolybdenum groups are almost invariably *cis* to each other, and second, bonds *trans* to oxo groups are lengthened relative to comparable bonds in a *cis* geometry. Combining these facts with the EXAFS results, it appears conceivable that in one case (xanthine oxidase), the medium length sulfurs are *trans* to the oxo groups, while in the other case (sulphite oxidase) the medium sulfurs are *cis* to the oxo's, thus:

xanthine oxidase   sulfite oxidase

In this picture we have assumed six-fold coordination of the Mo, but have as yet little clue as to the nature of Z or Y. Furthermore, in this picture it is hard to explain why Mo-S_ℓ in xanthine oxidase (2.84 Å) is the same as Mo-S_ℓ in sulfite oxidase, despite

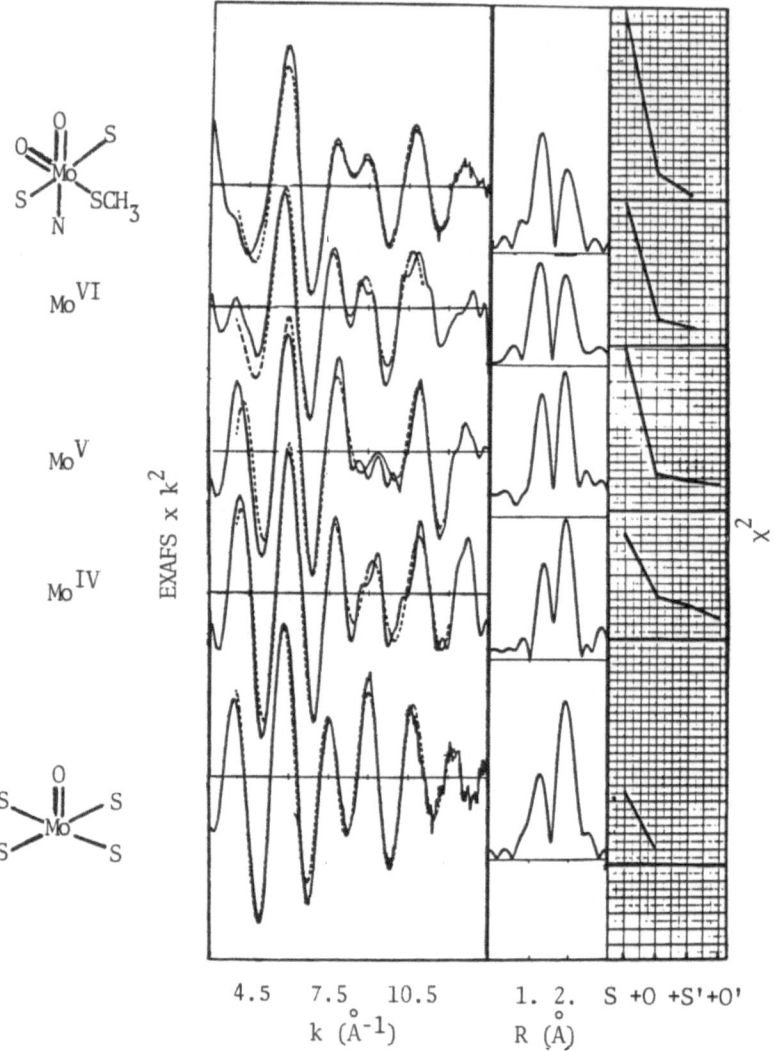

4.5   7.5   10.5   1. 2.   S +O +S'+O'
        k (Å⁻¹)   R (Å)

Figure 3.5 - Sulfite Oxidase and Model Fits and Fourier Transforms

the fact that in the first case, S$_\ell$ is assumed *cis* to Mo=O, while
trans in the second. The picture becomes still more complicated
when we consider data on the reduced forms of sulphite oxidase.

For both reduced forms of sulphite oxidase, the EXAFS fits
indicate not just two but three medium distance sulfurs, at shorter
distances than in the oxidized form. Furthermore, there now

Table 3.3   Curve-Fitting Results for Sulfite Oxidase and Model Compounds

Sample	Mo-S	Mo=O	Mo-S'	Mo-O'	Mo-N	χ^2
oxidized	2.430					4.364
sulfite oxidase	(2.3)					
	2.421	1.708				0.752
	(2.1)	(1.7)				
	2.416	1.707	2.840			0.567
	(2.0)	(1.8)	(0.7)			
sulfite-reduced	2.375					5.040
sulfite oxidase	(2.6)					
	2.367	1.710				1.098
	(2.5)	(1.8)				
	2.368	1.711	2.907			0.935
	(2.5)	(1.8)	(0.7)			
	2.362	1.719		2.008		0.764
	(2.9)	(1.8)		(1.0)		
	2.363	1.722			2.008	0.740
	(2.9)	(1.9)			(1.6)	
	2.364	1.723	2.897		2.010	0.561
	(2.9)	(1.9)	(0.7)		(1.6)	
dithionite-reduced	2.382					3.204
sulfite oxidase	(3.0)					
	2.378	1.688				1.336
	(2.9)	(1.2)				
	2.378	1.690	2.839			1.063
	(2.9)	(1.3)	(0.8)			
	2.379	1.682		2.058		0.974
	(3.3)	(1.1)		(1.0)		
	2.380	1.686			2.063	0.937
	(3.2)	(1.0)			(1.6)	
	2.378	1.691	2.849		2.056	0.651
	(3.3)	(1.0)	(0.9)		(1.6)	
$MoO_2[(SCH_2CH_2)_2$	2.406					5.359
$NCH_2CH_2SCH_3]^{2-}$	(2.0)					
	2.404	1.694				0.312
	(1.7)	(2.0)				
	2.399	1.694	2.799			0.200
	(1.7)	(2.1)	(0.5)			
$MoO(S_2CNEt_2)_2$	2.426					2.307
	(3.9)					
	2.429	1.663				0.501
	(3.7)	(1.2)				

appears to be a single oxygen or nitrogen ligand at about 2.05 Å. Finally, there seem to be two oxo groups in the sulphite-reduced state, but only a single Mo=O in the dithionite-reduced (Mo(IV)) state. There are several ways to explain these data, but instead of discussing the relative merits of each, I will concentrate on the one I think best.

With regard to the appearance of an extra S ligand, it seems possible that this ligand was present in the Mo(VI) state, but at a distance unresolvable from the shorter Mo-S bonds. Under conditions where the separation in Mo-S bond lengths is about 0.05-0.15 Å, 2 shorter and 1 longer Mo-S bonds might be seen as 2 bonds with an intermediate bond length. Thus, we might really have 2 S at about 2.4 Å and 1 S at about 2.5 Å in the Mo(VI) state.

The nature of the oxygen or nitrogen ligand observed at about 2.05 Å in the reduced data is of considerable interest. Stiefel has proposed that molybdenum enzymes possess a ligand, most likely nitrogen, that has a dissociable proton which is responsible for the protein splitting observed in the Mo(V) EPR. Although such a ligand is one possibility, I think an even better assignment is that Mo-X represents an Mo-O interaction in which the oxygen comes from bound sulfite or sulfate. Sulfite oxidase is known to bind anions, and since about a 50-fold excess of sulphite (and then dithionite) was used to reduce the enzyme, the binding of sulphite or product sulphate should be near saturation. Of course, we still need to explain the proton-splitting of the EPR, but I think it is reasonable to assume protonation of one of the oxo groups.

We also need to explain the shortening of the Mo-S bond length upon going from Mo(VI) to Mo(V), which is contrary to the general trend of longer bond lengths for lower oxidation states. Both this and the 7-coordinate nature of the Mo(V) state can be explained by postulating a pentagonal bipyramidal geometry about the molybdenum, with the only significant *trans* effect being on the single long sulfur. Finally, the loss of an oxo-group in going to the Mo(IV) state can explain the Mo(IV) data. We thus arrive at a self-consistent scheme for the sulfite oxidase data in terms of an oxo-molybdenum site with four sulfur ligands and the capacity to bind anions by assuming a 7-coordinate geometry.

The fact that our scheme assumes that an oxo group is regained between the Mo(IV) and Mo(V) states has interesting repercussions on the interpretation of the EPR data, since there should be a transient Mo(V) intermediate with a single Mo=O. This may well be the cause of the so-called "very rapid" EPR signal seen by Bray, et al., in their freeze-quench studies of xanthine oxidase. We have already assumed that it is protonation of one oxo group which leads to the observed hyperfine splitting. The "very rapid" EPR signal does not exhibit such splitting, perhaps because a single $Mo^V=O$ group is not basic enough to protonate at pH 7-9.

Although I hope you will find this model appealing, I will be the first to admit that it is rather a delicate house of cards at the moment. We have as yet said nothing about the chemical nature of the various S ligands, and a lot of work needs to be done on chemical analysis of the Mo cofactor. It is not yet clear whether there is a ligand difference between xanthine oxidase and sulfite oxidase, or whether it is purely a matter of geometrical isomerism. Finally, we really need more EXAFS data on 1) the protein at pH 7 (instead of 9), 2) the reduced forms in the absence of SO_3^{2-}/SO_4^{2-}, 3) the oxidized form in the presence of SO_4^{2-}, and 4) all forms in the presence of other anions such as Br^- and SeO_3^{2-}. Perhaps I will be able to return in a couple of years and tell you the results of such experiments.

REFERENCES

1. T. K. Eccles, Ph. D. Thesis, Stanford University (1977).
2. P. A. Lee and G. Beni, Phys. Rev. B, 16, 2862 (1977).
3. B.-K. Teo, P. A. Lee, A. L. Simons, P. Eisenberger, and B. M. Kincaid, J. Am. Chem. Soc., 99, 3854 (1977).
4. P. A. Lee, B.-K. Teo, and A. L. Simons, J. Am. Chem. Soc., 99, 3858 (1977).
5. S. P. Cramer, H. B. Gray, and Z. Dori, submitted to J. Am. Chem. Soc.
6. For a good review of nitrogenase, see W. H. Orme-Johnson and L. C. Davis, in Iron-Sulfur Proteins (W. Lovenberg, ed.) vol. 3, 15-60, Academic Press, N. Y. (1977).
7. V. K. Shah and W. J. Brill, Proc. Natl. Acad. Sci. USA, 74, 3249 (1977).
8. J. Rawlings, V. K. Shah, J. R. Chisnell, W. J. Brill, R. Zimmerman, E. Munck and W. H. Orme-Johnson, J. Biol. Chem., 253, 1001 (1978).
9. G. Schrauzer, Agnew. Chem., 87, 579 (1975).
10. a) T. E. Wolff, J. M. Berg, C. Warrick, K. O. Hodgson, R. H. Holm, and R. B. Frankel, J. Am. Chem. Soc., 100, 4630 (1978).
    b) G. Christou, C. D. Garner, and F. E. Mabbs, Inorg. Chem. Acta., 28, L189 (1978).
11. a) J. L. Johnson, H. P. Jones, and K. V. Rajagopalan, J. Biol. Chem., 252, 4994 (1977).
    b) P. T. Pienkos, V. K. Shah, and W. J. Brill, Proc. Nat. Acad. Sci. USA, 74, 5468 (1977).

SELECTED READINGS

A. EXAFS Theory

1. E. Stern, Phys. Rev. B, 10, 3027 (1974).
2. P. A. Lee and J. B. Pendry, Phys. Rev. B, 11, 2795 (1975).
3. C. A. Ashley and S. Doniach, Phys. Rev. B, 11, 1279 (1975).

B. Nitrogenase and Mo EXAFS

4. S. P. Cramer, K. O. Hodgson, E. I. Stiefel, and W. E. Newton,
    J. Am. Chem. Soc., 100, 2748 (1978).
5. S. P. Cramer, K. O. Hodgson, W. O. Gillum, and L. E. Mortenson,
    J. Am. Chem. Soc., 100, 3398 (1978).
6. S. P. Cramer, W. O. Gillum, K. O. Hodgson, L. E. Mortenson, E.
    I. Stiefel, J. R. Chisnell, W. J. Brill, and V. K. Shah, J. Am.
    Chem. Soc., 100, 3814 (1978).

C. EXAFS Reviews

7. S. P. Cramer and K. O. Hodgson, Prog. Inorg. Chem., 25 (1979).
8. R. G. Shulman, P. Eisenberger, and B. M. Kincaid, Ann. Rev.
    Biophys. Bioeng., 7, 559 (1978).
9. P. Eisenberger and B. M. Kincaid, Science, 200, 1441 (1978).

Note added in proof: The original spectra on sulfite-reduced sulfite oxidase were interpreted as indicating two oxo groups on the Mo^V species. Subsequent quantitative EPR titrations have shown that only 40 - 50% of the enzyme is converted to Mo^V under these conditions. Thus, further EXAFS analysis of the Mo^V state is required, using difference spectrum analysis on samples with known Mo^{IV}, Mo^V, and Mo^{VI} concentrations.

THE LAYOUT OF X-RAY OPTICS AND INSTRUMENTS FOR THE USE OF SYNCHROTRON RADIATION AT THE OUTSTATION OF THE EUROPEAN MOLECULAR BIOLOGY LABORATORY IN HAMBURG

H.B. Stuhrmann

European Molecular Biology
Laboratory, Outstation at Hamburg
Notkestraße 85, 2000 Hamburg 52.

1. GENERAL CONCEPT

The number of instruments which can simultaneously receive radiation is given by the ratio between the beam cross section and the size of the optical elements (mirrors, monochromators). The beam has a horizontal width of 25cm. As the vertical divergence of X-ray synchrotron radiation is about 0.4 milliradian, the vertical width of the beam at the entrance of the hall is about 1.0cm.

The spectrum of the radiation depends on the energy of the accelerated electrons (or positrons) and on the radius of curvature of the orbit. At the synchrotron (DESY) the particle energy is very often 7 GeV, whereas at the storage ring (DORIS) our laboratory receives radiation from positrons with energies of 4 GeV (Fig 1). The wavelength spectra are, however, almost identical because the radius of the orbit at DORIS (12.12m) is almost three times smaller than that of the synchrotron (31.7m). The intensity of synchrotron radiation per wavelength unit reaches a maximum at about 0.6A.

The design of instruments must be adapted to the given circumstances that the Outstation is mainly a parasitic user of synchrotron radiation produced by high energy physics experiments. This means the optical elements should be designed for a wavelength region from 0.3 to 3A. As the incident angles may

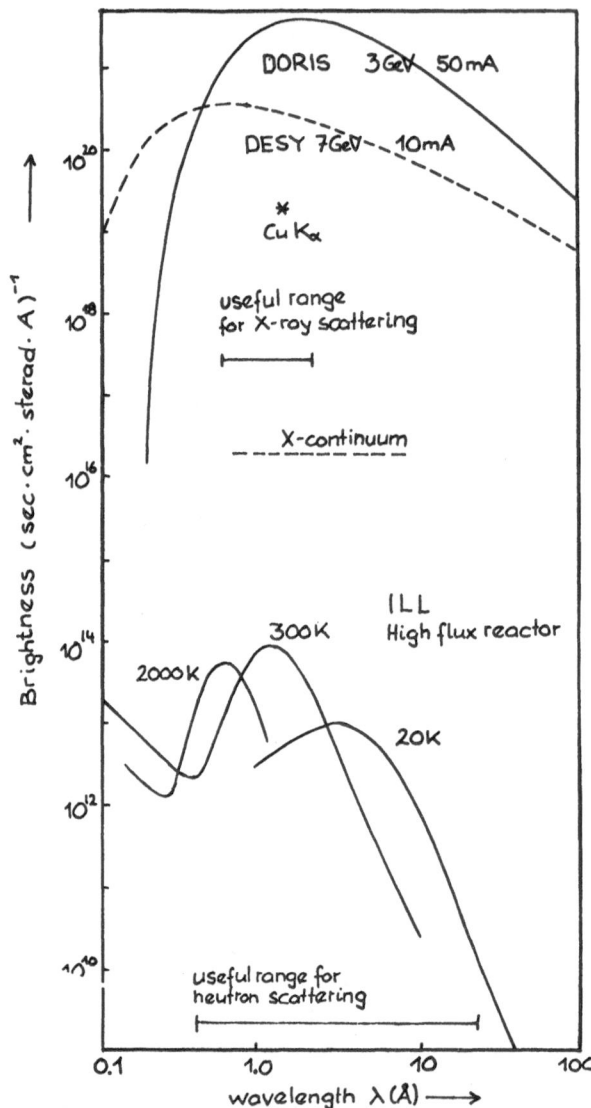

Fig 1   Brightness of synchrotron radiation and thermal
        neutron flux at the accelerators DORIS and DESY
        and the high flux reactor of the Institute
        Max von Laue respectively.

become rather small, very large mirrors and
monochromators would be needed in order to accept a
major part of the beam. In practice, mirrors 160cm
long are used for vertical focussing, and bent
monochromators of 7cm length demagnify the beam in the
horizontal plane. Therefore not more than 5cm
horizontal width are allotted to each instrument, and
since 25cm total horizontal beam width are available,
five instruments can be operated simultaneously in
each laboratory. The installation of five or more
independent instruments in the hall has been guided
by the following principles. In order to minimize
spatial interference, the instruments should be
distributed homogeneously over the area or, preferably,
the volume of the hall. Each user should have
independent access to his instrument, either by way of
an elaborate remote control system for alignment of the
camera and sample control, or by introducing radiation
shielding. For obvious reasons we have adopted the
second alternative whenever possible, after
discussing the design of the shielding with the DESY
radiation protection group.

2. THE OPTICAL LAYOUT

The line-shaped beam, 25cm long in the
horizontal plane, is divided into five sections. The
central part and the extreme outer parts are
reflected by quartz mirrors at grazing incidence. The
eight quartz mirrors in each of the three sections are
20cm long, 5cm wide, and 2cm thick; they are aligned
to form tangents of an ellipse in order to achieve
focussing in the vertical direction. Depending on the
angle of incidence only wavelengths beyond a critical
value will be reflected. The outer beams are
deflected by the monochromator crystals to the right
and to the left respectively (Ge_{111} reflexion); as
the crystals are cut asymmetrically and bent, a point-
line monochromatic beam, useful for diffraction
experiments, is obtained. The central beam has a
broad wavelength distribution; it is used for
diffuse small angle scattering and spectroscopic
experiments.

The beams between the mirrors are deflected
upwards by crystal monochromators. Second
monochromators, 130cm and 190cm above the central
beam respectively, direct the beam back into the

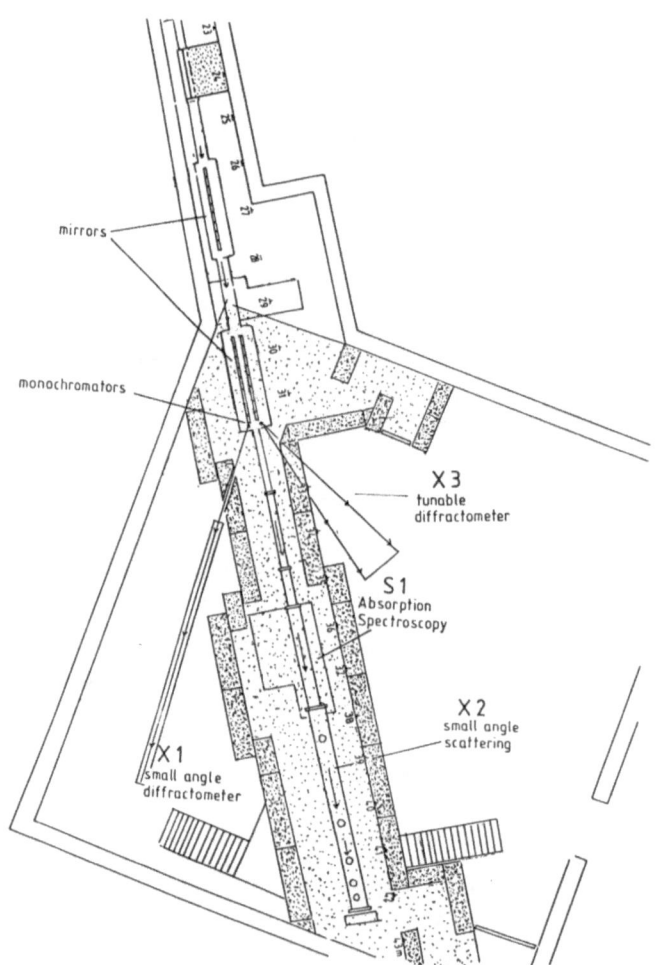

Fig 2 The laboratory of EMBL at the synchrotron DESY.

horizontal plane. The second monochromators are
fixed to H-section girders, 6m long and parallel to the
central beam; the displacement of the crystal along a
girder with concomitant rotation permits change of
wavelength without changing the direction of the beam
emerging from the second monochromator. The upward
deflection of the beam does not lead to any losses
because of its polarization, even at large angles,
where horizontal deflection of the almost linearly
polarized beam would lead to a dramatic decrease of
intensity.

3.   SHIELDING OF INSTRUMENTS AND RADIATION PROTECTION

Instruments along the central beam line are
remotely controlled. This is also true for all
mirrors and monochromators. We preferred to build
tunnels around the main beam. Instruments in the
tunnel can be reached through small holes in the
concrete for sample control and exchange. The tunnels
cover a part of the beam path; they are 2m high, have
an interior width of 1.8m with 0.4m thick walls. The
roofs of the tunnels form an upper floor on which
instruments using the upward deflected beams are
installed. Instruments on this first floor need no
remote control, nor do the instruments on the ground
floor outside the tunnel.

4.   INSTRUMENTS AT THE SYNCHROTRON (LAB 2)

The first diffraction experiments were carried out
in LAB 2 in the early 1970's using a small angle
diffractometer. This instrument was remotely
controlled and has produced diffraction patterns of
muscle and collagen. When the new instruments at the
storage ring became available this instrument was
removed. The shielding was redesigned and built up
in spring 1978 (Fig 2). The construction of the new
optical system follows essentially the ideas outlined
above. At present there is installed an EXAFS
instrument in the tunnel, which is remote controlled,
and a small angle instrument.

The new optical bench (X1), which replaces the
prototype instrument, has been ordered and most of the
parts have been delivered. It will be installed in
autumn 1978.

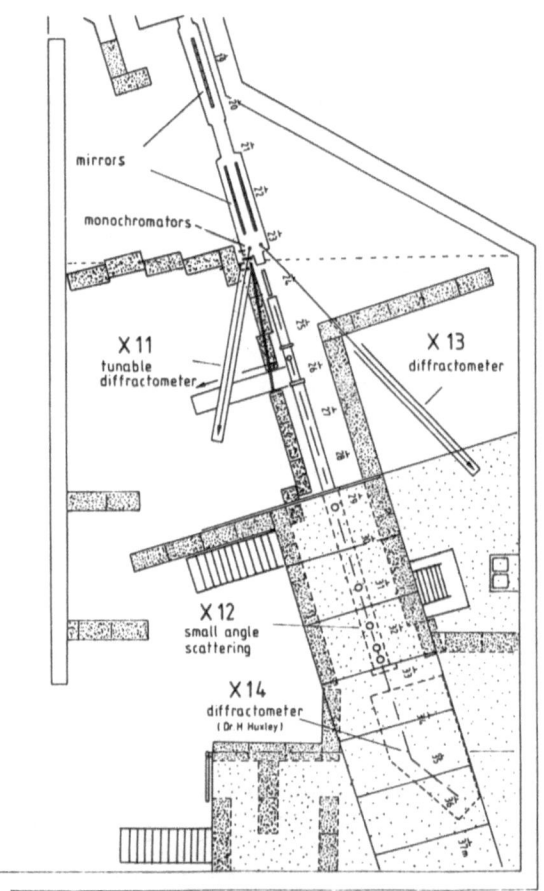

Fig 3   The laboratory of EMBL at the storage ring DORIS.

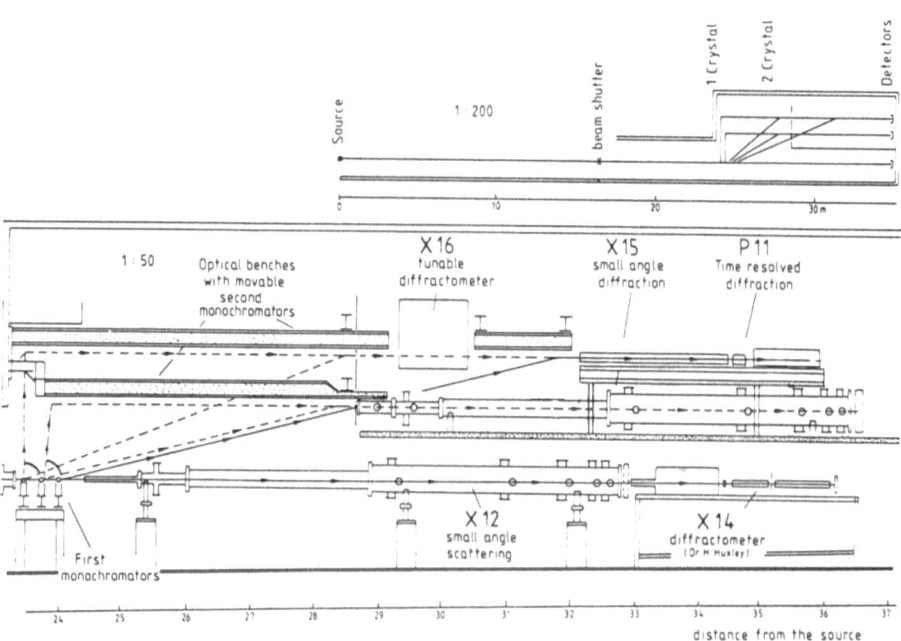

Fig 4   Vertical cross-section of the laboratory of
EMBL at DORIS along the beam axis.

TABLE 1:   INSTRUMENTS AT THE SYNCHROTRON

NAME	INSTRUMENT	PURPOSE
X1 until 20.2.1978	Small angle diffractometer; double focussing system with two mirrors and Ge-monochromator. Constant wavelength (1.54 Å)	1) Diffraction experiments on muscle, collagen, membranes
from Oct. 1978	Double focussing system with eight mirrors and Ge-monochromator. Constant wavelength (1.54 A) corresponds to X13 in LAB 4	3)
X2 * from April 1978 test till Summer 1978	Small angle scattering with "white" radiation, with energy dispersive counters and position-sensitive Gabriel detectors	2) Short time small angle scattering experiments on solutions and suspensions; stopped flow techniques
X3 from Dec 1978	(Movable) optical bench (length 5m) Double focussing system with two mirrors and Ge-monochromators, Wavelength 1.0 to 2.0 Å.	3) Diffraction experiments (for test and development of instruments).
S1 * ready	X-ray absorption spectrograph with two "locked" Ge-double monochromators Energy resolution about 1 eV. (constructed at Daresbury, U.K.)	1) X-ray absorption fine structure of metal proteins

* X2 and S1 can not run simultaneously

INSTRUMENTS AT THE STORAGE RING DORIS

NAME	INSTRUMENT		PURPOSE
Grnd Floor			
X11	Small angle diffractometer, double focussing system with eight mirrors and bent Ge-monochromator. Movable optical bench. Wavelengths:1.0 to 2.0 Å	3)	Diffraction expts. on muscle, collagen & other connective tissues, cryst. of proteins, viruses etc.
X12 ✱	Small angle scattering with "white" radiation. Position-sensitive area detectors of A. Gabriel (20x20cm). Distance sample-detector:25,45,90, 180,360 & 720cm. Use of energy dispersive counters	2)	Short time small angle scattering expts. Stopped flow technique
X13	Small angle diffractometer, double focussing system with 8 mirrors &(bent) Ge-monochromator. Fixed wavelength (1.5 Å).	3)	Mainly diffraction expts. on proteins, viruses
X14 ✱	Diffractometer, double focussing system installed in X14 (brought from Daresbury UK by H.E. Huxley MRC Cambridge)	1)	Diffraction expts on muscles

INSTRUMENTS AT THE STORAGE RING DORIS   (contd.)

NAME	INSTRUMENT	PURPOSE
Upper Floor		
X15	Small angle diffraction; parallel upwards shift of the beam by 130cm 3) with two Ge-monochromators. Wave-lengths: 0.7 to 4.5 A. Position-sensitive area counter. Distance sample-detector: 25,45,90,180,360 and 720cm	Diffraction from connective tissue membranes, solutions
X16 after Sep 1978	Place for a diffractometer. Wave-lengths 1.3 to 4.5 A 3)	Anomalous dispersion
P11 Test	Small angle diffraction. Detectors with a gating frequency adapted to 3) the pulse structure of the radiation (one pulse of 0.1ns each 1 ns)	Mössbauer diffraction

✱   X13 and X14 can not run simultaneously

Footnote to Table
1) Instrument not accessible when beam is on. Completely remote controlled
2) Instrument partly accessible when beam is on. Sample exchange possible while the main beam shutter is open. Completely remote controlled
3) Instrument accessible when beam is on. Mirrors and monochromators remote controlled

Further instruments are planned. They are dedicated mostly to test experiments, because the intensity which is available at the synchrotron is considerably lower than that of the storage ring. However, the synchrotron is running very often and therefore meets ideally our requirement of beam time for test experiments.

5.   INSTRUMENTS AT THE STORAGE RING (LAB 4 AT DORIS)

The instruments on the ground floor of LAB 4 at DORIS are shown in Fig 3. The construction of the optical system is nearly finished. It includes two double focussing systems with eight mirrors and bent Ge monochromators. Another set of mirrors will be installed this autumn.

The double monochromator system (Fig 4) feeding a small angle instrument on the upper floor has been tested with flat asymmetrically-cut germanium crystals of small mosaic spread, and the stability of the beam leaving the second monochromator is very satisfactory. A list of existing and forthcoming instruments is given in Table 1. The instruments X11 and X13 may turn out to be more powerful than X14, and in this case we may envisage the installation of a spectroscopic instrument and/or a diffractometer for anomalous dispersion techniques in the area of X14. Moreover, we plan to focus the central beam also in horizontal direction by using cylindrical mirrors.

SMALL ANGLE SCATTERING OF SOLUTIONS

H.B. Stuhrmann

European Molecular Biology
Laboratory, Outstation at Hamburg
Notkestraße 85, 2000 Hamburg 52

The measured small angle scattering is proportional
to: the flux of the incident beam
      the scattering cross-section of the particles
      the sensitivity of the detector

1.   THE INTENSITY OF THE PRIMARY BEAM

The divergence of the synchrotron radiation received
by the instruments at about 25m distance from the source
is small and therefore ideally suited for small angle
scattering experiments. A nearly point-like primary
beam with respect to the half-width of the central peak
of small angle scattering can be achieved without
severe loss of intensity.

Let us assume a sample with a cross-section of
5✳5mm. At 25m distance from the source about half of
the vertical intensity profile X-ray radiation is
accepted and the horizontal divergence is 0.2mrad.
About 10^{14} photons per wavelength unit (Å) and second
will arrive at this sample area. (DORIS: 3.5GeV, 50mA).
Absorption by windows and air will decrease this value
to about 3.10^{13}.

Rough monochromatisation of the beam is achieved
by total refraction of the X-rays at grazing incidence.
At a glancing angle of 2.5mrad quartz mirrors reflect
only wavelengths greater than 1Å. The absorption will
cut off the spectrum at the long wavelength side (Fig 1).

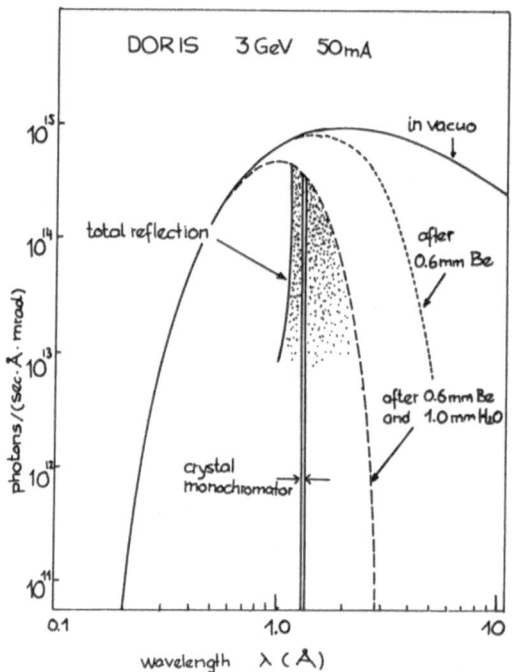

Fig 1   The wavelength spectrum of synchrotron radiation
as emitted from DORIS (3GeV, 50mA) after various
absorbers and total reflexion.

This results in a $\Delta\lambda/\lambda$ of about 0.3 to 0.5, which is still tolerable for the determination of the molecular weight and the radius of gyration from the central peak of small angle scattering (1). A flux of about 10^{13} photons per second would cross the sample under the conditions mentioned above.

A much better monochromatisation is achieved by a crystal monochromator. The wavelength distribution $\Delta\lambda/\lambda$ will then be of the order of 10^{-4}. Though this means a drastic decrease of intensity, the resulting flux of some 10^9 photons is considerably higher than with classical small angle X-ray equipment.

Furthermore we assume a distance of 10m between the sample and the detector. The angular resolution is then 0.5mrad and the smallest angle at which the scatter can be measured would be 1mrad. This means that periodic spacings up to 1500Å can be observed.

Very often biological macromolecules in solution have dimensions of about 100Å. At a resolution of 500Å the sample cross-section can be bigger. It is practical to increase the vertical and horizontal length by the factors 2 and 4 respectively, which enhances the intensity by a factor 8.

2. THE INTEGRAL OVER SMALL ANGLE SCATTERING

The probability W of an X-ray photon to be scattered into the small angle region is:

$$W = \frac{2\pi \int J(z) z \, dz}{P} \tag{1}$$

where J(z) is the small angle scattering and P the P intensity of the primary beam. z is the distance of a point in the detector plane from the primary beam. Assuming a two-phase system in electron density as a convenient model for solutions of macromolecules, we obtain:

$$W = 2 \tilde{\varrho}^2 \, v(1-v) \, \lambda^2 \, d \int_0^\infty H(x) \, dx \tag{2}$$

$\tilde{\varrho}$ is the difference in scattering density between the two phases (solute and solvent) and v is the volume of the solute. d is the thickness of the sample. The scattering density is the product of electron density and the scattering length b ($b=0.28 \cdot 10^{-12}$ cm, classical

radius of an electron). Its dimension is $cm/cm^3 = cm^{-2}$.
The characteristic function H(x) resembles the
Patterson function in crystallography (2).

As an example we take spherical particles. H(x)
is then the common volume of two spheres with the radius
r at a distance x. As H(o) is normalised to 1 we
obtain:

$$H(x) = 1 - \frac{3}{4}\frac{x}{r} + \frac{1}{16}\left(\frac{x}{r}\right)^3 \tag{3}$$

At low concentrations (about 1%) the volume fraction of
the solute is nearly one. After integration of H(x)
equation (2) is:

$$W = \frac{3}{2}\bar{\varrho}^2 \upsilon \, x^2 \, d \cdot r \tag{4}$$

As the scattered intensity will partially be absorbed
in the sample the probability of a scattered photon to
leave the sample is:

$$W = \frac{3}{2}\bar{\varrho}^2 \upsilon. \, \lambda^2 \cdot r \cdot d \cdot e^{-\mu d} \tag{5}$$

where μ is the absorption coefficient. For water μ
(cm^{-1}) is approximately $3\lambda^3$ (λ in Å). The optimal
thickness of the sample is given by the reciprocal of
the absorption coefficient.

Taking water as a typical biological sample d
becomes $1/3\lambda^3$ and the absorption of the sample adds
another factor of $e^{-1} = 0.37$

$$W = 0.17 \, \bar{\varrho}^2 \upsilon \cdot r/\lambda \tag{6}$$

As an example we calculate the contrast $\bar{\varrho}$ of a
protein in solution. The electron density of the
protein ($0.43Å^{-3}$) and that of water ($0.33Å^{-3}$) differ by
$0.1Å^{-3}$. Multiplication of the excess electron density
$0.1Å^{-3} = 0.1 \cdot 10^{24} cm^{-3}$ with the scattering length of one
electron ($=0.28 \cdot 10^{-12}$ cm) yields the contrast
$\bar{\varrho} = 2.8*10^{10} cm^{-2}$. The scattering probability for
proteins therefore becomes:

$$W = 1.33.10^{-4} \cdot v \cdot r / \lambda \tag{7}$$

For myoglobin (c=13.6mg/ml, i.e. v=0.01) we obtain
with r~20Å and λ=1.5Å:

$$W = 1.33_{10}{-4} \cdot 0.01 \cdot 20 / 1.5 = 1.77_{10}{-5}$$

Ribosomes are about 5 times larger in size than myoglobin. There is also an increased contrast by a factor of two, due to the large RNA content. The probability of a photon to be scattered by a 1.7% solution (v=0.01) is therefore:

$$W = 5 \cdot (2)^2 \cdot 1.77_{10}{-5} = 3.5_{10}{-4}$$

It is worth while noting that the dependence of the scattering probability on the wavelength is quite different for neutrons. Due to the fact that the absorption increases only slightly with the wavelength ($\sim \lambda^{1/2}$) a considerable increase of W at longer wavelengths is absorbed. As the contrast of proteins in H_2O is nearly the same for X-rays and neutrons, we obtain:

$$W = 1.33_{10}{-4} \cdot v \cdot r \cdot \lambda^{3/2} \tag{8}$$

Still higher values of W are encountered in neutron scattering of proteins in D_2O. The absorption coefficient is about ten times smaller and remains almost constant. Therefore the λ^2 dependence of W (equation 4) is conserved.

$$W = 3_{10}{-3} \cdot v \cdot r \cdot \lambda^2 \tag{9}$$

3. THE SENSITIVITY OF THE DETECTOR

We restrict the discussion to position sensitive counters with two-dimensional resolution as constructed by A. Gabriel (3). The sensitive area is given by 100 parallel conducting wires of 20mm thickness, 200mm length and 2mm spacing. There are two cathode planes with an area of 200*200mm in front and behind the anode plane. The wires of the cathode planes are connected to delay lines (coils). The distance between the cathode planes is 10mm. The detector chamber is filled with a mixture of argon and carbon dioxide (7:3).

The sensitivity of the detector is given by the maximum path (1) of counter gas which is offered to the photon. 10mm argon diluted with 30% CO_2 has a rather

low efficiency of about 13% for 1.5Å photons. We may
describe the absorption A of X-ray photons by this
counter gas in the following way:

$$A = 1 - e^{-0.042 \lambda^3} \cdot \ell \qquad (10)$$

λ in Å, ℓ is the distance between the cathode planes
in cm. For the purposes of small angle scattering the
sensitivity ($\sim$A) can be increased by adding a flat
drift chamber of convenient depth, in order to convert
a major part of the incident photons into electrons,
the signals of which can be processed.   (3)

RESULTS AND DISCUSSION

First small angle scattering experiments have been
made at various synchrotron radiation facilities:
SSRL Stanford, LURE Paris, VEPP Novosibisk, EMBL Hamburg).
Though the expected intensity of small angle scattering
in many cases should be high enough in order to saturate
position sensitive counters (limit$\sim 10^5$ Hz), this does
not yet appear to be the case at least as far as small
angle instruments in the western countries are
concerned (Fig 3). However, this is not surprising,
as this technique has been adapted to synchrotron
radiation quite recently and considerable improvements
in the monochromator and detector system are still in
progress. It is generally assumed that small angle
scattering with X-ray synchrotron radiation is
interesting in those cases where a drastic reduction
of measuring time is possible. This opens the way to
the observation of structural changes of macromolecules
in solution which occur in seconds or milliseconds
time. A first approach towards this aim has been made
by M. Moody at EMBL quite recently. With a rapid
mixing device it was found that the dissociation of
haemoglobin into its subunits is nearly finished at
50ms after the pH jump to low values.

Using monochromatic radiation ($\sim 10^{10}$ photons/sec)
the integral small angle scattering can be as high as
10^5 to 10^6 photons/sec, and with "white" radiation
($\Delta\lambda/\lambda$ = 0.3) 10^7 to 10^8 scattered photons are
expected , without taking into account the unwanted
background (fluorescence and scattering from metal
slits and monochromators). The minimisation of the
latter effect is an important point in the construction
of small angle instruments, and only little experience
with synchrotron radiation is gathered so far in this
domain.

Once a reasonable compromise is found between wavelength distribution $\Delta\lambda/\lambda$ and the tolerable distortion of small angle scattering the total small angle scattering will saturate the position sensitive photon counting devices known so far. There are, however, new detector systems under construction which finally will allow a counting rate of 10^7 to 10^8 counts per second.

REFERENCES

1. H.B. Stuhrmann, Quarterly Review of Biophysics 11, 71-98 (1978)

2. G. Porod, Kolloidzeitschrift 124 83-114 (1951)

3. A. Gabriel & G. Dupont, Rev. Sci. Instr. 43 1600-1602 (1972)

TWO FORTHCOMING METHODS FOR THE DETERMINATION OF MACRO-MOLECULAR STRUCTURES IN SOLUTION: MÖSSBAUER SCATTERING & SPATIAL CORRELATION OF SCATTERING FLUCTUATIONS

H.B. Stuhrmann

European Molecular Biology
Laboratory, Outstation at Hamburg
Notkestraße 85, 2000 Hamburg 52

The structure analysis from small angle scattering of solutions is by no means unambiguous. This is also true if we assume for the following considerations a dilute solution of equal macromolecules.

The small angle scattering intensity $J(k)$ of a dissolved particle described by its excess scattering density distribution $\rho(\vec{r})$ as an expansion of spherical harmonics $Y_{\ell m}$

$$\rho(\vec{r}) = \sum_{\ell=0}^{\infty} \sum_{m=-\ell}^{\ell} \rho_{\ell m}(r) \, Y_{\ell m}(\omega) \tag{1}$$

is given by:

$$J(k) = \sum_{\ell=0}^{\infty} \sum_{m=-\ell}^{\ell} |A_{\ell m}(k)|^2 \simeq S_{oo}(k) \tag{2}$$

where $A_{\ell m}(k)$ and $\rho_{\ell m}(r)$ are related by Hankel transforms of the ℓ-th order with spherical Bessel functions as a kernel. Inspection of equations (1) and (2) immediately shows that each partial structure $\rho_{\ell m}(r) Y_{\ell m}(\omega)$ results its own scattering function $(A_{\ell m}(k))^2$. $J(k)$ is invariant under rotation of $\rho(\vec{r})$. This is also true for each partial structure. Therefore independent rotation of partial structures changes the total structure but leaves $J(k)$ invariant. This is a simple way to construct different structures $\rho(\vec{r})$ which belong to the same $J(k)$.(1)

The family of possible models related to a scattering function is usually reduced by various assumptions (e.g. symmetry, homogeneity).

Though the condition of random orientation of
macromolecules in solution is most frequently
encountered one can still imagine a situation where an
accurate small angle scattering pattern measured in a
time which is short compared to the rotational
movement of the particles. For reasons of plausibility
we make the (unnecessary) assumption that only one
molecule becomes irradiated and a number of snapshots
of this molecule are taken. The scattering patterns
will all be different as they result from different
orientations of the molecule. The situation will not
change very much if two or more particles are present
in the irradiated sample volume. The fluctuation of
the momentary scattering intensity gives rise to
additional terms in equation (2):

$$D(\vec{k_1}, \vec{k_2}) = \langle J_t(\vec{k_1}) J_t(\vec{k_2}) \rangle = J(k_1) J(k_2) + C(\vec{k_1}, \vec{k_2}) \equiv D(k_1, k_2, \psi) \qquad (3)$$

K_1 and K_2 are vectors of momentum transfer and ψ is the
angle between K_1 and K_2. As $J(k)$ and $D(\vec{K}_1, \vec{K}_2)$ are
measurable quantities, we can see that one can extract
from such an experiment not only the angular average
of the scattering pattern (equation 2), but also the
average of its products for two arbitrary scattering
directions. (2) In terms of an expansion of spherical
harmonics one obtains (2)

$$C(\vec{k_1}, \vec{k_2}) \sim \sum_{\ell=0}^{\infty} P_\ell(\cos \psi) \left[\sum_{m=-\ell}^{\ell} S_{\ell m}(k_1) S_{\ell m}^*(k_2) \right] \qquad (4)$$

The contribution each term, 1, has a well-defined
angular dependence, given by the 1-th **Legendre**
polynomial P_ℓ. This allows to a fair extent the
calculation of $A_{\ell m}(k)$ from the experimental data (2).

What are the experimental requirements for the
observation of fluctuations of the scattering pattern?
The observation time has to be kept short enough in
order to avoid an averaging of the fluctuations due to
the rotation of the dissolved particles. The reciprocal
of the rotational diffusion coefficient D_r can be
chosen as the duration of the snapshot. If a resolution
of about 1/4 of the particle size (or equivalent we
match with spherical harmonics up to 1=4), it can be
shown that the signal to noise ratio is about one, if
a few scattered photons are detected from each particle
in $1/D_r$ seconds (2). A total of 10^{11} scattering
events per second from about 10^9 molecules and
$D_r \sim 10^2 sec^{-1}$ results a signal to noise ratio of one.
A concentration of 0.17 mg/ml of a 10^6 molecular weight

particle in 0.1*0.1*1.0mm scattering volume contains 10^9 particles and the scattering from the solution is about twice that of the solvent. 10^{11} scattered photons require an incident flux of 10^{16} photons per second in the primary beam in a wavelength range $\Delta\lambda/\lambda$ of about 0.3.

The measurement of the fluctuations of small angle scattering can provide a lot more information about the macromolecular structure in solution than any other method under similar conditions. Another approach to get unambiguous models is given by the measurement of defined intramolecular distances. As a successful example we cite the determination of protein-protein distances in E. coli ribosomes by neutron small angle scattering. The proteins of interest are marked by heavy deuteration and are clearly distinguished from the remaining proteins of the ribosomal particle. Originally this method was proposed for the determination of distances between metal atoms in complex biological structures by X-ray scattering (4).

We are now going to consider proteins which contain several iron atoms. This class of proteins deserves special interest because of the scattering properties of the iron atom. X-ray radiation is scattered by electrons, but at certain wavelengths absorption of X-rays by nuclei can be observed as well. The absorbed photons from an excited transition state of the Fe nucleus are emitted with a half time of about 100 nanoseconds. This emission time is by many orders of magnitude greater than the scattering process of photons by electrons.

The delayed emission of X-ray photons by ^{57}Fe (and some other muclei known in Mößbauer spectroscopy) finds a most exciting application with pulsed X-ray sources. The time structure of the synchrotron radiation emitted by the storage ring DORIS can be such that a burst of photons of 0.14ns duration is produced each μs. If the detector is gated off during the arrival of the photons scattered by electrons, the delayed scattering pattern from the iron nuclei can be observed separately. As iron proteins contain only a few iron atoms, the small angle scattering pattern can be simple enough in order to allow a straight forward interpretation in terms of iron-iron distances.

There are different methods which have been proposed for the isolation of the photons scattered by

iron nuclei (3). One of them consists of the use of a
fast scintillation counter, which has to be inactivated
at Megahertz frequency for a time of a few nanoseconds.
This method is being tried at the moment by various
groups.

For both the measurement of fluctuations of small
angle scattering and the measurement of photons emitted
from transition states of Mössbauer nuclei there is one
solution which appears to be most promising. This is
the use of special drift chambers proposed by
J. Hendrix, which can convert the property of the
pulsed synchrotron radiation into spatial resolutions
or follow the time dependence of radiation in the
nanosecond region. The principle of the detector is
simple: The cylindrical detector consists of a central
cathode, which is surrounded by 360 anode wires
equally spaced at 5cm distance from the centre. The
detector is filled with a mixture of Argon and CO_2 and
a field of some thousand volts is applied. It works
as follows: A bunch of photons with a length of 5cm
is travelling along the optical bench and scattered by
the sample. The scattered photons will become absorbed
by the counter gas of the detector, which converts them
into electrons. These small electrons which are
created in various regions of the cylindrical detector
at practically the same time will drift slowly to the
outer anode wires following the electric field. After
a microsecond's time all the created electrons will
have arrived at the outer anode wires and the game
starts again. The time of arrival of each electron
depends on the distance of its "place of birth" from
the circle of anode wires. The time slot for the
classification of the electrons can be as small as
10ns. The construction of both the detector and the
signal processing system is progressing. The main
advantages of the detector are:

- it is a fast counter (more than 10^7 cps)

- the localization of events in a polar coordinate
  system is well-suited for the observation of
  fluctuations of small angle scattering

- it is an ideal detector for all types of kinetic
  investigations in the millisecond region without
  losing the two-dimensional resolution

- it is easily converted into a detector to follow the
  time dependence of scattering patterns in the
  nanosecond region at the sacrifice of radial resolution

REFERENCES

1. H. Stuhrmann, Acta Cryst. A26 297-306 (1970)

2. Z. Kam, Macromolecules, in press

3. G.T. Tramell, J.P. Hannon, S.L. Kuby, Paul Flinn,
    R.L. Mößbauer & F. Parak, to be published

4. W. Hoppe, Israel Journal of Chemistry 10 321-333 (1972)

PROTEIN CRYSTALLOGRAPHY WITH SYNCHROTRON RADIATION

I. GENERAL DISCUSSION AND HIGH RESOLUTION DATA COLLECTION

R. FOURME

Laboratoire de Physicochimie Structurale, Université
Paris XII, 94000 Créteil and LURE (CNRS, Université
Paris-Sud), B 209C, 91405 Orsay, France

I - INTRODUCTION

I will discuss in these lectures the present and possible applications of X-synchrotron radiation to protein crystallography.
Basic underlying physics can be briefly summarised : when
X-radiation impinges on a free electron, the electron is forced
into oscillations of the same frequency as the incident wave and
scatters radiation. Scattered rays have the same wavelength as the
incident radiation and a definite phase relationship to the incident
beam ; hence, scattering is coherent. The scattering pattern of the
electrons in the sample is the Fourier transform of the electron
density. If the sample is a single crystal - e.g. a periodic three-
dimensional array of scatterers - the Fourier transform is a
weighted lattice ; the discrete values of the transform are called
structure factors $F(hkl)$. Except for centric zones, structure factors are complex numbers because protein structures are not centrosymmetric. Usual detectors record the intensities of diffracted
rays, which are proportional to $|F(hkl)|^2 = F(hkl).F^*(hkl)$. In this
process, the phases of structure factors are lost ; this information
has to be recovered by indirect ways from the magnitudes of scattered amplitudes only. Then the data can be back Fourier transformed to obtain the electron density in the crystal ; finally, from
this map and from the aminoacid sequence, the tertiary structure
of the protein is derived.

Even nowadays, the investigation of even a moderately complex
protein is a time-consuming job with many tedious steps. But it is
the only technique which reveals the complete three-dimensional

structure, thus giving an **invaluable** frame in any attempt to
understand functional aspects of the macromolecule.

At first glance,one might think that there are no really com-
pelling reasons to use synchrotron radiation (SR) in protein crystallography : i) Data-collection is only a step among many others,
and the biochemical aspects, including crystal growth, are generally
the most difficult and time-consuming part of the work ii) Contrary
to many VUV or EXAFS experiments, which simply cannot be done
without the SR source, powerful and stable standard X-ray sources
are available iii) There are presently only four facilities in the
world where high intensity X-SR is available (see table 1) and none
of them is dedicated for more than 25 % of the storage ring operation to the production of radiation ; protein crystallography would
require careful, unhurried operation and comparatively long beam
time iiii) To use the full potential of these facilities, many new
technical developments are necessary ; many interesting possibilities
are now rather speculative, due to the lack of suitable electronic
area detectors. In spite of these remarks, the potential of this
new source is high and, in my opinion, it is now somewhat underesti-
mated ; it is likely that within a few years, an appreciable fraction
of the diffraction data will be recorded using SR sources, and
especially those related to the most challenging problems of protein
crystallography.

II - X-RADIATION FROM STORAGE RINGS ; RELEVANT PROPERTIES
AND RELATED PROBLEMS

II - 1) The Most Important Property of SR for Protein Crystallo-
        graphy is the Availability of a <u>Continuous Spectrum</u> with
        a <u>High Brightness</u>

The <u>high brightness</u> is due to the fact that i) the intense
emission of relativistic electrons is essentially confined in the
orbit plane and ii) the electron beam cross-section is small. The
much lower brightness of commercial rotating sources is now near
technical limits set by the mechanical strength of anode materials.
From table 1, one has the flux for operating conditions which are
routinely - if not frequently - attained by relevant storage rings ;
numbers are rather similar near 1.5 Å ; machines with a shorter
critical wavelength are superior below $\sim$ 1 Å. In fact, the practical
results obtained from a given setup depend on many other factors
(and notably source size, source-to-monochromator distance, optics
and collimator, beam-line transmission) and may widely differ.

Table 1

Storage ring (laboratory, location)	Energy (GeV)	Current (mA)	Flux at 1.54 Å (photons.sec^{-1}.mrad^{-1}) $\Delta\lambda = 1$ Å
DCI (LURE, Orsay-France)	1.84	200	1.5×10^{15}
DORIS (EMBL Outstation, Hamburg, BRD)	3.7	35	1.5×10^{15}
SPEAR (SSRL, Stanford, USA)	3.7	35	1.4×10^{15}
VEPP-3 (Novosibirsk, USSR)	2.25	100	1.1×10^{15}

Substantial gains in brightness are expected either for dedicated new rings which are being built (SRS, Photon Factory, Brookhaven II) or for presently available machines optimised for dedicated operation (SPEAR, DORIS). Using a wiggler to decrease locally the critical wavelength, a machine like ADONE could become a good X-ray source.

For a standard beam-transport line with thin beryllium windows, the wavelength can be selected at will throughout the whole range of interest (0.7-3 Å). The best compromise for a given crystal — the optimum thickness is roughly given by the reciprocal of the linear absorption coefficient of the sample - and a given resolution can be achieved. A proper choice of wavelength will permit to use anomalous scattering with greater efficiency and flexibility. As discussed hereafter, a good or very good energy resolution is then necessary ; but even in most demanding cases, enough photons will be available at the output of the monochromator to make any experiment in reasonably good conditions.

II - 2) Other Properties of the Radiation Are Either Useless Now or Unwanted

The time structure of the emission has not yet been used in such experiments. The low duty cycle is rather a disadvantage for detectors which count photons, such as multiwire proportional chambers. The effect of short intense bursts on the protein lifetime with respect to a continuous source of the same average intensity is not known. The polarisation of the radiation could be used in selected cases ; it dictates the unusual setting of the four-circle diffractometers with the equatorial plane vertical rather than horizontal (Phillips et al, 1978). Theoretical expressions of the polarisation characteristics of the light emitted by a single electron on a circular orbit are available (Sokolov and Ternov, 1956). The polarisation ratio at the sample level has to be measured for accurate data collection ; a reasonable estimate can be obtained by

computing, provided that the source characteristics are known
(Lemonnier et al, 1978). The beam should be stable in position and
orientation, because any motion of the beam with respect to the
crystal will modify the polarisation ratio and will introduce errors.
Lastly, the flux of synchrotron radiation is not constant ; it
decreases at a typical rate of 3.10^{-3}/minute ; corrections for this
effect depend on the detection system which is used.

The unusual environment at the synchrotron radiation
facility has also to be mentioned, and especially since we discuss
biological applications. Many pioneering works were done in rather
poor conditions. Those difficulties are going to be gradually attenuated with better sources, more beam time, more flexible planning
and well-equipped biochemical facilities near the SR source. In contrast, speeding-up the experiment is an important advantage.

The high absorption cross-section of matter for X-rays
is the most limiting factor of X-ray scattering. The problem is
specially severe for biological specimens. The absorbed photons
produce heat effects and radiation damage (Barrington-Leigh and
Rosenbaum, 1976 ; Sturhmann, 1978 a).

In the case of high fluxes, the result is impressive : assume a
myoglobin crystal with dimensions of $\sim$ 0.5 mm and a linear absorption
coefficient of $\sim$ 10 cm^{-1} at 1.54 Å ; with a setup using a line-
focussing monochromator as described **hereafter,** the flux will be
at least 2×10^{10} photons.sec^{-1}.mm^{-2} at the sample location ; the
power dissipated in the crystal is then $\sim$ 3μW ! There are
$\sim$ 1.25 10^{15} molecules in this myoglobin crystal and $\sim$ 2.5 10^9photons.
sec^{-1} will be absorbed ; the absorption of each photon induces the
disruption of many molecules, so that the whole crystal is destroyed
within a few hours (6 hours if one assumes that each photon disrupts
25 molecules). Cooling the sample will always be useful to slow-
down the formation of reactive radiation products and evacuate heat.
Of prime importance also is the design of electronic multireflexion
diffractometers with as high collected information per scattered
photon as possible (Schulz and Rosenbaum, 1978).

III - HIGH RESOLUTION DATA-COLLECTION

III - 1) Introduction

Modern protein crystallography is faced with the need for an
improved description of tertiary structures. In most cases, discussion of functional aspects requires a detailed and accurate analysis
of selected molecular fragments. Diffraction data must be collected

to the highest possible resolution for a given sample. Another goal
is tackling more and more complex structures such as complexes like
t-RNA/t-RNA synthetase, multi-component systems, viruses and so on ;
the average integrated intensity of one reflexion is inversely proportional to the unit-cell volume ; at the same time, the angular
resolution has to be high enough to avoid overlapping of reflexions.
Using synchrotron radiation, suitable experimental requirements for
high resolution and/or large structures are fulfilled rather simply ;
they are summarised in table 2 which we will briefly comment, taking
as a working example the setup at LURE (Orsay) which has been routinely used by several groups since 1976 (Fourme et al, 1977).

III - 2) Experimental

    Horizontal focussing and monochromatisation is obtained by
reflexion of the synchrotron radiation on a plane parallel germanium
beam cut at an angle of 5-10° to (111) planes (fig. 1) ; the germanium crystal is shaped to an elongated triangle and mounted as a
cantilever so that a continuously adjustable cylindrical curvature
is obtained using a remotely controlled differential micrometer
screw (Lemonnier et al, 1978). The monochromator is tuneable to any
wavelength between 0.7 and 3.5 Å (fig. 2). An intense line focus
is obtained with a wavelength spread of 4.10^{-3} Å at 1.54 Å. The beam
is slightly convergent in the horizontal plane (horizontal angular
aperture $\sim$ 1.8 mrad) ; in the vertical plane, the low divergence
of the synchrotron radiation is preserved ($\sim$ 0.3 mrad). Two sets of
(X-Y) slits are incorporated. The whole system is connected to the
beam line and is evacuated by a separate pump. The rotation camera
is clamped to an optical bench fixed on a rigid frame rotating about
the same vertical axis as the crystal bender, which greatly simpli-
fies camera alignment when the wavelength is tuned (fig. 3). As
previously mentioned, the intensity of the synchrotron radiation
steadily decreases ; to get diffracted intensities at essentially
the same relative scale for all spots on a given rotation photograph,
the duration of an elementary oscillation has to be a small fraction
of the beam lifetime (say 1 % or less) ; the speed selector of the
commercial Nonius camera was purposely modified so that repeated
oscillations are now possible at any selected speed. In addition
the collimating system is also modified, with an enlargement of the
source aperture of the collimator. Finally, an helium-filled cone
with mylar windows and centred with cross-slides is placed between
the film and the sample. A cooling device will be shortly installed.

    The flux measured behind a 300 μm collimator was 1.4×10^9 pho-
tons.sec^{-1}, i.e. $\sim 2.10^{10}$ photons.sec^{-1}.mm^{-2} ; the storage ring was
operated at 1.72 GeV and 120 mA (the "design" parameters of this
machine are 1.84 GeV and 400-500 mA, so that a five-fold increase
in flux will be ultimately possible). For an Elliott GX6 rotating

Fig. 1 - Crystal bender of the focussing monochromator. An adjustable cylindrical curvature is obtained by displacing the tip of the germanium beam with a remotely controlled micrometer screw.

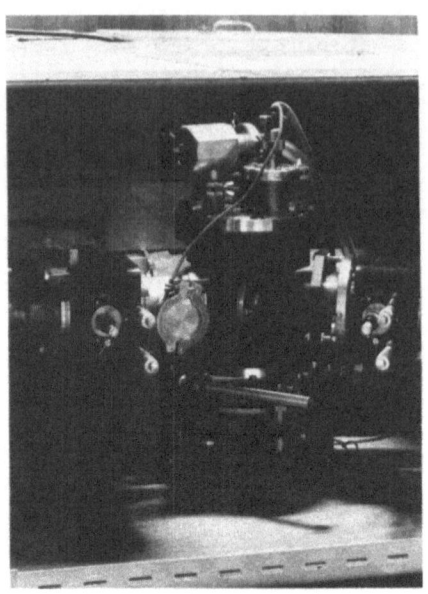

Fig. 2 - Focussing monochromator and slit systems.

Fig. 3 - Experimental setup for protein crystallography at LURE (the monochromator shielding has been partly removed for clarity).

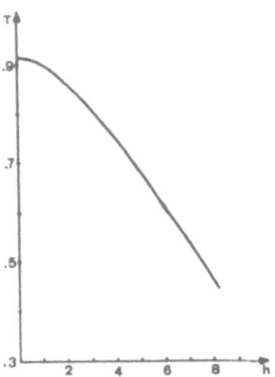

Fig. 4 - Polarisation of the synchrotron radiation at 17.1 m from the source, plotted vs the distance to the average plane of orbit (storage ring : DCI ; energy of stored electrons : 1.72 GeV) ($\mathcal{P} = (I_{//} - I_{\perp})/(I_{//} + I_{\perp})$)

Table 2

Requirements for normal and high-resolution protein crystallography
using photographic films

* source

    . high brightness
    . continuous spectrum
    . positional stability
    . polarisation must be known at the sample location
    . good beam lifetime

* monochromator/collimation system

    . $\Delta\lambda/\lambda \ : \ < 5.10^{-3}$
    . convergent beam   ⎫
                 ⎬ focussing optics
    . intensity   ⎭
    . wavelength may be tuned easily (0.7 - 2.6 Å) and
      displayed
    . remotely controlled focussing distance
    . < 0.1 % harmonics at film level

* camera

    . oscillation system with short oscillations ;
      oscillation speed and number of cycles must be adjus-
      table in a wide range
    . large solid-angle (non-planar cassettes)
    . crystal-to-film distance must be accurately known
      and adjustable between $\sim$ 4-25 cm
    . automatic film changer
    . low-temperature device

* miscellaneous

    . good processing unit for many films
    . accessories to mount and align crystals.

anode source (Elliott Brothers LTD, London) operated at 1.6 kW
(Ni filtered CuKα radiation), Harmsen et al (1976) have reported
9.10^8 photons.sec^{-1}.mm^{-2} for the same collimation system, that is
a factor of about 22 less ; they measured $\sim 10^8$ photons.s^{-1}.mm^{-2}
at the focus for a more powerful rotating anode source (Elliott
GX13) with a double focussing optics. This comparison is much more
significant because angular apertures of the beams are rather similar. The superiority of the synchrotron source is then quite striking, more than two orders of magnitude !

In this setup, the radiation scattered by the sample is essentially free from higher-order harmonics, for $\lambda < 1.6$ Å. This would
be no longer true for a storage ring with a critical wavelength much
shorter than DCI ; hard radiation is in such cases usually removed
by total external reflexion of the beam on a mirror.

The use of this setup is very simple. An interlock system prevents access to the monochromator when the beam shutter is opened.
Access to the camera is in contrast possible at any time. The
alignment of the camera is made practically in the usual way, with
the monochromatic beam attenuated by metal foils. During exposures,
a movable shield encloses the camera. Constraints are then minimi-
sed, which is of prime importance for routine data collection.

Standard flat cassettes are normally used ; "vee" shaped as well
as cylindrical cassettes were found quite useful in the case of
structures with large unit cells. Films are processed with an automatic film scanner. Spots on the photographs are smaller than usual
so that a fine grid scanning is necessary (50 µm steps) ; in such
conditions, the R-factors calculated using the standard formula
$R = \Sigma |I - \bar{I}|/\Sigma I$ (where I and $\bar{I}$ are respectively the intensity of a
reflexion, and the average intensity of symmetry related reflexions)
were comparable to usual values. A modified polarisation correction
was derived for the oscillation geometry (Kahn and Fourme, 1976)
and inserted in our data-reduction programs. The numerical value of
the rate of polarisation was computed with a program written by
De Bergevin taking into account the source characteristics and
assuming that the sample is bathed in the central part of the
synchrotron emission (fig. 4).

Tests and data collection were done with the setup at LURE
during sparse periods between mid-1976 and mid-1978. The operating
conditions of the storage ring DCI during dedicated shifts gradually
improved from 1.56 GeV, 50 mA to 1.8 GeV, 200 mA immediatly after
the injections. Between the injections, the current steadily decreases with a decay of 1/e in 8-11 hours. The positional stability of
the beam during one cycle is good, as well as the reproducibility
from one injection to another. The adjustment of the camera was
found quite stable, although routinely checked at the beginning of

each dedicated shift. Most experiments were done near 1.54 Å, but various tests from 1.2 Å to 2.5 Å gave equally good results.

III - 3) Results

First <u>tests</u> were performed on crystals of phycocyanin, a complex blue protein which is a photosynthetic pigment. The samples obtained from Institut Pasteur (Paris) had a very large unit cell (space group $P2_1$ a = 110 Å, b = 110 Å, c = 238 Å, β = 120°, asymmetric unit ~ 500,000 daltons), a high stability in the X-ray beam and satisfactory dimensions (~ 400 μm). 3 Å resolution, 1° oscillation photographs (fig. 5) were obtained in 25 minutes, in spite of poor operating conditions : 1.56 GeV, average current 40 mA (such photographs could be recorded now in 3 minutes). All spots were nicely resolved for a crystal-to-film distance of 100 mm. The overall quality of the pictures (signal/noise, spatial resolution) was at the level of what is obtained with mirror-mirror focussing optics and a rotating anode generator, as pointed out by Phillips et al (1976).

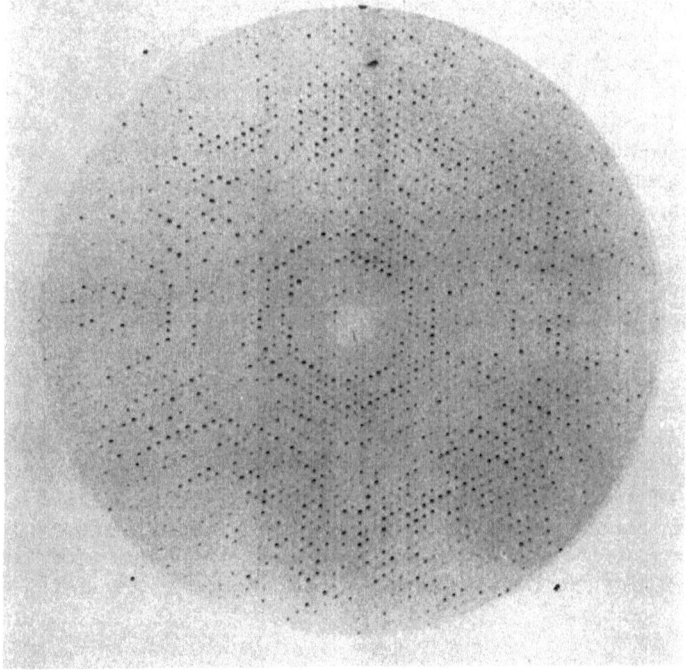

Fig. 5 - Oscillation photograph of a crystal of phycocyanin (1° oscillation, crystal-to-film distance : 10 cm, λ = 1.55 Å, exposure time 25 minutes, DCI storage ring operating at low energy and current : 1.56 GeV, 40 mA) (1976).

Data collection was performed for native arginine phosphokinase from
lobster (space group P2₁, a = 75 Å, b = 88.5 Å, c = 57.2 Å, β = 102°,
asymmetric unit 74,000 daltons) ; crystals are small (at best 400 x
150 x 100 μm³), have a stability of ∿ 30 hours for a standard X-ray
source and diffract usually to ∿ 2.7-2.9 Å (Berthou et al, 1975).
In that case, the resolution was markedly better with data extending
to ∿ 2 Å for fresh crystals ; with operating conditions of 1.72 GeV,
100 mA, the gain in exposure time was 20 with respect to the filtered radiation of a GX6 generator operated at 2.4 kW and each crystal had a lifetime of ∿ 5 hours. The decomposition is thus proceeding
at an increased rate, but the decay process is non-linear in dose
rate. Two isomorphous derivatives (platinum, mercury) were still
more sensitive to radiation, as already observed with the standard
source. But, in all cases, more useful data could be collected from
one sample. Photographs collected with the two sources were evaluated
and the data are currently used for electron density calculations.

A third example will be given with the collection to 1.9 Å resolution of a complete set of data for tts (tyrosyl-t-RNA synthetase
from Bacillus Stearothermophilus, space group P3₁2₁2, a = b =
64.4 Å, c = 238.8 Å, asymmetric unit ∿ 90,000 daltons). Hexagonal
plates (400 μm thick, 200 μm wide) oscillating about the long c
axis were used with a crystal-to-film distance of 100 mm. In that
case, 6° oscillation could be obtained from each crystal-better
than usual- with high quality photographs and exposure times of
∿ 25 minutes.deg⁻¹. The data has now to be evaluated and possibly
used to compute an improved electron density of this important
protein (Monteilhet, Blow and Fourme, 1978).

III - 4) Conclusions

It has been demonstrated that high resolution data collection is
possible now in a rather routine way with X-synchrotron radiation,
using essentially standard techniques and photographic films. Main
advantages are : a) drastically reduced exposure time, especially
for crystals with unit-cells which would normally require a mirror-
mirror focussing optic ; b) better resolution and better signal-to-
noise ratio ; c) high angular resolution ; d) more data will generally be collected from any sample ; e) the optimal wavelength can
be selected according to the sample size ; f) somewhat smaller and/
or poorly diffracting crystals can be used, as a result of preceding
points.

As a rule of thumb, the interest of the new source is roughly
proportional to the unit cell volume of the crystal. But, even when
the unit-cell is not very large, going to a synchrotron radiation
will become a normal practice either to get higher resolution data
or for problems which cannot be tackled with rotating anode genera-
tors. Tests on various medium or large-sized proteins have been

reported at Stanford (Phillips et al, 1976), Hamburg (Harmsen et al, 1976 and Sturhmann, 1978 b) and Novosibirsk (Mokulskii, 1978). The results are similar in many respect to ours, although the gain in exposure time seems to be less than what could be expected.

Last, it will be interesting to make a series of tests using short wavelengths (1 Å and below) to evaluate possible advantages in terms of crystal lifetime, reduced absorption, etc...*

*References will be found at the end of the following paper.

PROTEIN CRYSTALLOGRAPHY WITH SYNCHROTRON RADIATION

II. ANOMALOUS SCATTERING AND THE PHASE PROBLEM

R. FOURME

Laboratoire de Physicochimie Structurale, Université
Paris XII, 94000 Créteil and LURE (CNRS, Université
Paris-Sud), B 209 C, 91405 Orsay, France

I - INTRODUCTION

The synchrotron source is uniquely suited to anomalous diffraction experiments since the wavelength of the X-ray beam can be tuned
around the wavelength of the K or L absorption-edges for many heavy
atoms. With usual sources, only characteristic emission lines of
anode targets are available so that nearly all experiments where
anomalous scattering was used to assist in the phasing of protein
diffraction data were performed at a single wavelength, usually
Cu Kα radiation.

For a diffraction experiment at a single wavelength, anomalous
scattering is manifested in the Bijvoet differences. Blow (1958)
and Blow and Rossmann (1961) proposed the use of these differences
in the phasing of structure factors. Although multiple isomorphous
replacement has been the basis of most protein structure determinations to date, several structures have been solved from only one
isomorphous derivative plus anomalous differences. They include
rebredoxin (Herriott, Sieker, Jensen and Lovenberg, 1970), ferro-
doxin (Sieker, Adman and Jensen, 1972), flavodoxin (Watenpaugh,
Sieker, Jensen, Le Gall and Dubourdieu, 1972), phosphoglycerate
kinase (Blake and Evans, 1974), hemerythrin (Stenkamp, Sieker,
Jensen and Loehr, 1976) and penicillinase (Knox, Kelly, Moews and
Murthy, 1976). Improved Patterson maps for locating heavy atoms
can be obtained from combined isomorphous derivative differences
and Bijvoet differences. Matthews (1966) has shown how to scale
these two types of differences and extended the method to difference-
Fourier techniques.

II — THE PHASE PROBLEM

(Elementary background and usual methods for macromolecular structures.)

The electron density in a crystal is expressed as :

$$\rho(\underset{\sim}{r}) = \frac{1}{V} \underset{h}{\Sigma} \underset{k}{\Sigma} \underset{l}{\Sigma} \ F(\underset{\sim}{H}) \ \exp - 2\pi i \ \underset{\sim}{H}.\underset{\sim}{r}$$

($\underset{\sim}{r}$ is a vector in the unit cell and $\underset{\sim}{H}$ is a vector in the reciprocal space with integer coordinates h, k, l).

Structure factors $F(\underset{\sim}{H})$ are complex numbers with phases missing from the experimental data. The methods adopted in circumventing this phase problem are specific of protein crystallography.

A simple way of looking at the various techniques used to elicit phase   information from structure amplitudes was given by Kartha (1975). The principle is shown in **figure 1** ; assume that $\underset{\sim}{OA}$ and $\underset{\sim}{OB}$ are vectors of known magnitude but unknown phase, whereas $\underset{\sim}{AB}$ (label vector) is known both in magnitude and direction ; then :

$$\cos \psi = (OA^2 + AB^2 - OB^2)/2 \ OA.OB$$

The phase of $\underset{\sim}{OA}$ can take one of the values :

$$\alpha = \Theta \pm \psi$$

because the phase triangle can be constructed equally well on either side of $\underset{\sim}{AB}$. This result means that the amplitude change that has occured on addition of a known vector gives information about the component of the unknown vector in the direction of the label vector.

By use of two <u>non-collinear</u> label vectors instead of one, it is obvious that an unambiguous solution can be found to the phase problem.

II — 1) The Multiple Isomorphous Replacement (MIR) Method

In this method, one prepares in addition to the native protein an isomorphous heavy atom derivative ; assume that the heavy atom site(s) have been determined, so that the scattering vector for the heavy atom (s) F_H can be calculated. From the intensity data, we obtain $|F_L|$ and $|F_{LH}|$, the magnitudes of the structure factors respectively for the light atoms and the labelled protein. The relationship among those vectors is shown in fig. 2. As seen previously, resolution of phase ambiguity of a single derivative is obtained if at least <u>another</u> derivative is available. A graphical

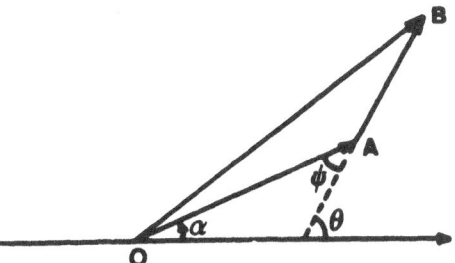

Fig. 1 – Principle of the derivation of phase information from structure amplitudes

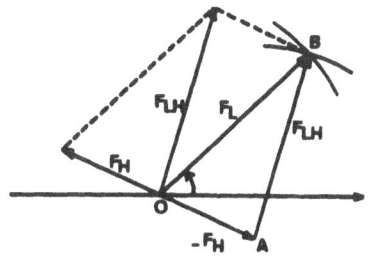

Fig. 2 – Phase ambiguity as a result of single isomorphous replacement. F_H is known (magnitude and phase) as well as the magnitudes of F_L and F_{LH}. Circles with radius $|F_L|$ and $|F_{LH}|$ and centres at O and A intersect in two points B and B', giving two possibilities for F_L (OB or OB')

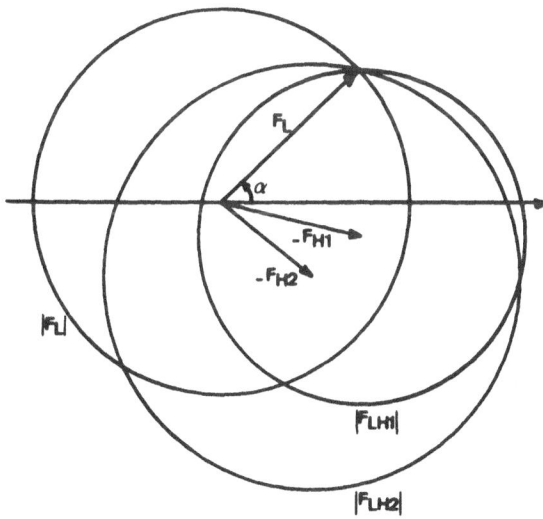

Fig. 3 – The phase ambiguity of figure 7 is solved with the data from two heavy atom derivatives (labelled 1 and 2) (MIR)

construction (Harker, 1956) shows how the phase can thus be determined, assuming for sake of simplicity that there are no experimental
errors **(fig. 3)**. Similar determinations are made for all reflexions
hkl in the experimental data set.

II - 2) Single Isomorphous Replacement plus Anomalous Scattering (SIRAS) Method

Assume that one heavy atom is an anomalous scatterer. Its
scattering factor is then $f + f' + if''$; let F_H be the real part
of the heavy atom contribution to the reflexion hkl and δ the imagi-
nery part ; F_L and F_{LH} are defined as above ; in a similar way,
we define $\overline{F}_H$, $\overline{\delta}$, $\overline{F}_L$ and $\overline{F}_{LH}$ for the reflexion $\overline{hkl}$. Those vectors
are shown in **fig. 4**. Since the imaginary part of the heavy atom
contribution is always phase advanced by $+ \pi/2$, the heavy atom
contributions are no longer symmetrical with respect to the real
axis. This is more clearly shown in **fig. 5** where the lower part
of fig. 4 has been reflected accross the real axis and superimposed
on the upper part as suggested by Blow and Rossmann (1961). (That is,
we use the complex conjugates of relevant contributions). The
Harker (1956) construction for this case is shown in **fig. 6.** The
native circle and the two from the derivative intersect at a single
point, which removes the ambiguity of the single derivative in a
way entirely analogous to the case for two derivatives.

Using synchrotron radiation, a well-chosen wavelength will give
enhanced differences with respect to usual sources.

III - MULTIWAVELENGTH METHODS

III - 1) Two-Wavelength Method Without Bijvoet Pair Measurement

In this method, the experimental problems of which have been
discussed by Arndt (1978), the intensity of each reflexion is measured
at two wavelengths on either side of an absorption edge. In the case
of a K-edge, the variation of f' is roughly symmetrical with respect
to the absorption edge and the two wavelengths are chosen so that
f' is the same in both cases. It can be shown that the phase angle
can be determined with one ambiguity if f'' and the phase angle of
the heavy atom contribution are known.

The average percentage change in intensity as a result of a
change f'' can be estimated as :

$$\frac{\Delta I}{I} \sim (\frac{2N_H}{N_L})^{1/2} \frac{f''}{f_L}$$

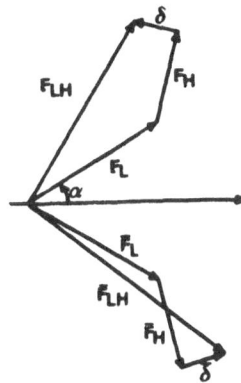

Fig. 4 - Intervector relations for a Friedel pair of reflexions when the label atom exhibits significant anomalous scattering

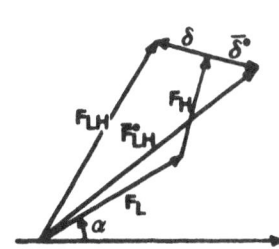

Fig. 5 - Same as figure 9, except that vectors associated to the second reflexion of the Friedel pair have been replaced by their complex conjugates

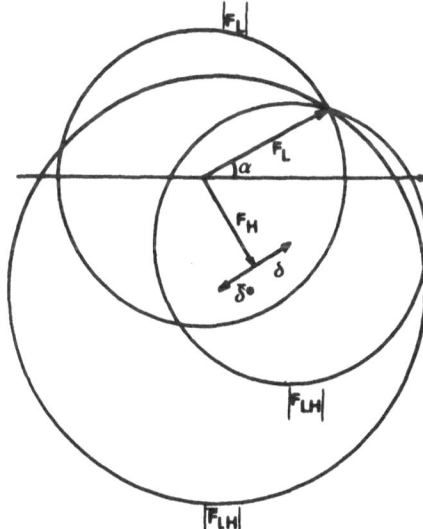

Fig. 6 - Harker construction showing unique solution for protein phase angle from single isomorphous replacement and anomalous scattering (SIRAS)

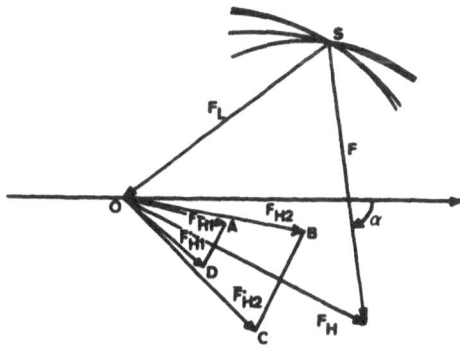

Fig. 7 - Unique solution for protein phase angle from single isomorphous replacement with Bijvoet pair measurements at two wavelengths (labelled 1 and 2)

where N_H and N_L are respectively the number of anomalous scatterers and the number of light atoms in the protein ; f" is the imaginary component of the scattering factor of the anomalous scatterer and f_L is the average scattering factor of light atoms.

Example :

For a protein with 2000 C, O, N atoms and one anomalous scatterer
at the K-edge $\frac{\Delta I}{I} \sim 1.8$ %. It will be necessary to make measurements
with a precision of $\sim 0.5-1$ % for an average reflexion and to ensure
that the systematic errors are small compared with the statistical
ones.

Using data from the native protein, the phase ambiguity can be
solved by combining this two-wavelength method and single isomorphous
replacement.

III - 2) Two-Wavelength Method With Bijvoet Pair Measurements

Here, all that is in principle required is a protein with a
label atom which exhibits anomalous scattering at λ_1 and λ_2. Fig. 12
is a composite diagram which shows the principle of the unique determination of the phase ; F, F_1 and F_2 are respectively the structure
factors for λ (an hypothetical wavelength where the anomalous scattering is negligible), λ_1 and λ_2 ; F_H, F_{1H} and F_{2H} are the heavy
atom contributions at those wavelengths ; the definition of $\overline{F}_{1H}^*$ and
$\overline{F}_{2H}^*$ is then obvious. Starting from the origin of the complex plane,
we draw F_{1H}, F_{2H}, $\overline{F}_{1H}^*$ and $\overline{F}_{2H}^*$. Circles drawn with origins at A, B,
C, D intersect at the point O, which represents the origin of the
gaussian plane for this reflexion. The phase angle of F is then
readily determined.

Between points A, B, C, D, six vectors can be drawn which form
with the corresponding structure factors F_{1H}, F_{2H}, $\overline{F}_{1H}^*$, $\overline{F}_{2H}^*$ six
phase triangles. This information is redundant since three of the
six phase triangles are in principle sufficient for the determination of the phase. It is further necessary that the phase-determining
vectors are inclined to each other (e.g. $\underset{\sim}{AB}$, $\underset{\sim}{DC}$ or $\underset{\sim}{AB}$, $\underset{\sim}{BC}$). The phase
ambiguity cannot be solved if there is no change in the real part
of the atomic scattering factor changes ; for this reason, the choice
of two wavelengths just below and just above an absorption edge may
not be the optimum choice ; contrary to the previous case, a substantial wavelength change may be necessary, as discussed here after.

Let us assume that accurate information exists on the values
of f'. and f" close to the absorption edges of the heavy atom scatterer. This information is unfortunately not known for most elements
used to make isomorphous derivatives ; f' and f" are related to the

absorption coefficient μ ; in the energy region just above the absorption edge, high values of μ have been frequently observed (Launois, 1978) ; abnormal values for f' and f" should then be observed, as above the L_{III} edge of cesium (Phillips, Templeton, Templeton and Hodgson, 1978). With such data, one optimum wavelength pair can be selected using the following expression (Bartunik, 1978) which gives the wavelength-dependent phase error for a relative accuracy on all intensity measurements $\Delta I/I$

$$\Delta\alpha \sim \frac{\sqrt{2}}{4} K f_L (\frac{N_L}{N_H})^{1/2} \frac{\Delta I}{I}$$

where : N_L and N_H are the numbers of normal and anomalous scatterers, f_L is the average scattering factor of normal scatterers

$$K = \frac{[2 + (f'_1/f''_1)^2 + (f'_2/f''_2)^2]^{1/2}}{|f'_1 - f'_2|}$$

(subscripts 1 and 2 are related to λ_1 and λ_2)

In practice, the choice of the wavelength pair is also influenced by the experimental conditions ; for instance, the wavelength difference should be kept as small as possible, together with a reasonably small value for K.

Example : for a protein with 2000 light atoms and 1 samarium atom, a possible choice could be λ_1 = 1.54 Å, λ_2 = 1.84 Å (below and above the L_{III} absorption edge of Sm) giving K $\sim$ 0.3. For $\Delta I/I$ = 2 %, $\Delta\alpha$ = 38° (Bartunik, 1978).

Very few x-ray diffraction studies of a protein crystal structure have used data collected at several wavelengths. Hoppe and Jakubowski (1975) have demonstrated that is was possible to measure a reflexion and its Bijvoet pair at two wavelengths with sufficient accuracy to obtain a useful estimate of the phases. They used NiKα and CoKα X-ray tubes with a four-circle diffractometer. For a small iron protein (erythrocruorin), they found an expected deviation of $\sim$ 50° from the true phases. The limit of this method is data-collection time and it was strongly pointed that a synchrotron source combined with a multireflexion diffractometer would be ideal for such an experiment.

Experiments at several wavelengths have been reported at SSRL (Stanford) (Phillips, Wlodawer, Goodfellow, Watenpaugh, Sieker, Jensen and Hodgson, 1977). A series of equatorial precession photographs of the Fe-containing protein rebredoxin were recorded at several wavelengths above and below the FeK absorption edge. The data showed intensity changes due to f' varying with wavelength as well as from Bijvoet differences. The iron site was located from

different Patterson and Fourier maps based either on f' or f" or
combined f' and f" maps. Phases were obtained with an average difference of 60° when compared to the known phases for rubredoxin,
which indicates that some information was available even from the
relatively poor data. More recently, a single-channel four-circle
diffractometer has been used to measure the values of f' and f"
above and below the L$_{III}$ edge of cesium (Phillips et al, 1978) ;
those results will be useful to assist in the phasing of low resolution data for gramicidin A, recorded with the same apparatus
(Hodgson, 1978).

III - 3) Three-Wavelength Methods

As discussed by Herzenberg and Lau (1967), phases can also in
principle be determined by measuring structure factors of each
reflexion at three wavelengths. This scheme has not yet been applied
to a real analysis.

IV - EXPERIMENTAL REQUIREMENTS FOR MULTIWAVELENGTH METHODS

The solution of the phase problem for small or medium protein
structures is essentially a technical problem. In my opinion, a
suitable experimental setup should have characteristics listed in
table 3.

Table 3

Requirements for multiwavelength phasing

* source : same as in table 2

* monochromator/collimation system

  . good $\Delta\lambda/\lambda$ ($\sim 10^{-3}$)
  . fast, accurate and reproductible tuning (0.7 - 2.6 Å)
  . fixed exit slit
  . low divergence of the exit beam
  . flux is monitored after collimation

* detector

  . photon counting
  . detection area with large solid angle
  . spatial stability
  . stability of quantum detection efficiency
  . "reasonable" quantum detection efficiency
  . energy resolution ~ 20 %
  . low-temperature device

A storage ring with a critical wavelength below 4 Å is a suitable source. The longer the beam lifetime, the better. The positional stability of the beam is of prime importance.

The monochromator is designed for high resolution and fast/accurate tuning rather than for highest intensity. The exit slit must be fixed : this can be achieved for instance with a separate two-crystal setup where one crystal is vertically translated by a cam system when the monochromator is rotated (Lemonnier et al, 1978 b). Germanium (silicon) crystals with an asymmetric cut to (111) planes will be preferred at wavelengths beyond (below) $\sim$ 1.2 Å (Kohra et al, 1978) ; such crystals do not produce second-order harmonics ; the third-order harmonics is weak and can be totally eliminated by a slight misorientation of one crystal with respect to the other. The intensity of the beam should be continuously monitored after the collimation system.

The detector is oriented towards accurate intensity measurements, since small differences are to be detected. Another crucial point is as high acquired information per photon scattered by the sample as possible (this point was introduced by Schulz and Rosenbaum, 1978). I will discuss now practical ways to fulfil these requirements.

V - CHOICE OF THE DATA-COLLECTION SYSTEM

Essentially four data-collection systems are used for protein crystallography.

- film instruments coupled with film scanners.

The oscillation camera (Arndt-Wonacott type) is a powerful recording system, especially for crystals with large unit-cells. But a film is a rather unefficient detector, in terms of acquired information per scattered photon ; the film processing is tedious ; the recording is discontinuous ; the picture is analog and has to be digitized via a cumbersome treatment. For multiwavelength phase determination, films can probably just be used for preliminary tests.

- four-circle diffractometers. The accuracy of these instruments is good and the diffraction data are readily obtained ; being a single-channel system, it is very unefficient even for a conventional X-ray source.

- area detectors with TV systems. Several such detectors have been described (Arndt and Ambrose, 1968 ; Reynolds, 1970 ; Minor, Milch and Reynolds, 1974).

The system is basically an oscillation goniometer where the film is replaced by a fluorescent screen coupled to an image intensifier and a TV camera. The image intensifier is delicate and difficult to operate. The detector area is small. The TV camera output is analog and suffers from noise, geometric distortion, saturation and instability, so that very precise measurement of integrated intensities has not yet been demonstrated.

- area detectors with multiwire proportional chamber

Multiwire proportional chambers (MWPC) have a digital output and the integrated intensity of a reflexion can be estimated by counting photons detected on a given area. A group at the University of California at San Diego (Cork et al, 1973, 1975) has designed and successfully used for several crystal structure determinations a system based on a flat MWPC with electronic readout into a large core memory. This multireflexion diffractometer gives on-line data with good accuracy ; the count rate is $\sim$ 30,000 photons.sec^{-1}, a good rate for protein crystallography with sealed X-ray tubes. These results were a strong incentive for us to go along the same line. A collaboration between LURE and the team of G. Charpak at CERN was initiated late 1974 to make a MWPC-based multireflexion diffractometer with improved capabilities.

VI - DESCRIPTION OF THE MULTIREFLEXION DIFFRACTOMETER

VI - 1) The detector

The detector itself includes a standard flat MWPC with a spherical drift gap (Charpak et al, 1974, 1977, 1978).

The MWPC has three planes of low-resistive wires ; the central plane (anode plane) has 2 mm spacing and is at a positive potential of a few kV ; the outer crossed planes (cathode planes) have 480 wires spaced by 1 mm. Electrons resulting from the conversion of the X-ray photons are accelerated toward one of the anode wires ; as a result of the avalanche of secondary electrons thus produced, a pulse is detected with an amplitude proportional to the energy of the incoming photon. The system is triggered by these anode pulses. Simultaneously, positive pulses are induced on a few nearby wires of cathode planes and the (X, Y) coordinates of the event are determined from an analysis of these pulses.

The drift gap is enclosed in a conical edge and two spherical electrodes at potentials V_1 and V_2 (the upper one is made of 30 µm aluminium and is hydroformed ; the lower one is made of a mesh). The conical edge is provided with conductive rings set at potentials

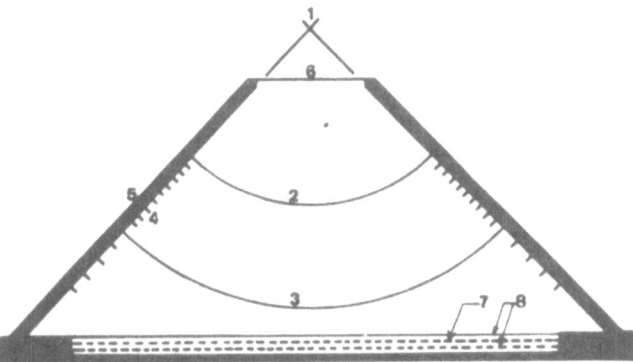

Fig. 8 — Cross section of the spherical drift chamber through central axis perpendicular to anode wires 1 — Focus 2 — Upper electrode 3 — Lower electrode 4 — Edge electrodes 5 — External shell 6 — Mylar window 7 — Anode plane 8 — Cathode planes

Fig. 9 — View of the prototype MARK I spherical drift chamber with the oscillation system. The source is a sealed X-ray tube

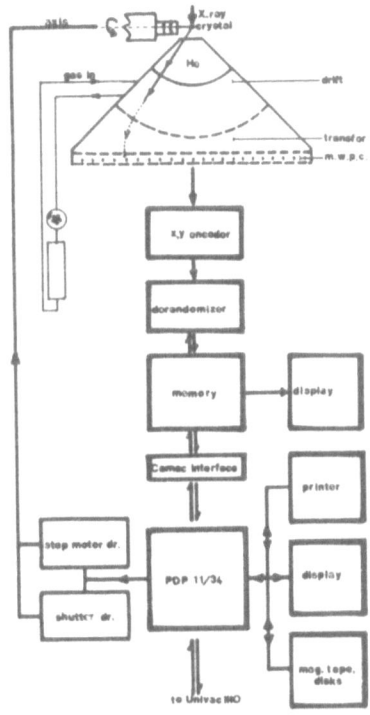

Fig. 10 — Major components of the multireflexion diffractometer

corresponding to a radial field, essentially to avoid edge effects
(fig. 9 and 10).

The drift space and the MWPC are filled with a mixture of xenon,
argon, ethane and carbon dioxyde circulating in a closed-loop circuit with a unit for gas purification.

The crystal is placed at the center of equipotential spheres.
X-ray photons are converted somewhere in the drift region and
electrons are drifted by the radial field and are transferred into
the MWPC through the lower electrode. The advantages of the drift
gap are :

- the drift gap is very thick, so that the probability for a photon
  to be converted is $\sim$ 100 % at a wavelength of 1.54 Å.

- for a flat MWPC, the angular resolution decreases at large angles.
  This effect is suppressed, provided all photons are converted in
  the drift region.

- as a result of the long drift path, the cloud of electrons is
  expanded so that the anode structure is lost. A comparable accuracy
  is then obtained for X and Y directions.

- since the speed of drifted electrons is very low, the time-struc-
  ture of the synchrotron radiation is partially lost, with a reduction of counting losses due to coincident events.

VI - 2) (X, Y) Cathode Readout

A fast (X, Y) encoding system was designed ; it is purely digital
with one amplifier-per-wire as in the system described by Parkman
et al. (1975) and Baru et al. (1978).

To a photon corresponds a distribution of pulses on several wires
in each cathode plane. Let X_1 and X_2 be the farthest channels of the
distribution along X, determined   with respect to a fixed threshold
amplitude ; X_1 and X_2 are found with priority encoders and the mid-
point with a digital hardware processor. A similar analysis is
simultaneously done along the Y direction, giving a couple (X, Y).
The total dead-time of the position encoder is now 350 ns and the
accuracy is $\sim$ 1 mm in both directions.

(X, Y) couples are buffered in a small shift-register which
acts as a derandomiser ; X and Y are adresses for a storage location
(cell) in a mass core memory (128 K 16-bit words memory with a read-
increment-write cycle of $\sim$ 1.2 μs, to be extended to 256 K). The
corresponding cell is then incremented by 1. Each cell can record

64,000 events. The number of effective pixels is then 480 x 240
(latter : 480 x 480). The maximum count-rate is $\sim$ 500 kHz. This
is a good rate for protein crystallography using either a rotating
anode source or a high-resolution non-focussing monochromator and
X-synchrotron radiation.

VI - 3) Computer Control and Software

A minicomputer (PDP 11/34), with 96 K of 16-bit words memory
and a floating point processor, controls the system and makes all
calculations necessary for on-line data collection. The external
buffer is included in a Camac crate with direct memory access (DMA)
to the central unit, so that high data-rate transfers to and from
the buffer are possible. The computer also controls the stepping
motor and the shutter of a modified Nonius oscillation camera and
an ion chamber which monitors the intensity of the collimated
incoming beam (fig. 10).

Experimental or predicted images are observed on a display ter-
minal which is used at various stages of the experiment, especially
to orient the protein single crystal.

A highly efficient data-collection method for the multireflexion diffractometer is to record a series of stationary diffraction
patterns (stills) rotating the crystal by a small fixed angle (0.05°)
between each exposure (frame). This method is similar to the electronic stationary picture described by Xuong et al. (1978) ; it is
equivalent to the step-scanning technique used in standard diffrac-
tometry except that here all "active" reflexions are measured. A
particular reflexion will occur in consecutive scan increments (or
frames). Since we can predict the frame in which each reflexion
will appear, the computer can read-out the counts accumulated at all
the reflexion coordinates of a particular frame and store them in a
data-file associated with each reflexion. Integrated intensities are
computed from the profile obtained from successive frames.

Stationary data collection has two advantages over moving crystal
methods : i) reflexion overlapping is eliminated, ii) the peak-to-
background counting ratio in enhanced. Both of these advantages are
very attractive for crystal with large unit-cells like proteins.

VI - 4) Conclusion

The MARK II version of this instrument will be installed at
LURE in early 1979 and will be extensively tested, for hardware as
well as software, using selected protein crystals and a sealed
X-ray tube. When these tests will be completed, it will be moved

to one of the ports of the synchrotron radiation beam line ; we
hope to use it as a general purpose facility for single crystal
data collection ; such a system, which is both accurate and efficient,
could be used for multiwavelength methods of phasing diffraction
data from small and medium-sized protein structures.

Acknowledgements

I would like to acknowledge the contributions of several scientists to the research in protein crystallography at LURE and especially R. Kahn, A. Gadet, Dr D. André, Dr J. Janin, Pr D. Blow,
Dr C. Monteilhet and Dr J.L. Risler. R. Kahn, assisted by B. Caudron,
R. Bosshard and G. Morel, played a central role in the multireflexion
diffractometer project, in close collaboration with the group of
Dr G. Charpak at Geneva including R. Santiard, R. Bouclier, R. Benoit,
R. Million and N. Greguric. Dr H.D. Bartunik contributed to fruitful
discussions. We are indebted to P. Marin and his collaborators from
the Laboratoire de l'Accélérateur Linéaire at Orsay who beautifully
operated the storage ring DCI.

REFERENCES

Arndt U.W. (1978) Nucl. Instr. and Meth., 152, 307..

Arndt U.W. and Ambrose B.K. (1968) IEEE Trans. Nucl. Sci., NS15 :
    Nb3, 92

Barrington-Leigh J. and Rosenbaum G. (1976) Ann. Rev. Biophys. and
    Bioeng., 5, 239.

Bartunik H.D. (1978) (to be submitted to Acta Cryst. A).

Baru S.E., Proviz G.I., Savinov G.A., Sidorov V.A., Feldman J.G.
    and Khabakhpashev A.G. (1978) Nucl. Instr. and Meth., 152, 195.

Berthou J., Rérat C., Rérat B., Gadet A., Fourme R., Renaud M.,
    Dubord C., Pradel L.A., Roustan D. and Thoai N.V. (1975)
    J. Mol. Biol., 95, 331.

Blake C.C.F. and Evans P.R. (1974) J. Mol. Biol., 84, 585.

Blow D.M. (1958) Proc. Roy. Soc., A247, 302.

Blow D.M. and Rossmann M.G. (1961) Acta Cryst., 14, 1195.

Charpak G., Hajduk Z., Jeavons A.P., Kahn R. and Stubbs R.J. (1974)
    Nucl. Instr. and Meth., 122, 1307.

Charpak G., Demierre C., Kahn R., Santiard J.C. and Sauli F. (1977)
    IEEE Trans. Nucl. Sci. 24 (1), 200.

Charpak G. Sauli F. and Kahn R. (1978) Nucl. Instr. and Meth., 122,
    185.

Cork C., Fehr D., Hamlin R., Vernon W. and Xuong N.H. (1973) J. Appl. Cryst., 7, 319.

Cork C., Hamlin R., Vernon W. and Xuong N.H. (1975) Acta Cryst., A31, 702.

Fourme R., Gadet A., Kahn R., André D., Janin J. and Risler J.L. (1977) European Congress of Crystallography, Oxford, UK and LURE Activity Report (1976-1977), LURE, Orsay, France.

Harker D. (1956) Acta Cryst., 9, 1.

Harmsen A., Leberman R. and Schulz G.E. (1976) J. Mol. Biol., 104, 311.

Herriott J.R., Sieker L.C., Jensen L.H. and Lovenberg W. (1970) J. Mol. Biol., 50, 391.

Herzenberg A. and Lau H.S.M. (1967) Acta Cryst., 22, 24.

Hodgson K.O. (1978) (private communication).

Hoppe W. and Jakubowsky U. (1976) Anomalous Scattering, edited by S. Ramaseshan and S.C. Abrahams, p. 437 - Copenhagen : Munksgaard.

Kahn R. and Fourme R. (1976) (unpublished results).

Kartha G. (1975) Anomalous Scattering, edited by S. Ramaseshan and S.C. Abrahams, p. 363 - Copenhagen : Munksgaard.

Knox J.R., Kelly J.A., Moews P.C. and Murthy N.S. (1976) J. Mol. Biol., 104, 865.

Kohra K., Ando M., Matsushita T. and Hashizume H. (1978) Nucl. Instr. and Meth., 152, 161.

Launois H. (1978), (Private communication.)

Lemonnier M., Fourme R., Rousseaux F. and Kahn R. (1978) Nucl. Instr. and Meth., 152, 173.

Lemonnier M., Collet O., Depautex C., Esteva J.M. and Raoux D. (1978) Nucl. Instr. and Meth., 152, 109.

Matthews B.W. (1966) Acta Cryst., 20, 230.

Minor T.C., Milch J.R. and Reynolds G.T. (1974) J. Appl. Cryst., 7, 323.

Mokulskii M.A. (1978) European Physical Society Meeting, York, UK.

Monteilhet C., Blow D. and Fourme R. (1978) European Physical Society Meeting, York, UK.

Parkman C., Hajduk A., Jeavons A., Ford N. and Lindberg B. (1975) CERN report DD/75/14, CERN, Geneva, Switzerland.

Phillips J.C., Wlodawer A., Yevitz M.M. and Hodgson K.O. (1976) P.N.A.S., 73, 128.

Phillips J.C., Wlodawer A., Goodfellow J.M., Watenpaugh K.D., Sieker L.C., Jensen L.H. and Hodgson K.O. (1977) Acta Cryst., A33, 445.

Phillips J.C., Templeton D.H., Templeton L.K. and Hodgson K.O. (1978) Science, 201, 257.

Reynolds G. (1970) IEEE Trans. Nucl. Sci., NS 17 : Nb 3, 310.

Schulz G.E. and Rosenbaum G. (1978) Nucl. Instr. and Meth., 152, 205.

Sieker L.C., Adman E. and Jensen L.H. (1972) Nature, Lond. 235, 40.

Sokolov A.A. and Ternov I.M. (1956) J. Exp. Theor. Phys. (USSR) 31, 473.

Stenkamp R.E., Sieker L.C., Jensen L.H. and Loehr J.S. (1976) J. Mol. Biol., 100, 23.

Stuhrmann H.B. (1978 a) Quart. Rev. Biophys., 11(1), 71.

Stuhrmann H.B. (1978 b) EMBL Outstation Activity Report, Hamburg, BRD.

Watenpaugh K.D., Sieker L.C., Jensen L.H., Le Gall J. and Dubourdieu M. (1972) P.N.A.S. 69, 3185.

Xuong N.H., et al., (1978) Acta Cryst., A34 (2), 289.

X-Ray Lithography and Microscopy

Eberhard Spiller, Ralph Feder and David Sayre

IBM T. J. Watson Research Center

Yorktown Heights, N.Y. 10598

X-rays have considerably shorter wavelengths than visible light and offer, therefore, the potential of drastically improved resolution. However, due to the fact that all materials are absorbing and that all refractive indices are very close to one in the x-ray region, conventional lenses cannot be fabricated for x-rays and magnification and demagnification cannot be obtained by refraction of x-rays. The oldest x-ray microscopy technique is the x-ray contact microscopy introduced by Goby.[1] Goby produced an x-ray shadowgraph of an object by bringing it into contact with a photographic film and irradiating it with x-rays. He viewed the developed unmagnified image with an optical microscope. Despite the fact that the resolution of this method cannot be better than that of the optical microscope, the technique has been widely used in biology and medicine, mostly for the differences in contrast and resolution compared to optical microscopy. A detailed discussion of the method can be found in several books and review articles.[2-4] There have also been several attempts to improve the resolution of the method, either by producing a magnified image by projection[5] or by the use of grain-free recording media coupled with an electron microscope for magnified viewing.[6,7] Drastically improved resolution close to the diffraction limit has been obtained recently.[8,9] In this case an organic x-ray resist has been used as the recording medium for the x-ray image and a scanning electron microscope provided the magnification. Organic resist materials are the radiation sensitive materials used in the fabrication of semiconductor devices. They are chemically modified by radiation in such a way that they, for example, can be dissolved in the areas that have been exposed, leaving the unexposed areas of a device wafer covered with a protection film which resists all the subsequent processing steps used for the fabrication of a device. Nearly all electronic microcircuits produced today are fabricated by photolithography (exposure by light) and have smallest linewidths around 2.5μm. It is expected that in the future, when linewidths around or below 1μm will be used, optical lithography will be replaced by electron beam or (and) x-ray lithography. X-ray lithography as it is used today in many laboratories is a 1:1 magnification shadowgraph technique like Goby's contact microscopy. A first mask of a microcircuit pattern is produced by

Fig. 1   Resist pattern made from an e-beam mask using x-ray lithography. SEM
picture shows part of a bubble pattern with 1 micron line widths.

an electron beam system, and this pattern is then transferred to a device wafer using x-rays (see Fig. 1). X-ray lithography has been discussed in detail in several review articles[10-13]; therefore, we will summarize here only the most important results and recommend these review articles for further details.

Despite the successes of the contact microscopy method, it is still very desirable to develop imaging devices for x-rays. Focussing devices are obviously required for x-ray astronomy and considerable efforts went into the development of grazing incidence telescopes.[14] The most important application of a high resolution focussing element for x-ray microscopy would be its use in a scanning x-ray microscope. Zone plates[15-18] or normal incidence mirrors using multilayer coatings[19-22] might provide the required focussing.

Wavelength Selection

The wavelengths used for x-ray lithography and microscopy are selected to adjust contrast and transmission. Fig. 1 shows that except for the jumps at absorption edges, the absorption coefficient increases with increasing wavelength. Therefore, small features will produce higher contrast for larger wavelengths. On the other hand, the material surrounding the feature or the substrate that supports the pattern, has higher transmission for shorter wavelengths. For features around or below $1\mu m$, the 5-100Å wavelength region is suitable. In x-ray lithography, the circuit pattern is usually defined by a gold absorber pattern. Fig. 2 shows that the absorption coefficient is larger than $1\mu m^{-1}$ in this wavelength region, such that gold films of less than $1\mu m$ thickness have sufficient contrast. A thin membrane fabricated from light elements is used to support the gold pattern. Fig. 2 shows that this membrane can be several μm thick and still have sufficient transmission. The jumps in the absorption edges of elements can be used to highlight the distribution of different elements in an x-ray micrograph. The distribution of a specific element in a specimen can be obtained from the difference of two micrographs, one obtained at a wavelength just above, and other just below an absorption edge of this element. Some absorption edges which are of importance for biological studies are the edges of $O(\lambda=23.3Å)$, $N(\lambda=31Å)$, $C(\lambda=43.6Å)$ $Ca(\lambda=31.1Å)$ and $P(\lambda=81Å)$. For wavelengths above $\lambda=23Å$ water has very little absorption, and the interior of a relatively thick, wet specimen can be made visible using those wavelengths. A wavelength above 44Å will be transmitted by the carbon atoms in an organic structure and allow easy observations of features which contain heavy elements, while the carbon structure can be imaged with high contrast at wavelengths shorter than 44Å. These few examples show that the x-ray wavelength can be selected to analyze the chemical composition of a microscopic sample and to suppress or highlight certain features in a specimen. This possibility is an important advantage of x-ray microscopy over electron microscopy.

X-ray Resists and Resolution

All presently used resist materials are organic polymers which are modified by the incident radiation in such a way that a subsequent development process can distinguish between unexposed and exposed regions removing one and leaving the other intact.

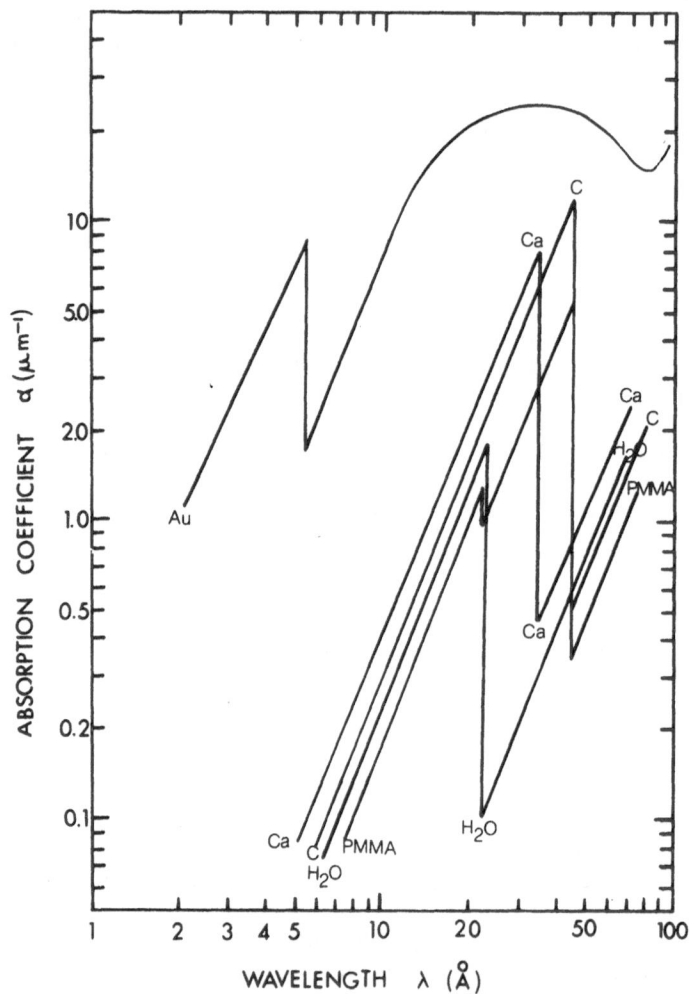

Fig. 2   Absorption coefficients of selected materials in the soft x-ray region.
Compounds of light elements can serve as mask substrates, gold is
usually   used   to   define   the   circuit   pattern.   **PMMA**
(polymethylmethacrylate) is a widely used x-ray resist.

The exposure of a resist film by x-rays is very similar to the exposure of the resist by an electron beam. An absorbed x-ray photon produces a shower of secondary electrons and these secondary electrons interact strongly with the resist material and are responsible for most of the chemical changes in the resist.

The effective range of these secondary electrons represents a limit for the resolution of a resist material; the highest resolution of all resists has been obtained with polymethylmethacrylate (PMMA).[9] Figure 3 shows that the effective range of the secondary electrons generated by x-rays in PMMA increases with the photon energy.[11] Diffraction effects, on the other hand, are less limiting for higher photon energies, and Fig. 3 shows that the highest resolution of about 50Å can be expected with ultrasoft x-rays with wavelengths around $\lambda=50$Å. It is a very favorable coincidence that about the same wavelength region also gives the highest contrast for very high resolution replication.

The sensitivity of a resist material is theoretically limited by the shot noise of the absorbed photons. A minimum number of photons has to be absorbed per resolution element to produce a certain signal to noise ratio. Therefore, resolution and sensitivity are correlated; the resist of higher resolution requiring a higher exposure. Similarly, contrast and sensitivity are correlated. (A high contrast resist can record an image of low contrast with high fidelity). The actual exposure of a resolution area that is exposed to an average exposure of $\bar{n}$ photons is described by the Poisson distribution with parameter $\bar{n}$; that of an area exposed to a lower exposure $T_{abs} \cdot \bar{n}$, with T_{abs} being the transmission of an absorbing feature in a specimen, is described by the corresponding Poisson distribution with parameter $T_{abs} \cdot \bar{n}$. The overlap area of the two Poisson distributions defines the probability of being unable to distinguish the two areas in an image.[11] The requirement that the two distributions overlap only outside their respective 3σ-limits allows us to calculate a minimum exposure (in absorbed photons per resolution element of the resist) for the replication of a feature which attenuates the incident radiation by a factor T_{abs}:

$$\bar{n} = \frac{9}{(1-\sqrt{T_{abs}})^2} \tag{1}$$

Features of low contrast ($T_{abs} \to 1$) require considerably higher exposure levels than features producing high contrast ($T_{abs} \to 0$).

PMMA resist shows the characteristics described by Eq. 1; the slope of its dissolution rate versus exposure curve becomes larger at higher exposure level allowing for large differences in dissolution rate for small differences in exposure.

Many resists of higher sensitivity but lower resolution and often lower contrast than PMMA are available. They are most useful in x-ray lithography for the fabrication of circuit patterns with a resolution around 1μm and have been reviewed elsewhere.[11]

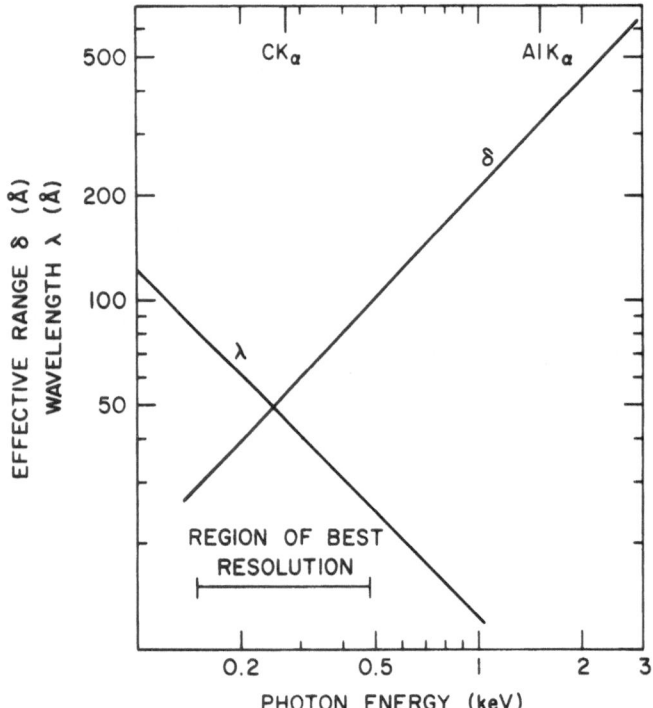

Fig. 3   Effective range of secondary electrons generated in x-ray resist
(PMMA) by x-ray photons of energy E and corresponding wavelength λ
with the region of optimum resolution around the crossover point of the
two curves.

Applications

X-ray lithography is presently used in research and development laboratories for the fabrication of exploratory devices and represents together with electron beam lithography the technologies which will replace optical lithography when the linewidths of devices drop below $1\mu m$. However, it is generally not expected that this replacement will occur on a commercial scale within the next few years. Commercial electronic devices which really utilize the resolution of around $50\mathring{A}$ are not expected within the foreseeable future.

On the other hand, biological specimens contain structural information on all scales down to the atomic level and the extension of the resolution capability of x-ray microscopy will provide valuable data on many biological structures, once the technology has been transferred to the biologist as a viable working tool. Synchrotron radiation sources are most suitable for high resolution work due to their collimation and high brightness; they will allow the studying of wet living specimens with high resolution and tolerably short exposure times.

Scanning X-ray Microscopes

The scanning x-ray microscope will be the most useful instrument for x-ray microscopy when high resolution focussing elements for soft x-rays become available. It cannot be expected that focussing elements will obtain a better resolution than that already obtained by the contact method with PMMA as a recording medium. There are, however, two other advantages which make the development of a scanning x-ray microscope worthwhile. 1. The electronically recorded and digitally stored image obtained from a photon counter behind the object, which is scanned by a small spot, is much more suitable for quantitative measurements than the resist method. For instance, two micrographs, obtained with two wavelengths at different sides of an absorption edge of an element, can easily be subtracted digitally to obtain the distribution of this element in the specimen. Similarly, digital image processing could be used to reconstruct a three-dimensional image from a series of micrographs obtained at different angles. 2. The photon counter behind the specimen can have a very high quantum efficiency and practically count all the photons that are transmitted by the specimen. This feature allows the highest possible image fidelity consistent with the resolution and only limited by the random nature of the photon counting process and allows one to obtain images with the lowest possible radiation damage. It has been shown that soft x-rays can produce comparable images in the resolution range above $100\mathring{A}$ at a lower radiation damage than electrons,[23] as shown in Fig. 4. This is surprising at first sight because electrons can be elastically (i.e. non damaging) scattered to some extent, while the main interactions of x-rays with matter is absorption, which is always damaging. The reason for the lower possible radiation damage is due to the better contrast obtainable by selecting an optimum wavelength for each feature. This allows a lower exposure (see Eq. 1) than can be obtained with electrons of sufficient energy to penetrate comparable material thickness. We mentioned before that PMMA has a performance (sensitivity, resolution and contrast) close to the theoretical limit, if one counts the number of photons absorbed per resolution element for adequate exposure levels. However, for a very high resolution close to the diffrac-

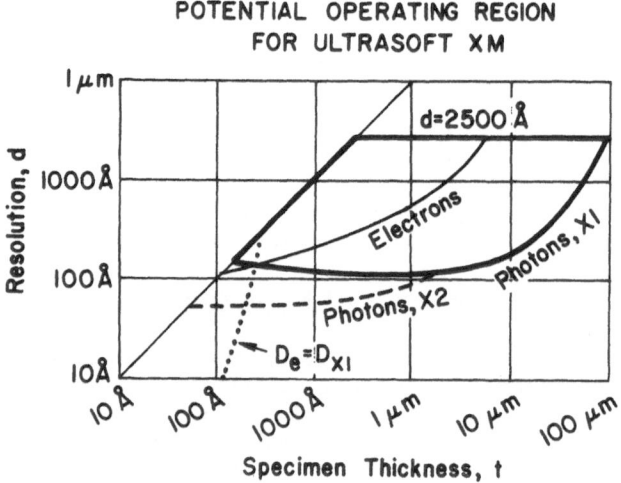

POTENTIAL OPERATING REGION
FOR ULTRASOFT XM

Fig. 4   Contours in the (t, d)-plane (where t is the specimen thickness and d is
              the resolution) of the radiation dose that must be undergone by a speci-
              men consisting of protein features in a water background. Contours are
              at 10^9 rads.   The area bounded by heavy solid lines is the potential
              suboptical operating region (for the protein-water specimens) for bright-
              field transmission microscopy with ultrasoft x-ray photons. Other curves
              are described in the text.

tion limit, only very thin resist films can be used which absorb only a small fraction of the incident radiation (between 10^{-3} and 10^{-2} depending on wavelength). The required exposure, measured by the number of photons incident on the resist, i.e. the number of photons that have gone through the specimen is about 100 times higher with the corresponding higher radiation level. In a scanning microscope, the resolution is determined by the spot size of the illuminating beam. No spatial resolution is required for the x-ray counter; therefore, this counter can be made sufficiently thick to absorb all the incident photons and therefore produce an image at the lowest possible radiation damage.

The focussing element which produces the illuminating spot is the key element of a scanning x-ray microscope. In a first instrument, that spot was simply produced by a pinhole illuminated with synchrotron radiation and a resolution of around 1μm has been obtained.[24] Grazing incidence x-ray telescopes have been improved and fabricated during the last twenty years; the fabrication tolerances obtained up to now are, however, not sufficient to build an x-ray microscope with a resolution better than 1μm. The presently most promising focussing elements for soft x-rays are precision zone plates and normal incidence mirrors.

The fabrication of zone plates for soft x-rays has been reviewed recently,[18] and a resolution better than 1μm has been obtained. For the fabrication of a zone plate, a fabrication process is required that has the same resolution capability that one attempts to obtain when one uses the zone plate to produce a diffraction limited focussed spot. Because electron beam[25] and x-ray lithography have a resolution capability of around 100Å, a similar resolution should be obtainable from zone plates fabricated by these techniques. The main disadvantage of zone plates is the fact that they require monochromatic radiation. A highly collimated beam as obtainable from synchrotron radiation sources is desirable for the scanning microscope to produce a small spot size. However, the wide spectral range of synchrotron radiation requires selecting only a very narrow spectral range with a corresponding loss in intensity.

Normal incidence mirrors as focussing elements have nearly spherical surfaces and can be fabricated with considerable higher precision than grazing incidence mirrors. The reflectivity for normal incidence can be raised to acceptable levels by the use of multilayer coatings.[19-20] Extremely smooth films with stable boundaries are required in these coatings. Multilayer coatings consisting of amorphous layers of ReW and C have been fabricated with surface roughness $\sigma < 5$Å, suitable for reflecting coatings with a short wavelength limit below $\lambda = 100$Å.[26]

References

1.   P. Goby, C. R. Acad. Sci Paris, *156*, 686 (1913).

2.   P. Kirkpatrick, H. H. Pattee, Jr., X-Ray Microscopy, in Handbuch der Physik, ed. by S. Flügge, Vo. XXX (Springer, Berlin, Heidelberg, New York (1957).

3.   V. E. Cosslett, W. C. Nixon, X-Ray Microscopy, University Press, Cambridge (1960).

4.   T. A. Hall, H. O. E. Röckert, R. L. de C. H. Saunders, X-Ray Microscopy in Clinical and Experimental Medicine, C. C. Thomas, Springfield, Ill, (1972).

5.   W. C. Nixon, Proc. Roy. Soc. A*232*, 475 (1955).

6.   W. A. Ladd, M. Hess, M. W. Ladd, Science *123*, 370 (1956).

7.   S. K. Asunmaa, X-Ray Optics and X-Ray Microanalysis, ed. by H. H. Patte, V. E. Cosslett, A. Engstrom (Academic Press, New York, 1963).

8.   E. Spiller, R. Feder, J. Topalian, D. Eastman, W. Gudat, D. Sayre: Science *191*, 1172 (1976).

9.   R. Feder, E. Spiller, J. Topalian, A. N. Broers, W. Gudat, B. J. Panessa, Z. A. Zadunaisky, J. Sedat, Science *197*, 259 (1977).

10.   H. I. Smith, Proc. IEEE 62, IEEE 62, 1361 (1974).

11.   E. Spiller and R. Feder, X-Ray Lithography in X-Ray Optics, ed. by H.-J. Queisser, Topics in Appl. Phys. Vo. 22, *35*, (1977).

12.   J. H. McCoy, SPIE Vol. 100, Semiconductor Microlithography II, p. 162 (1977).

13.   H. I. Smith, Fabrication Techniques for Surface Wave Devices in Acoust. Surface Wave Devices, ed. by A. A. Oliner, Topics in Appl. Phys. Vol. 24 305 (1978).

14.   J. H. Underwood, J. E. Milligan, A. C. deLoach, R. B. Hoover, Appl. Opt. *16* 859 (1977).

15.   G. Schmahl, D. Rudolph, Optik *29*, 577 (1969).

16.   J. Kirz, J. Opt. Soc. Am. *64*, 301 (1974).

17.   B. Niemann, D. Rudolph, G. Schmahl, Appl. Opt. *15*, 1882 (1976).

18.   D. Rudolph and G. Schmahl, J. de Physique (1978), in press.

19.   E. Spiller, Appl. Phys. Lett. *20*, 365 (1972).

20.   R.-P. Haelbich, C. Kunz, Opt. Commun. *17*, 187 (1976).

21.   E. Spiller, Appl. Opt. *15*, 2333 (1976).

22.   A. V. Vinogradov, B. Ya. Zeldovich, Appl. Opt. *16*, 89 (1977).

23.   D. Sayre, J. Kirz, R. Feder, D. M. Kim, E. Spiller, Science *196*, 1339 (1977) and Ultramicroscopy *2*, 337 (1977).

24.   D. Horowitz and J. A. Howell, Science *178*, 608 (1972).

25.   A. N. Broers, W. W. Molzen, J. J. Cuomo, N. D. Wittels, Appl. Phys. Lett. *29*, 596 (1976).

26.   R.-P. Haelbich, A. Segmüller, E. Spiller, to be published.

INDEX